BIOENGINEERING:

Biomedical, Medical and Clinical Engineering

BIOENGINEERING:
Biomedical, Medical and Clinical Engineering

A. Terry Bahill
Carnegie-Mellon University

PRENTICE-HALL, INC., *Englewood Cliffs, New Jersey 07632*

Library of Congress Cataloging in Publication Data

Bahill, Terry.
Bioengineering—biomedical, medical, and clinical engineering.

Includes bibliographies and index.
1. Biomedical engineering. 2. Human physiology
–Mathematical models. I. Title. [DNLM:
1. Biomedical engineering. 2. Models, Biological.
QT34 B151b]
R856. B33 610'.28 80-18945
ISBN 0-13-076380-2

Editorial/production supervision and interior design: Nancy Moskowitz
Manufacturing buyer: Anthony Caruso
Cover design: Edsal Enterprises

Printed in the United States of America

10 9 8 7 6 5 4 3 2 1

Prentice-Hall International, Inc., *London*
Prentice-Hall of Australia Pty. Limited, *Sydney*
Prentice-Hall of Canada, Ltd., *Toronto*
Prentice-Hall of India Private Limited, *New Delhi*
Prentice-Hall of Japan, Inc., *Tokyo*
Prentice-Hall of Southeast Asia Pte. Ltd., *Singapore*
Whitehall Books Limited, *Wellington, New Zealand*

Dedicated to Karen and the Jabberwock.[1]

[1] Lewis Carrol said "that the Anglo-Saxon word 'wocer' or 'wocor' signifies 'offspring' or 'fruit.' Taking 'jabber' in its ordinary acceptation of 'excited and voluble discussion,' would give (Jabberwock) the meaning of 'the result of much excited discussion.'" Dodgson, C. L. (Lewis Carrol), *The Annotated Alice*, introduction and notes by Martin Gardner, New American Library, New York, 1960, p. 195.

CONTENTS

PREFACE

There are dozens of books on bioengineering, there are scores of books on biomedical engineering, and there are numerous books on clinical engineering. However, I have not been able to find a single book which covers this broad range of topics in a manner suitable for an undergraduate engineering textbook. In my initial efforts at teaching a senior-level course at Carnegie-Mellon University, I had to use five textbooks: one for membrane biophysics, one for instrumentation, one for systems control theory, one for biological modeling, and one for clinical engineering. These areas, basically, became the chapters of this text. Unfortunately, one cannot present the detail in one chapter which can be presented in an entire book. I tried to discuss certain important areas in depth, and to discuss the rest of the field broadly.

Most textbooks on physiological systems modeling discuss many different physiological systems, but none in enough detail to really explain the principles of good bioengineering modeling. If I have erred in this book, I have erred on the opposite extreme: Chapters 3 and 4 present only a few physiological systems, but cover them in depth. If the reader thinks that too much detail is provided, he or she may skip certain sections.

No rules of thumb or clever shortcuts are used in this engineering textbook because the student will probably forget them unless they are used constantly. And I hope that no bioengineer will be continually doing the same thing. Methods, techniques, and basic principles, not specific results, are stressed.

The text was written with seniors in biomedical or electrical engineering in mind. However, the course was taught to students with many different engineering and science backgrounds.

The order in which the chapters are covered is unimportant, except that Chapter 3 must precede Chapter 4. I use Chapters 3, 4, and 1 for a one-semester graduate course in physiological systems modeling. For a one-quarter course taught to engineers or a one-semester course taught to life scientists, Chapters 3 and 4 could be used.

For a comprehensive senior-level course on bioengineering, I use the whole book, but skip certain sections, such as the Einstein relationship (Sec. 1-1-3), the Hodgkin–Huxley equations (Sec. 1-9), the predictions of the reciprocal innervation model (Sec. 3-7-4.4), and thermoregulation (Sec. 4-5). These sections may be omitted without affecting comprehension of later material. Also, I presume the students have a good working knowledge of systems theory and only rapidly review (Secs. 3-1, 3-2, and 3-6). Reasonable prerequisites for a one-semester course using the whole book are a systems or control theory course where the use of Laplace transforms and Bode diagrams are taught, an electrical engineering or physics course where electrical circuits are discussed, and a physiology or biology course where the action of nerve and muscle cells is presented.

This text should not overlap a typical course in physiology, which would probably cover the physiological properties of the systems covered in this text and in addition would cover cardiovascular, renal, respiratory, and endocrine systems.

The author is indebted to Drs. J. Robert Boston, Mark B. Friedman, and Oscar M. Reinmuth, who gave advice and support; to Jose R. Latimer and the other graduate students who took the course and made helpful comments; to Anita Nebiolo, who typed the text into the computer; and to Carolyn E. Stewart, who edited the manuscript.

A. Terry Bahill

Pittsburgh, Pennsylvania

BIOENGINEERING:

Biomedical, Medical and Clinical Engineering

0

INTRODUCTION–MODELING

Modeling is an important facet of most bioengineering studies. Models are simplified representations of objects and systems. For this reason, models are also an important part of everyday living. For example, if you wanted to drive from Pittsburgh to San Francisco, you could use a model of the highway system called a road map. This map allows you to study and understand the highway system of the United States without having to drive on every road. If you wanted a model of the life of southern aristocrats at the time of the Civil War, you could use the book *Gone with the Wind*. Highway engineers create models to see how traffic-light synchronization or lane obstructions will affect traffic flow. Ohm's law is a good model of a resistor. $F = ma$ is a simple model for slow-speed movement of a mass. Models allow us to transfer mathematical notation to real situations. Whenever something is too big, too complicated, too far away, or when mathematical precision is necessary, we use models.

Models are used to help configure and evaluate computer systems. For example, if a university is going to purchase a new computer system, it will initially build a model of its user community. This model will specify the users; for example, 50 students on computer terminals, a physics experiment analyzing data in batch mode, and the university's accounting system. It will specify the type of load each user will put on the system; for example, a student using a text editor may enter five characters per second but require little CPU time, while another student running an economic simulation program will enter very few characters but may require a large amount of CPU time. When this model of the university's user community is finished, university personnel will go to computer manufacturers and ask them which of their computers would best satisfy the university's needs. The computer manufacturers would then use models of their computer systems to determine which computer system and which peripherials would be best suited for the university. The model of the university community would then be run on either a real computer system or on a model of this proposed system to provide quantitative data on computer system performance. The final outcome of all this modeling would be a multimillion dollar purchase order for a new computer system.

Animal models are frequently used in biomedical research. As a means of understanding such models, let us examine the following example from Kandel (1979).

Early in this century it was reported that children who were discovered being raised in social isolation (e.g., an attic or a closet) exhibited permanent loss of social responsiveness and speech capability. It was difficult to tell what was wrong with them or even to know if they were mentally disturbed before the isolation.

In the 1940s René Spitz compared the development of infants raised in an orphanage with the development of infants raised in a nursing home attached to a women's prison. Both institutions were clean and provided adequate food and medical care. The babies in the nursing home were cared for by their mothers. Because they were in prison and away from their families, the mothers tended to pour affection on their infants in the time allotted each day. By contrast, in the orphanage the infants were cared for by nurses, each of whom was responsible for about seven infants. As a result, the children in the orphanage had much less contact with other human beings than did those in the nursing home. This produced social deprivation in the orphanage children. Furthermore, these children suffered sensory deprivation, because the bars of their cribs were covered with sheets which prevented the infants from seeing outside. Spitz found that in the second and third years of life the nursing home children were walking and talking, whereas the orphanage children were not. The orphanage children were withdrawn and were more susceptible to infection. These studies showed a cause and an effect, but did not explain the problem.

The next step in understanding the problem was the development of an animal model of infant social isolation. This step was taken by Margaret and Harry Harlow in the 1950s. They separated newborn monkeys from their mothers a few hours after birth. These monkeys were fed by remote control and observed through one-way mirrors. Monkeys raised in isolation for a year were seriously impaired socially and psychologically. When returned to the monkey colony they did not play or interact with the other monkeys. Many would crouch in a corner and rock back and forth in a manner reminiscent of autistic children. When they reached sexual maturity they did not mate. These profound deficits resulted from only 6 months of total isolation during the first year of life. Such isolation in later life had little effect on social behavior. These studies suggested that there is a *critical period* for social development.

To find out what happens during this critical period, we once again use animal models and assume that the structure, organization, and development of the human cortex is similar to that of animals. In the 1960s and 1970s Hubel and Wiesel (references are in chapter 3) and many others examined the effects of visual deprivation on nerve cells in the visual cortex of newborn kittens and monkeys. Most cells in the visual cortex respond to an appropriate stimulus presented to either eye. However, if a monkey is raised from birth to 3 months with one eyelid sewn closed, that animal will be permanently blind in that eye, although there is nothing wrong with that retina. Recordings made of single cells in the visual cortex after the stitches were removed and the eyelid opened showed that no cortical cells were driven by stimuli presented to that eye. This deficit occurs because cells in the lateral geniculate nucleus that receive input from the closed eye regress and lose their connections with cortical cells, and geniculate cells from the open eye sprout and take their place. These studies show that sensory deprivation early in life can alter the structure of the cerebral cortex.

At this stage we must examine the adequacy of our animal models. Monkeys with an eyelid sewn closed for 1 week sometime in the first 2 months of life will be permanently blind in that eye. For kittens the critical period is 3 weeks to 3 months, but some of the effects can be reversed by several months of subsequent experimental manipulation of the visual environment. Similar visual deprivation in human beings (caused by cataracts) has less dramatic effects on vision. This could be telling us that there are significant differences in the visual cortex of kittens, monkeys, and man, or that the critical period is longer or later in man (perhaps lasting until 6 years of age). However, the models have served their purpose. They have answered some questions and have posed some new ones.

There are many possible models for any system. The simplest is usually the easiest to use, but the more complex models are more accurate. An engineer usually selects the simplest model that is adequate for the study. For example, Fig. 0-1 shows three valid models of a capacitor. I usually use the first one, the most uncomplicated.

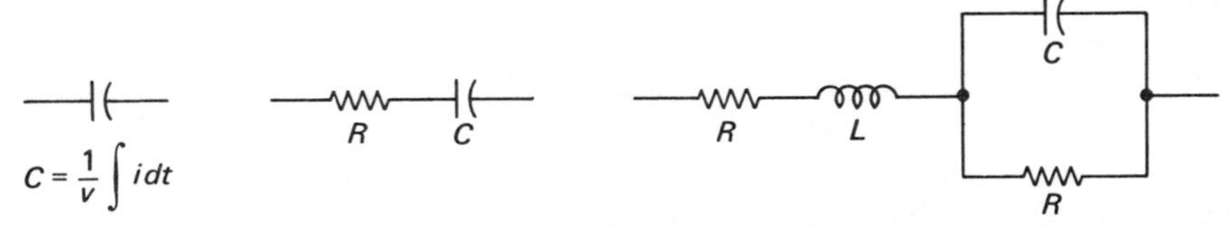

Figure 0-1 Three models of a capacitor. The first is the simplest and the most common.

Many types of models have been used to study biological systems. We list here several types of models, with a few examples of each. This list is necessarily brief. The book by Gordon (1978) is an extensive study of simulation[1] techniques, and the texts by Stark (1968), Talbot and Gessner (1973), and Kline (1976) give many examples of physiological systems models.

I. Physical homolog models where the model structure is similar in form and structure to the biological system being modeled: prosthetic limb, glass eye, DNA double helix, alpha helix, animal models, iron-wire neuronal model, or a wind tunnel.
II. Physical analog models where some quantity is said to be analogous to the quantity being simulated: analog computer, automobile speedometer, or a clock.
III. Schematic, verbal and symbolic models: novel, road map, poem, painting, or circuit diagram.
IV. Mathematical models
    A. Discrete systems: sampled data, Z transforms, difference equations, or Monte Carlo simulations
    B. Continuous systems
        1. Nonlinear systems
            a. Wiener kernels
            b. Nonlinear differential equations: for example, the Hodgkin–Huxley equations
        2. Linear systems
            a. Equations
                (1) Differential equations (time domain), (ζ, ω, τ)
                (2) Algebraic equations (frequency domain), Laplace transforms
                (3) Transfer functions (including modulation transfer functions)

[1] The terms *modeling* and *simulation* are often used interchangeably, although they have different meanings. Modeling means constructing a model that emulates or explains a physical system. Simulation means implementing or running this model, usually on a digital computer.

(4) Weighting function (impulse response)
(5) State-variable models
(6) First-order Wiener kernels

b. Graphs
(1) Bode diagrams
(2) Pole–zero plots and root locus
(3) Nyquist plots
(4) Power spectrum density (time domain), autocorrelation (frequency domain)
(5) Cross-correlation and cross spectrum functions (may also be analytic)

Models codify, simplify, teach and explain. They are heuristic. Models should enable predictions to be made about the future behavior of systems. Biomedical engineering models should be able to do all these things, and to suggest new physiological experiments.

This textbook is primarily about modeling linear, continuous, deterministic, constant-coefficient, lumped-parameter systems. We will study neural models, membrane models, muscle models, neuromuscular control system models, and electrical circuit models. The study of bioengineering models is the common feature of the diverse material presented in this book.

When you finish this book you should understand most of these terms and should realize how widespread modeling is. But most important, when you finish this book you should also understand how to validate, or assess the goodness, of a model.

REFERENCES

GORDON, G., *System Simulation*. Englewood Cliffs, N.J.: Prentice-Hall, Inc., 1978.

KANDEL, E. R., "Psychotherapy and the Single Synapse, the Impact of Psychiatric Thought on Neurobiologic Research," *The New England Journal of Medicine*, 30 (1979), 1028–37.

KLINE, J., *Biological Foundations of Biomedical Engineering*. Boston: Little, Brown and Company, 1976.

STARK, L., *Neurological Control Systems, Studies in Bioengineering*. New York: Plenum Press, 1968.

TALBOT, S. A., and U. GESSNER, *Systems Physiology*. New York: John Wiley & Sons, Inc., 1973.

MODELS OF NEURONS

Physiological systems are composed of muscle and nerve cells which act electrically. The torpedo, an electric ray, uses electricity to paralyze its prey. Aristotle described this behavior of the torpedo in 341 B.C. The Romans used this fish therapeutically by having the patient stand upon it and receive its 1-kW electric jolt (Bennett 1968). In 1791, Luigi Galvani stated that electricity was responsible for nerve and muscle behavior. However, it was not understood how this electricity was generated until 1952, when Hodgkin and Huxley published their model of the electrical behavior of the squid giant axon. Their model, four nonlinear differential equations, is one of the most famous models of our time. In this chapter, we will gradually work our way up to this model. We must first understand the behavior of ions in solutions separated by biological membranes.

1-1 BASIC BIOPHYSICS TOOLS

Four basic mathematical tools are needed to calculate voltages across membranes.

1. Fick's law for diffusion of particles:

$$J = -D\frac{d[\mathrm{C}]}{dx} \tag{1-1}$$

2. Ohm's law for drift of charged particles in electric fields:

$$J = -\mu z\frac{dv}{dx}[\mathrm{C}] \tag{1-2}$$

3. The Einstein relationship:

$$D = \frac{kT\mu}{q} \tag{1-3}$$

4. Space-charge neutrality:

$$\sum \mathrm{C}^{+} = \sum \mathrm{A}^{-}$$

We will discuss these tools and use them to derive the Nernst, Goldman, and Donnan equilibrium equations. The parameters and variables of these equations will be defined when the equations are derived. The analogies to the tools of semiconductor physics are obvious and will not be elaborated upon.

1-1-1 Diffusion of Uncharged Particles in Aqueous Solutions

Classical membrane theory is based on two processes: diffusion of particles caused by concentration differences and drift of ions caused by potential differences. This section discusses the pure diffusion of uncharged particles (e.g., perfume in air or glucose in water). In it, we mathematically analyze the diffusion of the sugar glucose, a nonionic substance, dissolved in pure water.

Motion of the water molecules is imparted to the glucose molecules by collisions. The glucose molecules begin to move freely in the aqueous medium. If the body of water is stationary, the individual water molecules have no preferred direction of motion, and their translational velocities are distributed equally in all directions. The velocities that the water molecules impart to the glucose molecules are also distributed equally in all directions. Therefore, the glucose molecules must move randomly in the aqueous medium: their velocities must be uniformly distributed in all directions. If the concentration of glucose molecules was not uniform, the

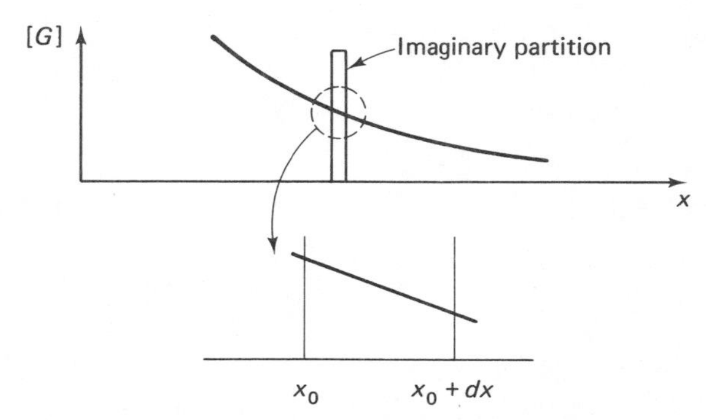

Figure 1-1 If the molecular concentration varies as a function of x, a diffusion force will result which will cause molecules to move from the region of high concentration to the region of lower concentration.

random motion of these molecules would result in a diffusion of glucose from regions of higher concentration to regions of lower concentration.

To illustrate this, consider the situation shown in Fig. 1-1, where the glucose concentration, [G], varies along the x coordinate. Imagine a very thin partition in the solution (whose thickness is small compared to the mean free path of glucose molecules), which is perpendicular to the x coordinate. At any instant of time, a certain number of glucose molecules will be situated at the left-hand edge (x_0) of the partition. This number will be proportional to the glucose concentration at x_0. Because of their uniform velocity distribution, approximately half of these molecules will have a velocity component to the right, into the partition. The flow of glucose molecules, therefore, will be proportional to their concentration at x_0. For electric circuits we calculate electric current, or the rate of flow of electrons, with units of coulombs per second. For diffusion of particles we want a similar rate of flow measure that incorporates time. For a molecule moving with constant velocity

$$\mathrm{v} = \frac{dx}{dt}$$

So, the time required for the molecule to cross the partition will be directly proportional to the partition's thickness (dx). Therefore, the rate of flow of glucose molecules across the partition from the left to the right is proportional to

$$\frac{[\mathrm{G}]_{x_0}}{dx} \tag{1-4}$$

where $[G]_{x_0}$ is the concentration[1] of glucose at x_0. Similarly, the rate of flow in the other direction (from right to left) is proportional to

$$\frac{[G]_{x_0+dx}}{dx} \tag{1-5}$$

The *net* flow from left to right across the partition is proportional to the difference between Eqs. (1-4) and (1-5), which is, in the limit as dx approaches zero, the derivative of the concentration.

$$\frac{[G]_{x_0}}{dx} - \frac{[G]_{x_0+dx}}{dx} = \frac{-d[G]}{dx} \tag{1-6}$$

Thus, because all glucose molecules are moving randomly and because there are more molecules on the left-hand side than there are on the right, there is a net diffusion of glucose from left to right across the imaginary partition. The rate of flow is directly proportional to the slope (gradient) of the concentration.

Thus the flow of molecules from left to right across the membrane becomes

$$J_G(x) = -D_G \frac{d[G]}{dx} \tag{1-7}$$

Note that J is the rate of flow of particles, not electric current density.

The diffusion flux (glucose molecules/cm^2/sec) is $J_G(x)$, and D_G is a constant of proportionality called the diffusivity (cm^2/sec). The diffusivity is a function of position, concentration, solute type, temperature, molecular weight, molecular shape and size of the hydrated ion. We will neglect variations in diffusivity and treat it as a constant for each particle studied. Equation (1-7) can be generalized to three dimensions.

Equation (1-7) is *Fick's first law*. It is generally considered an empirical rather than an axiomatic law. It states that diffusion takes place *down* the concentration gradient and is everywhere directly proportional to the magnitude of that gradient.

[1] Theoretically, activity rather than concentration should be used, but in most physiological investigations concentrations may be used without undue error.

An explanation of the difference in these two terms can be illustrated with the following analogy. Suppose that you wish to walk across a campus mall filled with 1000 students playing football, soccer, and frisbee. It would be difficult to do this quickly without getting trampled. However, if 990 of the students were chained to the benches and trees, movement through the mall would be facilitated. The concentration of the students is the same in each case, but their activity is different.

1-1-2 Drift of Charged Particles in Aqueous Solutions

Charged particles in a solution will experience an additional force: the force resulting from the interaction of their charges and any electrical fields present in the solution. When sodium chloride is dissolved in water, the salt becomes dissociated into positively charged sodium ions (Na^+) and negatively charged chloride ions (Cl^-). In the presence of an electric field, **E**, the force on an individual sodium ion is

$$\mathbf{F} = q\mathbf{E} \tag{1-8}$$

and the force on an individual chloride ion is in the opposite direction,

$$\mathbf{F} = -q\mathbf{E} \tag{1-9}$$

Where q is the magnitude of charge on each individual ion which is equal to the magnitude of the charge on an electron.

$$|q| = 1.60186 \times 10^{-19} \text{ coulomb}$$

According to Eqs. (1-8) and (1-9), either type of ion would be accelerated indefinitely in an electric field with

$$\mathbf{a} = \frac{\mathbf{F}}{M}$$

where M is the mass of the ion. However, this infinite acceleration does not occur because the ion soon collides with a water molecule and (ideally) loses all its kinetic energy. Hence the ion drifts through the solvent with a uniform drift velocity. If the electric field is small enough, the average drift velocity, $\mathbf{v}_d$, is proportional to the electrical field:

$$\mathbf{v}_d \propto \mu\mathbf{E}$$

where μ is the constant of proportionality. This is consistent with *Stokes' law*, which states that in a viscous medium a steady force acting on a particle produces a constant velocity proportional to this force. The collisions of the ions in the solvent act as a drag on their motion.

To further explore this proportionality of velocity and electric field, let us consider an isolated sodium ion in water at thermal equilibrium. We imagine the idealized trajectory of the ion in the solvent as a series of straight lines between collision points, as shown in Fig. 1-2. These lines have lengths that are directly proportional to the times between collisions, $t_1, t_2, \ldots, t_n$. The average is the mean free time, $\bar{t}$. The collisions are

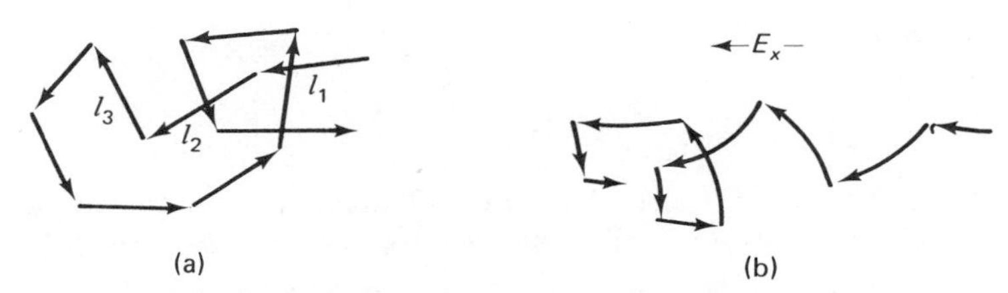

Figure 1-2 Trajectory of a positively charged ion (a) in thermal equilibrium and (b) in the presence of an electric field.

entirely random in nature; that is, after many collisions, the expected displacement of a given particle will be nearly zero. Between collisions, the ion behaves as a free particle moving with constant velocity. The exact velocity over any given path depends on the momentum and energy transferred from it to the solvent during the last collision. These collisions are such that the ion has an average drift velocity along each path.

Under the influence of an applied electric field, $\mathbf{E}_x$, the paths become parabolic. A particular ion is considered to start each path with the velocity it would have had in the absence of the applied field. During free flights, the ion gains momentum and energy from the electric field, and then transfers them to the solvent at the end of its free flights. Thus energy is transferred from the electric field to the solvent (which we interpret externally as joule heating), and the ion undergoes a net displacement (see Fig. 1-2b).

Under normal conditions, the added displacement caused by the electric field, in a free time t_i, is small compared to the corresponding free path length l_i. And therefore, the electric field does not disturb the basic thermal motion. The motion in the presence of the field consists of a slow drift of the equilibrium position, around which the ion moves rapidly and randomly. Once again we conclude that the average drift velocity is proportional to the electric field.

$$\mathbf{v}_d \propto \mu \mathbf{E}$$

We now wish to formulate an equation for the motion of sodium ions dissolved in water and maintained in an electric field. First, this flow of ions is proportional to the drift velocity, $\mathbf{v}_d$.

$$\mathbf{J}_{\text{Na}} \propto \mathbf{v}_d \propto \mu_{\text{Na}} \mathbf{E}$$

Next we can see that the flow depends upon the concentration of sodium ions. The more there are, the more will cross the imaginary barrier in each instant of time.

$$\mathbf{J}_{\mathrm{Na}} \propto \mu_{\mathrm{Na}}[\mathrm{Na}^+]\mathbf{E}$$

And finally, because $\mathbf{F} = q\mathbf{E}$, the flow will depend upon the number of electronic charges carried by the ion, +1 for sodium. We call this quantity z.

$$\mathbf{J}_{\mathrm{Na}} = \mu_{\mathrm{Na}}[\mathrm{Na}^+]z\mathbf{E}$$

The electric field is the gradient of the voltage potential. So for our problems,

$$E = \frac{-dv}{dx}$$

and our statement of *Ohm's law* becomes

$$J_{\mathrm{Na}} = -\mu_{\mathrm{Na}} z[\mathrm{Na}^+]\frac{dv}{dx} \tag{1-10}$$

The mobility, μ_{Na}, will be treated as a constant, although it actually depends upon concentration, temperature, electric field, electronic charge, and physical properties of both the water and the sodium ion. Ohm's law written for a chlorine ion is

$$J_{\mathrm{Cl}} = -\mu_{\mathrm{Cl}}[\mathrm{Cl}^-]z\frac{dv}{dx}$$

In this case z equals -1. Therefore,

$$J_{\mathrm{Cl}} = +\mu_{\mathrm{Cl}}[\mathrm{Cl}^-]\frac{dv}{dx}$$

Note that J_{Cl} is not current due to chlorine flow, but rather is the flow of chlorine ions. These equations are statements of Ohm's law. Although this law can be derived from theoretical considerations, it, too, is generally considered an empirical law rather than an axiomatic law. It states that drift of *positively* charged particles takes place *down* the electric potential gradient and is everywhere directly proportional to the magnitude of that gradient. And it states that *negatively* charged particles drift *up* the potential gradient in proportion to the gradient's magnitude.

The magnitude of z becomes important for divalent ions, such as calcium (Ca^{2+}) and sulfate (SO_4^{2-}). In general, if an ion has valence z, where the magnitude of z is the number of electronic charges carried by the ion, and the sign of z is the polarity of the ion, then Ohm's law can be written

$$J_C = -z\mu_C[C]\frac{dv}{dx} \tag{1-11}$$

where C is the ion species under consideration.

1-1-3 The Einstein Relationship

The mobility, μ, and the diffusion constant, D, both depend upon the ability of the ion to move through the solution. It is therefore not surprising to find that a relationship exists between D and μ. This was first recognized by Einstein in 1908. He generalized the concept of mobility to include drift in an electric field, as we have studied here, and also to movement in other force fields, such as those due to osmotic pressure and gravity.

The relation between the diffusion and drift processes is called the Einstein relationship. Our discussion of diffusion and drift is similar to that encountered in semiconductor physics. So it is fitting that we use the following derivation for the Einstein relationship from Gibbons (1966, p. 120). An alternative derivation using Stokes' law and viscous drag is given by Kittel (1958, pp. 154–56); and the original derivation using osmotic pressure and diffusion is given by Einstein (1956, p. 75).

The equation connecting D and μ, called the *Einstein relationship*, is

$$\frac{D}{\mu} = \frac{kT}{q}$$

where D is the diffusion constant, μ is the mobility, k is Boltzmann's constant, T is the absolute temperature, and q is the magnitude of the electrical charge. To see how this relationship is obtained, we use the expressions for D and μ for holes and electrons moving in a semiconductor lattice:

$$D = \frac{\bar{l}^2}{\bar{t}}$$

where $\bar{l}$ is the mean free path of the particle and $\bar{t}$ is the mean free time between collisions.

$$\mu = \frac{q\bar{t}}{m}$$

where q is the electronic charge, $\bar{t}$ is once again the mean free time, and m is the effective mass of the particle. Now we can form the ratio of D to μ.

$$\frac{D}{\mu} = \frac{\bar{l}^2}{\bar{t}} \frac{m}{q\bar{t}} = \left(m \frac{\bar{l}^2}{\bar{t}^2} \right) \frac{1}{q} \tag{1-12}$$

The ratio of $\bar{l}$ to $\bar{t}$ is a velocity. This velocity is the thermal velocity: the velocity the free electrons have because they absorb heat from their environment. We call this velocity $\mathbf{v}_T$. Using this in Eq. (1-12), we obtain

$$\mathbf{v}_T = \frac{\bar{l}}{\bar{t}}$$

$$\frac{D}{\mu} = m\mathbf{v}_T^2 \frac{1}{q}$$

Now $m\mathbf{v}_T^2/2$ is the energy the electron has as a consequence of being in thermal equilibrium with the lattice, which is at temperature T. From the principle of equipartition of energy, in the one dimensional case,

$$\frac{m\mathbf{v}_T^2}{2} = \frac{kT}{2}$$

Therefore, we have

$$\frac{D}{\mu} = \frac{kT}{q} \tag{1-13}$$

which is the Einstein relationship. Although we have only shown the relation to be valid for an electron moving in a semiconductor lattice in a one-dimensional example, it is also true for ions in solution in three dimensions.

Basically, the Einstein relationship states that the resistance to flow of ions will be the same whether the driving force is drift or diffusion.[1]

1-2 EQUILIBRIUM IN A ONE-ION SYSTEM

The simplest system for studying ionic flow would be a system where only one ion can flow. Such a system is shown in Fig. 1-3, where only potassium can pass through the membrane. For equilibrium there must be no net

[1] Some authors (e.g., MacGregor and Lewis 1977, and Plonsey 1969) use an alternative statement for the Einstein relationship, namely, $D = kT\mu$, and consequently use zq in place of z in their statement of Ohm's law, Eq. (1-10).

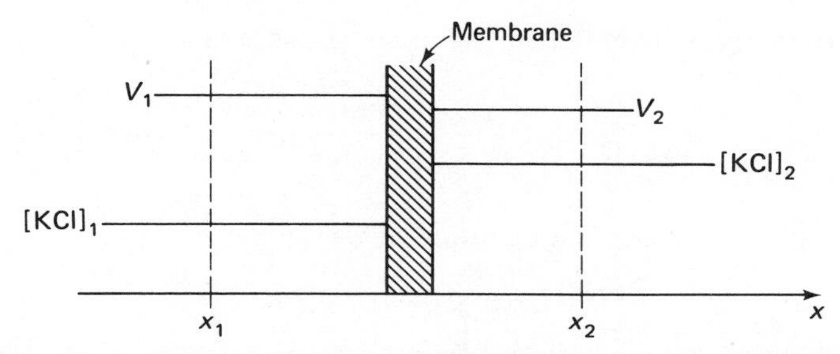

Figure 1-3 Tank divided by a membrane that is permeable only to potassium ions and water. The concentration of KCl is different on each side. This will produce a voltage potential across the membrane.

flow of either ion. Chlorine ions cannot flow under any circumstances, so we only have to consider potassium ions. The potassium flow has two components: drift and diffusion. The sum of the two must be zero. From Fick's law, Eq. (1-7), the diffusion flow from left to right (which is in the positive x direction) is

$$J_{\mathrm{K}}(\text{diffusion}) = -D_{\mathrm{K}} \frac{d[\mathrm{K}^+]}{dx}$$

This movement of charged particles will set up a voltage potential. And then, by Ohm's law, Eq. (1-10),

$$J_{\mathrm{K}}(\text{drift}) = -\mu_{\mathrm{K}} z \frac{dv}{dx} [\mathrm{K}^+]$$

The total potassium flow is

$$J_{\mathrm{K}}(\text{total}) = -\mu_{\mathrm{K}} z \frac{dv}{dx} [\mathrm{K}^+] - D_{\mathrm{K}} \frac{d[\mathrm{K}^+]}{dx}$$

From the Einstein relation, Eq. (1-13),

$$D_{\mathrm{K}} = \frac{kT}{q} \mu_{\mathrm{K}}$$

Thus, for equilibrium,

$$J_{\mathrm{K}}(\text{total}) = 0 = -\mu_{\mathrm{K}} z \frac{dv}{dx} [\mathrm{K}^+] - \frac{kT}{q} \mu_{\mathrm{K}} \frac{d[\mathrm{K}^+]}{dx}$$

$$-\mu_{\mathrm{K}} z \frac{dv}{dx} [\mathrm{K}^+] = \frac{kT}{q} \mu_{\mathrm{K}} \frac{d[\mathrm{K}^+]}{dx}$$

$$\frac{dv}{dx} = \frac{-kT}{zq} \frac{1}{[\mathrm{K}^+]} \frac{d[\mathrm{K}^+]}{dx}$$

Now if we integrate from left to right across the membrane, we obtain

$$\int_{x_1}^{x_2}\left(\frac{dv}{dx}\right)dx=\frac{-kT}{zq}\int_{x_1}^{x_2}\left(\frac{1}{[\mathrm{K}^+]}\frac{d[\mathrm{K}^+]}{dx}\right)dx$$

Changing the variables of integration yields

$$\int_{v_1}^{v_2}dv=\frac{-kT}{zq}\int_{[\mathrm{K}^+]_1}^{[\mathrm{K}^+]_2}\frac{d[\mathrm{K}^+]}{[\mathrm{K}^+]}$$

$$v_2-v_1=\frac{-kT}{zq}\{\ln[\mathrm{K}^+]_2-\ln[\mathrm{K}^+]_1\}$$

$$=\frac{-kT}{zq}\ln\frac{[\mathrm{K}^+]_2}{[\mathrm{K}^+]_1} \qquad (1\text{-}14)$$

This equation, called the *Nernst equation*, expresses the voltage difference required to balance potassium diffusion with equal but opposite potassium drift, thus making the net flow of potassium zero. The equation sometimes appears in the form

$$v_2-v_1=\frac{-RT}{zF}\ln\frac{[\mathrm{K}^+]_2}{[\mathrm{K}^+]_1}$$

where F is 1 Faraday (the charge in 1 mole of protons) and R is the universal gas constant.

$$R=kA$$
$$F=qA$$

where A is Avogadro's number.

$$\frac{kT}{q}=\frac{RT}{F}$$

At room temperature, 27°C, kT/q is approximately 26 mV.

EXAMPLE

A given membrane has small pores of such a size that chlorine ions can pass through but sodium ions cannot. A 100-millimole (mmol) per liter solution of NaCl is put on the right side and a 50-mmol/liter solution of NaCl is put on the left side. However, very few ions actually move before a potential difference is set up that prevents further diffusion. The final concentrations are approximately the same as the original concentrations.

Use the Nernst equation, Eq. (1-14), to find the potential difference.

$$E_{12} = \frac{RT}{-F} \ln \frac{[C_2]}{[C_1]} = -26 \times 10^{-3} \ln 2 = -18\,\text{mV}$$

Chlorine has an atomic weight of 35 and sodium has an atomic weight of 23; therefore, chlorine should be the larger atom. How can the membrane be permeable to chlorine ions but not to sodium ions?

Because the sodium ion is smaller, it exerts a greater electrostatic force on the surrounding polar water molecules. Thus it holds more water molecules in its shell, allowing the diameter of the water shell around the sodium ion to be greater than the diameter of the water shell around the chlorine ion. Therefore, it is possible for membranes to have pore sizes that will allow the hydrated chlorine ions to pass, but not the larger hydrated sodium ions (Katz 1966, pp. 50–51; Junge 1976, p. 44).

1-3 DONNAN EQUILIBRIUM

We have just found the voltage across a membrane if it is permeable to only one ion. Now let us complicate our membrane by making it permeable to K^+ and Cl^-, but not to a large organic ion called R^+ (Fig. 1-4).

For equilibrium,

$$J_K(\text{total}) = 0 = -\mu_K z \frac{dv}{dx}[K^+] - D_K \frac{d[K]}{dx} \tag{1-15}$$

$$J_{Cl}(\text{total}) = 0 = -\mu_{Cl} z \frac{dv}{dx}[Cl^-] - D_{Cl} \frac{d[Cl^-]}{dx} \tag{1-16}$$

$$J_{R^+}(\text{total}) = 0 \tag{1-17}$$

From Eq. (1-15),

$$v_2 - v_1 = \frac{-kT}{qz} \ln \frac{[K^+]_2}{[K^+]_1}$$

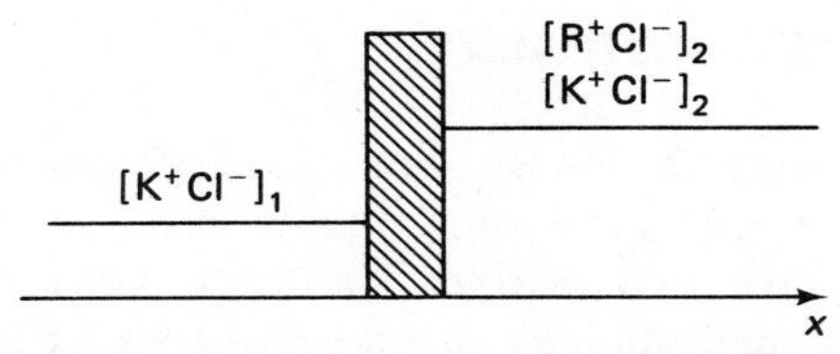

Figure 1-4 Tank divided by a membrane that is permeable to potassium and chlorine ions but not to a large organic ion called R^+. The concentration of KCl is different on each side. This situation will produce a voltage difference across the membrane.

and $z = +1$ for $[K^+]$, so

$$v_2 - v_1 = -\frac{kT}{q}\ln\frac{[K^+]_2}{[K^+]_1}$$

Now from Eq. (1-16),

$$v_2 - v_1 = \frac{-kT}{qz}\ln\frac{[Cl^-]_2}{[Cl^-]_1}$$

and $z = -1$ for $[Cl^-]$, so

$$v_2 - v_1 = +\frac{kT}{q}\ln\frac{[Cl^-]_2}{[Cl^-]_1} = \frac{-kT}{q}\ln\frac{[Cl^-]_1}{[Cl^-]_2}$$

We cannot tell by inspection whether $v_2 - v_1$ will be positive or negative, because we do not yet know the final concentrations of K^+, Cl^-, and R^+. But we do know that

$$v_2 - v_1\big|_{\text{from } K^+ \text{ equations}} = v_2 - v_1\big|_{\text{from } Cl^- \text{ equations}}$$

Therefore,

$$\frac{[K^+]_2}{[K^+]_1} = \frac{[Cl^-]_1}{[Cl^-]_2}$$

If another ion, such as the divalent calcium ion, were present, the equilibrium equation would become

$$\frac{[K^+]_2}{[K^+]_1} = \frac{[Cl^-]_1}{[Cl^-]_2} = \sqrt{\frac{[Ca^{2+}]_2}{[Ca^{2+}]_1}} \tag{1-18}$$

This is called the *Donnan equilibrium*, and it is a true equilibrium.

1-4 SPACE-CHARGE NEUTRALITY

The last of our four basic biophysical tools is *space-charge neutrality*, which says that the number of positive ionic charges in a given volume is equal to the number of negative ionic charges. This approximation is usually quite accurate, because a small number of uncompensated charges will produce very large potential differences.

For example, let us calculate how many ions would have to be removed from a 1-cm^3 cell in order to develop a potential difference of 100 mV across the cell membrane (based on MacGregor and Lewis, 1977).

The capacitance of cell membranes is about 1 $\mu F/cm^2$. Charge equals voltage times capacitance. Therefore, to create a potential difference of 100 mV, 10^{-7} coulomb (C) of charge must be separated by every square centimeter of membrane.

Each singly charged ion (K^+, Cl^-, Na^+) carries 1.6×10^{-19} C (the charge on an electron). Therefore, 10^{-7} C is equivalent to approximately 0.6×10^{12} singly charged ions. Thus 0.6×10^{12} singly charged ions must be uncompensated on each side of every square centimeter of membrane in order to provide 100 mV. The surface area of a 1-cm cube is 6 cm^2; therefore, 3.6×10^{12} ions must be uncompensated.

Considering only potassium, typical cytoplasm is about 0.3 molar (0.3 × Avogadro's number of ions per liter of cytoplasm). Therefore, there are $0.3 \times (6 \times 10^{23})$ ions per liter, which is equivalent to 1.8×10^{20} ions per cubic centimeter of cell volume. Thus 1.8×10^{20} ions are contained in the volume while only 3.6×10^{12} ions must be uncompensated to provide 100 mV. If only one ion in approximately 10^7 is uncompensated, sufficient charge will be present to create a 100-mV potential across this cell's membrane.

For a very small cell (with a radius of 10^{-5} cm), the volume is

$$\tfrac{4}{3}\pi r^3 = 4 \times 10^{-15}\ \text{cm}^3$$

And the surface area is approximately

$$4\pi r^2 = 12 \times 10^{-10}\ \text{cm}^2$$

Such a cell, or equivalently a long thin nerve axon, has the lowest volume/surface area ratio and sets the upper limit on the proportion of ions that must be uncompensated. The cell contains at least 7.2×10^5 potassium ions, of which only 7.2×10^2 would have to move across the membrane unaccompanied by negative ions in order to develop 100 mV. Thus only one ion in 1000 must be uncompensated. Even in the smallest cell, more than 99.9% of all ions are escorted by ions of the opposite charge.

Therefore, while counting ions, we can assume space-charge neutrality, which states that the summation of positive charges, called *cations*, equals the summation of negative charges, called *anions*, on either side of a membrane.

$$\Sigma C^+ = \Sigma A^- \tag{1-19}$$

The typical membrane capacitance of 1 $\mu F/cm^2$ is large compared to capacitors used in electronic equipment. The large value is made possible by the thinness of the biological membrane, about 10 nm. Lining one side of such a membrane with negative charges and the other side with positive charges will produce a very large electrical field. A voltage as small as 100 mV across such a membrane will produce an electric field of 10^7 V/m! (This is within one order of magnitude of the voltage needed to produce breakdown in Zener diodes.)

EXAMPLE

A membrane, shown in Fig. 1-5, is permeable to H_2O, K^+, and Cl^- but not to a large organic ion, R^+. Find V_m and give the polarity. Will there be any osmotic pressure? If so, in which direction?

By conservation of mass,

$$(1)\ [K_a] + [K_b] = 500$$
$$(2)\ [Cl_a] + [Cl_b] = 600$$

By space-charge neutrality,

$$(3)\ [K_a] + 100 = Cl_a$$

For equilibrium of drift and diffusion (Donnan equilibrium),

$$(4)\ \frac{[K_a]}{[K_b]} = \frac{[Cl_b]}{[Cl_a]}$$

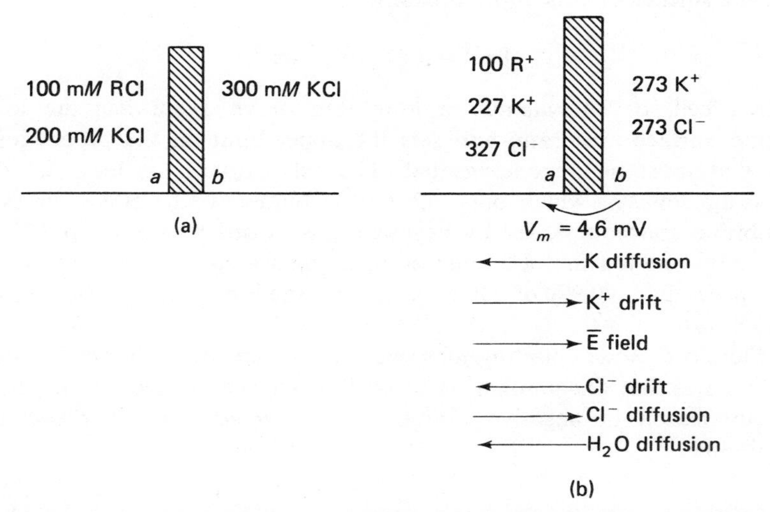

Figure 1-5 The solution is divided by a membrane that is permeable to potassium and chlorine but not to the large inorganic ion R^+. Initial ionic concentrations are indicated in (a) and final ionic concentrations are indicated in (b). A nonzero equilibrium membrane potential will exist. The ionic flows caused by drift and diffusion forces are shown.

Put (1) and (2) into (4):

$$(5)\quad \frac{[K_a]}{500-[K_a]}=\frac{600-[Cl_a]}{[Cl_a]}$$

Put (3) into (5):

$$\frac{[K_a]}{500-[K_a]}=\frac{600-[K_a]-100}{[K_a]+100}$$

Therefore,

$$[K_a]=227$$
$$[K_b]=273$$
$$[Cl_a]=327$$
$$[Cl_b]=273$$
$$V_m=25\text{ mV}\ln\frac{273}{227}=(25\text{ mV})\,(0.184)$$
$$V_{ab}=V_m=4.6\text{ mV}$$

or

$$V_{ba}=-4.6\text{ mV}$$

Water molecules will diffuse from the right side to the left side, creating an osmotic pressure difference.

1-5 VOLTAGE ACROSS A MEMBRANE WITH NONZERO PERMEABILITY FOR ALL IONS

In our previous examples we could only have an equilibrium voltage across a membrane if it were impermeable to at least one ion present in the solution. However, it is possible to have a voltage across a membrane with nonzero permeabilities for all ions. Consider the situation shown in Fig. 1-6.

Because $\mu_K > \mu_{Cl}$, let us suppose that potassium ions cross the membrane first. The initial movement of K^+ will set up an E field that will aid Cl^- drift. The equilibrium voltage is obviously zero, but if the tanks are large and $\mu_K \neq \mu_{Cl}$, quasi-static conditions can be reached when $J_K = J_{Cl}$.

Find V for these quasi-static conditions. By the Einstein relationship,

$$D_K=\frac{kT\mu_K}{q}$$

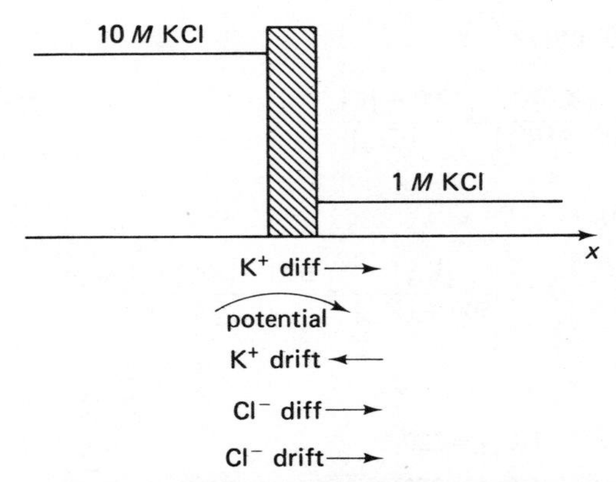

Figure 1-6 Tank divided by a membrane that is permeable to potassium and chlorine, with $\mu_K > \mu_{Cl}$. Concentrations are initially different. This will produce a voltage across the membrane that will approach zero as time approaches infinity.

Using Fick's and Ohm's laws yields

$$J_K = \mu_K z \frac{dv}{dx}[K^+] - \frac{kT\mu}{q}\frac{d[K^+]}{dx}$$

$$J_{Cl} = -\mu_{Cl} z \frac{dv}{dx}[Cl^-] - \frac{kT\mu_{Cl}}{q}\frac{d[Cl^-]}{dx}$$

and

$$z_K = +1$$
$$z_{Cl} = -1$$

Space-charge neutrality requires that after the initial transient,

$$[K^+] = [Cl^-] = [KCl]$$

We can take the derivative of this equation with respect to time to obtain

$$\frac{d[K^+]}{dt} = \frac{d[Cl^-]}{dt}$$

which is

$$J_K = J_{Cl}$$

Therefore,

$$-(\mu_K + \mu_{Cl})[\text{KCl}] = \frac{dv}{dx}\frac{-kT}{q}\frac{d[\text{KCl}]}{dx}(\mu_{Cl} - \mu_K)$$

$$\frac{dv}{dx} = \frac{kT}{q}\frac{\mu_{Cl} - \mu_K}{\mu_{Cl} + \mu_k}\frac{1}{[\text{KCl}]}\frac{d[\text{kCl}]}{dx}$$

Integrating left to right,

$$V_R - V_L = \frac{kT}{q}\frac{\mu_{Cl} - \mu_K}{\mu_{Cl} + \mu_K}\ln\frac{[\text{KCl}]_R}{[\text{KCl}]_L} \tag{1-20}$$

Note that if $\mu_{Cl} = \mu_K$, the voltage is zero. This results from the fact that if the mobilities are equal, each potassium ion will be accompanied across the membrane by a chlorine ion, and no charge imbalance will result.

This potential, of course, approaches zero as time approaches infinity: the concentration of each ion will become the same on each side of the membrane. To have a permanent potential across a membrane, the membrane must be either impermeable to at least one ion species, or else active ion pumps must be present.

1-6 THE GOLDMAN EQUATION

An old and famous equation in the study of biological membranes is the *Goldman equation*. It is sometimes called the constant-field equation, because this was one of the most important assumptions used in its derivation.

In deriving the equation for the drift motion of an ion in an electric field, we assumed a *constant* electric field (see Fig. 1-2). A biological membrane can be modeled as having a constant electrical field within it. This model is accurate if the permeable ions are univalent, if the total ionic concentrations on each side of the membrane are equal, and if the membrane is thin (Goldman, 1943). Therefore, by assuming that the electric field in the membrane is constant, we can use our previously derived relationships for movement of ions to derive the Goldman equation. The flow of potassium ions caused by diffusion and drift is

$$J_K = -\mu_K z\frac{dv}{dx}[\text{K}^+] - \frac{kT}{q}\mu_K\frac{d[K^+]}{dx}$$

Now assume a constant electric field in the membrane,

$$E = \frac{-dv}{dx} = \frac{V}{w}$$

where w is the width of the membrane. Let $z = 1$:

$$J_{\mathrm{K}} = -\mu_{\mathrm{K}}\frac{V}{w}[\mathrm{K}^+] - \frac{kT}{q}\mu_{\mathrm{K}}\frac{d[\mathrm{K}^+]}{dx} \tag{1-21}$$

Define a new "constant," P_K, the permeability coefficient, as

$$P_K = \frac{\mu_{\mathrm{K}}}{w}\frac{kT}{q} = \frac{D_{\mathrm{K}}}{w}$$

Solving for μ_{K} yields

$$\mu_{\mathrm{K}} = \frac{P_{\mathrm{K}}wq}{kT}$$

Substituting this into Eq. (1-21) yields

$$J_{\mathrm{K}} = -P_K\frac{q}{kT}V[\mathrm{K}^+] - P_K w\frac{d[\mathrm{K}^+]}{dx}$$

Solving for the concentration gradient yields

$$\frac{d[\mathrm{K}^+]}{dx} = -\frac{J_{\mathrm{K}}}{P_K w} - \frac{q}{kT}\frac{V}{w}[\mathrm{K}^+]$$

$$dx = \frac{d[\mathrm{K}^+]}{\dfrac{J_{\mathrm{K}}}{P_K w} - \dfrac{q}{kT}\dfrac{V}{w}[\mathrm{K}^+]} \tag{1-22}$$

In our quasi-static conditions J_{K} is independent of x, so we can use mathematical tables or Laplace transforms, or invoke the "it is intuitively obvious to the casual observer" technique to find a solution for differential equation (1-22). My math tables state that

$$\int \frac{dx}{a + bx} = \frac{1}{b}\ln(a + bx)$$

Therefore, the integration from the left edge of the membrane to the right edge will give us

$$x\Big|_0^w = \frac{-kT}{q}\frac{w}{V}\ln\left[-\frac{J_K}{P_K w} - \frac{q}{kT}\frac{V}{w}[K^+]\right]_{[K^+]_L}^{[K^+]_R}$$

Substituting the upper and lower limits yields

$$-\frac{qV}{kT} = \ln\frac{\dfrac{-J_K}{P_K w} - \dfrac{q}{kT}\dfrac{V}{w}[K^+]_R}{\dfrac{-J_K}{P_K w} - \dfrac{q}{kT}\dfrac{V}{w}[K^+]_L}$$

Using the definition of a logarithm, namely, if $a = \ln b$, then $e^a = b$, we obtain

$$e^{-qV/kT}\left(\frac{J_K}{P_K w} + \frac{q}{kT}\frac{V}{w}[K^+]_L\right) = \frac{J_K}{P_K w} + \frac{q}{kT}\frac{V}{w}[K^+]_R$$

Solving this equation for J_K, we arrive at

$$J_K = \left(\frac{qVP_K}{kT}\right)\frac{[K^+]_R - [K^+]_L e^{-qV/kT}}{e^{-qV/kT} - 1} \tag{1-23}$$

Repeating this procedure for chlorine yields

$$x\Big|_o^w = \frac{kT}{q}\frac{w}{V}\ln\left[\frac{-J_{Cl}}{P_{Cl} w} + \frac{q}{kT}\frac{V}{w}[Cl^-]\right]_{[Cl^-]_L}^{[Cl^-]_R}$$

Using the definition of a logarithm, we obtain

$$e^{+qV/kT}\left(\frac{-J_{Cl}}{P_{Cl} w} + \frac{q}{kT}\frac{V}{w}[Cl^-]_L\right) = \frac{-J_{Cl}}{P_{Cl} w} + \frac{q}{kT}\frac{V}{w}[Cl^-]_R$$

Now multiply both sides by $e^{-qV/kT}$ and solve for J_{Cl}.

$$J_{Cl} = \left(\frac{qP_{Cl}V}{kT}\right)\frac{-[Cl^-]_L + [Cl^-]_R e^{-qV/kT}}{e^{-qV/kT} - 1} \tag{1-24}$$

By space-charge neutrality for our quasi-static conditions,

$$J_K = J_{Cl}$$

Using Eqs. (1-23) and (1-24) yields

$$P_K([\mathrm{K}^+]_R - [\mathrm{K}^+]_L e^{-qV/kT}) = P_{\mathrm{Cl}}(-[\mathrm{Cl}^-]_L + [\mathrm{Cl}^-]_R e^{-qV/kT}) \tag{1-25}$$

Finally, solving this equation for V yields

$$V = \frac{-kT}{q} \ln \frac{P_K[\mathrm{K}^+]_R + P_{\mathrm{Cl}}[\mathrm{Cl}^-]_L}{P_K[\mathrm{K}^+]_L + P_{\mathrm{Cl}}[\mathrm{Cl}^-]_R} \tag{1-26}$$

When this equation is used, it normally includes the contribution of sodium current, and the subscripts o and i are used to denote concentrations outside and inside the cell, respectively. The equation becomes

$$V = \frac{kT}{q} \ln \frac{P_K[\mathrm{K}^+]_o + P_{\mathrm{Na}}[\mathrm{Na}^+]_o + P_{\mathrm{Cl}}[\mathrm{Cl}^-]_i}{P_K[\mathrm{K}^+]_i + P_{\mathrm{Na}}[\mathrm{Na}^+]_i + P_{\mathrm{Cl}}[\mathrm{Cl}^-]_o} \tag{1-27}$$

This is the Goldman equation. This equation is used extensively by electrophysiologists to explain variations in membrane-permeability changes. It is also used to determine relative permeabilities of ions.

1-7 ION PUMPS

There are active transport mechanisms in cell membranes that move ions from one side of the membrane to the other (Katz 1966, p. 67: Selkurt 1976, pp. 10–13, 28–31; Junge 1976, pp. 113–23; Erlij and Grinstein 1976). Their energy source is adenosine triphosphate (ATP). They can be blocked with the chemical ouabain and slowed by low temperatures. The reaction to low temperatures is due to a decreased utilization of ATP. This effect changes the resting membrane potential about 2 mV/°C. The ion pumps may be electrogenic, where there is a net transfer of charge, or nonelectrogenic, where no charge is transferred (e.g., a one-for-one exchange of internal sodium ions for external potassium ions).

Let us calculate the steady-state concentration difference created by the electrogenic pump of Fig. 1-7. Let the membrane be permeable to both K^+ and Cl^- and allow the pump to supply a constant flow (J_p) of potassium. Find the equilibrium concentration differences created by the pump.

For equilibrium, $J_{\mathrm{K}} = J_{\mathrm{Cl}} = 0$.

$$J_{\mathrm{K}} = J_{\mathrm{P}} - \mu_{\mathrm{K}} z[\mathrm{K}^+]\frac{dv}{dx} - D_{\mathrm{K}}\frac{d[\mathrm{K}^+]}{dx} = 0 \tag{1-28}$$

$$J_{\mathrm{Cl}} = -\mu_{\mathrm{Cl}} z[\mathrm{Cl}^-]\frac{dv}{dx} - D_{\mathrm{Cl}}\frac{d[\mathrm{Cl}]}{dx} = 0 \tag{1-29}$$

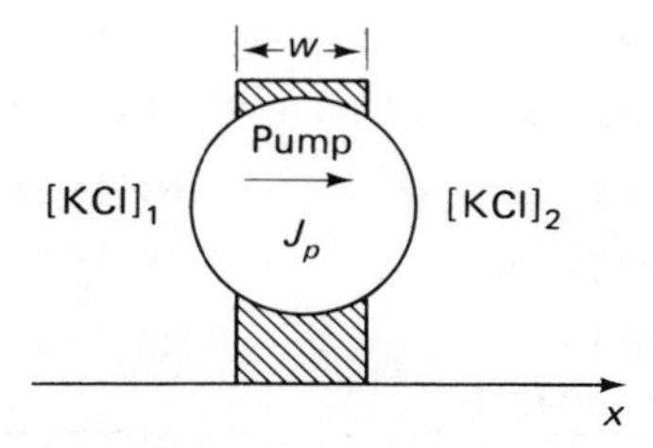

Figure 1-7 Tank divided by a membrane that is permeable to potassium and chlorine and also contains a pump that transports potassium from the left side to the right side. The action of this pump will establish different concentration on each side and will therefore produce a steady-state voltage difference.

Solving Eq. (1-29) for dv/dx, we obtain

$$\frac{dv}{dx}=\frac{kT}{q}\frac{1}{[\mathrm{Cl}^-]}\frac{d[\mathrm{Cl}^-]}{dx}$$

By space-charge neutrality, $[\mathrm{Cl}^-]=[\mathrm{K}^+]$.

$$\frac{dv}{dx}=\frac{kT}{q}\frac{1}{[\mathrm{K}^+]}\frac{d[\mathrm{K}^+]}{dx}$$

Substituting this into Eq. (1-28) yields

$$J_p=\mu_{\mathrm{K}}z[\mathrm{K}^+]\frac{kT}{q}\frac{1}{[\mathrm{K}^+]}\frac{d[\mathrm{K}^+]}{dx}+\frac{kT\mu_{\mathrm{K}}}{q}\frac{d[\mathrm{K}]}{dx}$$

$$J_p=2\frac{kT\mu_{\mathrm{K}}}{q}\frac{d[\mathrm{K}^+]}{dx}$$

Assume that J_p is independent of concentration, separate the variables, and integrate over the width of the membrane.

$$[\mathrm{K}^+]_R-[\mathrm{K}^+]_L=\frac{J_pwq}{2kT\mu_{\mathrm{K}}}$$

1-8 MEMBRANE POTENTIALS FOR BIOLOGICAL MEMBRANES

Armed with our four basic biophysical tools (Fick's law of diffusion, Ohm's law of drift, the Einstein relationship, and space-charge neutrality), we are now prepared to look at voltages across actual physiological membranes. The ionic concentrations on the two sides of two typical cell

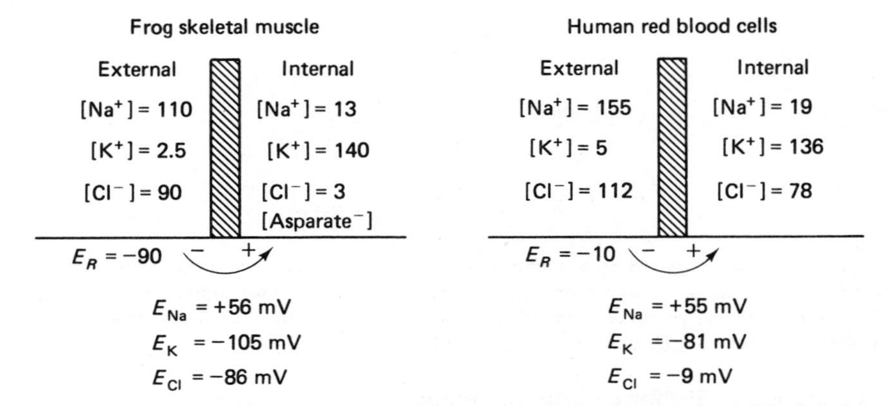

Figure 1-8 Approximate concentrations of the most important ion species and the resulting Nernst potentials for two types of cells (after Katz 1966, and Selkurt 1976). Voltage potentials of the inside with respect to the outside are given in millivolts. Concentrations are in milliequivalents per liter of water.

membranes are shown in Fig. 1-8. Membrane potential depends on the concentration ratios and the related Nernst potentials. To understand this relationship, we should examine Fig. 1-9, a model of a cell membrane patch. By writing Kirchhoff's voltage law around the loop containing the sodium elements, we obtain

$$+V_m - E_{\mathrm{Na}} + R_{\mathrm{Na}} i_{\mathrm{Na}} = 0$$

E_{Na} is the Nernst potential which is due to the concentration difference of sodium ions. So we can write

$$i_{\mathrm{Na}} = \frac{-V_m}{R_{\mathrm{Na}}} + \frac{kT}{q} \ln \frac{[\mathrm{Na}^+]_o}{[\mathrm{Na}^+]_i}$$

We can see that there is a drift component that depends upon the ionic concentration ratio. In general,

$$V_m \neq E_{\mathrm{Na}}$$

Hence the net sodium flow is not zero. The sodium–potassium pump is included in the model to remove the sodium that leaks into the cell. Therefore, the total flow of sodium should be

$$i_{\mathrm{Na}} = \frac{-V_m}{R_{\mathrm{Na}}} + \frac{E_{\mathrm{Na}}}{R_{Na}} - Jp_{\mathrm{Na}}$$

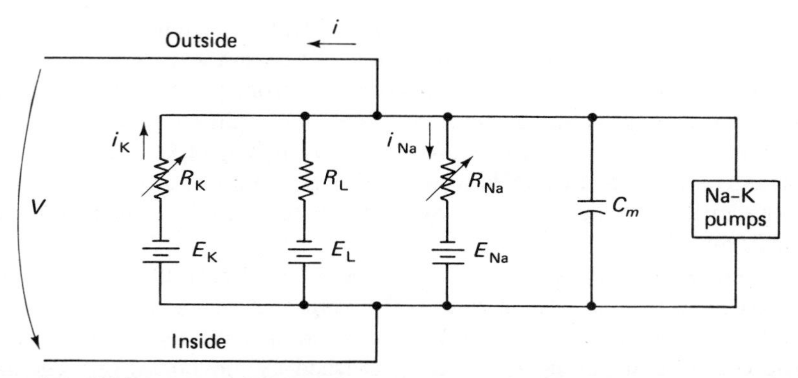

Figure 1-9 Model of a small patch of a nerve or muscle membrane. The batteries are the Nernst potentials for sodium, E_{Na}, potassium, E_K, and leakage, E_L, ions: E_K and E_L have negative values. The leakage channel is primarily that of chlorine ions. Current flowing out of the cell is defined to be positive current. The sodium and potassium pumps are very slow compared to the time course of a voltage spike.

For simplicity, we normally assume that

$$Jp_{Na} = Jp_K$$

and then we ignore the contributions that these pumps make to the membrane potential.

Normally, potassium leaks out of the membrane as shown by i_K in Fig. 1-9. It is forced out by the concentration gradient, (E_K), which is normally a negative number. And it is forced in by the membrane voltage, V_m, which is also normally negative. At the same time, sodium is leaking in, owing to both its concentration difference, (E_{Na}), and the membrane voltage. The sodium–potassium pump must counteract the leakage currents. The variable nature of each channel's resistance is explained in Katz (1966, Chap. 5), MacGregor and Lewis (1977, Chap. 7), and Keynes (1979). The elements E_L and R_L represent the leakage channel, which is due primarily to chlorine. The membrane potential is a measure of the driving force of drift: it is proportional to the electric field and to the ion flow rate caused by drift. The driving force of diffusion can be represented by the Nernst potential for the particular ion under consideration. This equation gives the ratio of the concentrations and therefore tells us how great the ion flow rate will be due to diffusion. The difference between the Nernst potential and the membrane potential is indicative of how hard the pump must work.

In cell membranes, the Nernst potential for sodium is quite different from the resting membrane potential. This means that the sodium pump must work very hard to maintain the concentration difference. The Nernst potential for chlorine is just about the same as the resting membrane potential. In this case, the Cl^- ions are in electrochemical equilibrium and no direct energy expenditure is required to maintain the steady-state concentration ratios. In other words, the inward flow of Cl^- ions down the concentration gradient is balanced by an equal but opposite outward flow of ions up the voltage gradient. We can, therefore, conclude that there is no chlorine pump.

As an example of how to use this model, let us look at the behavior of the model as adapted for a neuronal synapse. A *synapse* is the junction where one nerve cell excites another or where a nerve excites a muscle or gland cell. In some cases, this is accomplished by movement of a chemical transmitter substance from one cell to the cell membrane of the next. In others, the presynaptic and postsynaptic cells are electrically coupled. In a chemical synapse, an action potential invades the terminal of the presynaptic neuron, releasing the chemical transmitter into the space between the cells. This gap, or synaptic cleft, is about 15 nm (150 angstroms) wide. A short time after the arrival of the presynaptic impulse (about 0.5 msec in mammalian neurons, longer in invertebrates), the transmitter diffuses across the cleft and causes channels to open up, allowing sodium and potassium to flow through the membrane. This is the reason why R_{Na} and R_K are shown as variable resistors in Fig. 1-9. These resistors are composed of two parallel current pathways: one resulting from leakage through the membrane, and the second resulting from channels that are specific for sodium or potassium. These channels can be opened by certain chemicals, such as the neurotransmitter acetylcholine, ACH.

Observe how the model can be used to explain membrane potential variations. Use the model of Fig. 1-9, but ignore the leakage channel, R_L and E_L, and ignore the sodium–potassium pumps.

A representative motor endplate has

$$\text{area} = 10^{-5}\ \text{cm}^2$$

$$\text{capacitance} = 1\ \mu\text{F/cm}^2$$

$$E_K = -100\ \text{mV}$$

$$E_{Na} = 60\ \text{mV}$$

$$R_{K_{\text{leakage}}} = 10^8\ \Omega$$

$$R_{Na_{\text{leakage}}} = 1.5 \times 10^9\ \Omega$$

$$R_{K_{ACH}} = 10^5\ \Omega$$

$$R_{Na_{ACH}} = 10^5\ \Omega$$

Find $V_{m_{rest}}$ and find $V_{m_{reversal}}$ (i.e., the voltage with the ACH gates open). In steady state $i_K = i_{Na}$, by Kirchhoff's voltage law.

$$i_K = i_{Na} = \frac{-E_K + E_{Na}}{R_K + R_{Na}} = \frac{160 \text{ mV}}{1.6 \text{ G}\Omega} = 10^{-10} \text{ A}$$

$$V_{m_{rest}} - E_K - i_K R_K = 0$$

$$V_{m_{rest}} = (10^{-10})(10^8) - 100 \text{ mV} = -90 \text{ mV}$$

This is indeed a value that is typical of biological cells. Now, to find the reversal potential, we merely perform the same calculation with the ACH gates open.

$$R_{K_{new}} = R_{K_{leakage}} \text{ in parallel with } R_{k_{ACH}}$$

$$= 10^8 \text{ in parallel with} 10^5 = 10^5 \ \Omega$$

Assume steady state; then

$$i_K = i_{Na} = \frac{160 \text{ mV}}{100 \text{ k}\Omega + 100 \text{ k}\Omega} = 0.8 \ \mu\text{A}$$

$$V_{m_{reversal}} = (0.8 \times 10^{-6})(100 \text{ k}\Omega) - 100 \text{ mV} = -20 \text{ mV}$$

This is a typical value for the reversal potential at a neuromuscular junction. Physiological experiments to determine this value are common. In this particular case, this reversal potential value means that the ACH opened both the sodium and potassium gates.

1-9 THE HODGKIN–HUXLEY MODEL

The circuit model of Fig. 1-9 is similar to the Hodgkin–Huxley (1952) model. Because it is such an old and famous neuronal model, let us examine it in greater detail, following the development of MacGregor and Lewis (1977), and use it to explain neuronal voltage spikes.

The experiments that formed the basis for this model were performed on the squid giant axon. Because of the large diameter of that axon, a

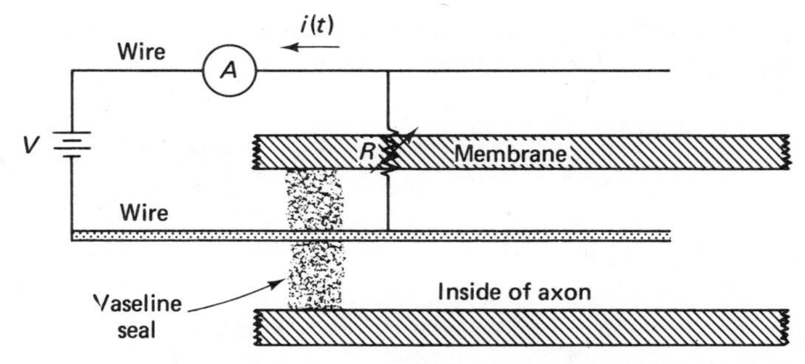

Figure 1-10 Model of the voltage-clamp condition. The voltage across the membrane is held constant. If the conductance of the membrane changes, then, by Ohm's law, the current will change. Measuring this current will yield a measure of the conductance.

silver wire could be pushed through the cut end of the axon and into its axoplasm. This electrode extended the entire length of the axon. The voltage potential on this wire, and thus inside the axon, was carefully controlled. This technique, developed by Cole (1949) and Marmont (1949), is called a voltage clamp. In 1952, Hodgkin, Huxley, and Katz tried to model the changes in the conductance of the membrane, and the voltage-clamp technique provided a convenient method for doing this. They could hold the voltage across the membrane constant while measuring the current passing through the membrane. This current is a good indicator of the conductance, as indicated in Fig. 1-10.

The voltage clamp technique included a guard system with cylindrical electrodes, which enabled the extracellular solution (seawater) to be kept isopotential as well. This eliminated all axial components of the current, leaving only radial components in the voltage-clamped portion of the axon. It allowed the voltage clamp to be applied to a large area of the membrane.

When Hodgkin and Huxley applied a step depolarization to a voltage-clamped membrane, they observed three phases of current, as shown in Fig. 1-11A. First, there was a current spike, lasting a few microseconds, which discharged the membrane capacitance. Next, they observed an inward current, which would have brought about further depolarization if the membrane were not clamped by a voltage source. The time course and magnitude of this inward current depended on the magnitude of the depolarizing step and on the initial voltage from which the step was made. Within 1 or 2 msec, the current changed direction from inward to outward; its magnitude and dynamics also depended upon the size of the voltage step and the membrane potential prior to it. This outward current persisted for tens of milliseconds when the clamped depolarization was maintained, but subsided quickly when the membrane was repolarized. Based on the model, we can hypothesize that the inward current was due to sodium ions and the delayed outward current was due to potassium.

To test this hypothesis, most of the sodium in the external seawater was replaced with the large ion choline, so that nothing would flow through the sodium channels whether they were opened or closed. Under these conditions, the inward phase of the current was abolished (Fig. 1-11B) and the late outward phase began almost immediately after the voltage-clamped step.

Thus, when sodium ions were present, there was a transitory inward phase followed by an outward phase; when sodium ions were not present, there was only an outward phase. The reversal potential of the outward phase was very close to the Nernst potential of potassium ions, whereas that of the inward current was very close to the Nernst potential of sodium ions. This would be expected if the outward current were carried predominantly by potassium ions and the inward current were carried predominantly by sodium ions, as Hodgkin and Huxley had assumed. These concepts and some other experimental details are summarized in the model shown in Fig. 1-9.

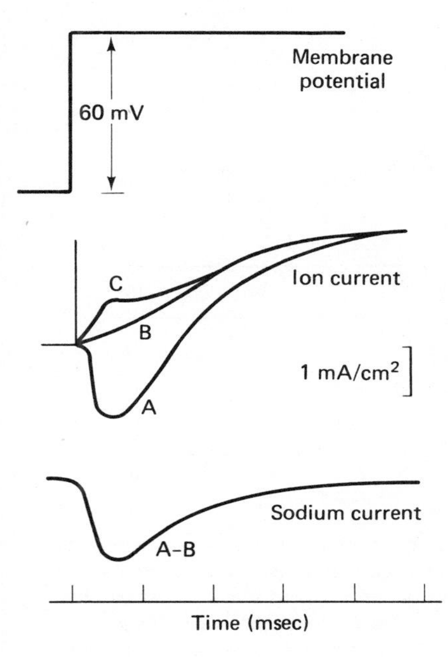

Figure 1-11 Analysis of ionic membrane current by the voltage-clamp technique. The membrane potential of a squid axon is suddenly depolarized from −60 mV to 0 (upper trace). The resulting ionic current (middle trace) is recorded under different conditions: (A) axon surrounded by sea water; (B) $\frac{9}{10}$ of external sodium replaced by choline (approximately equalizing the sodium activities on either side of the membrane; (C) external sodium totally replaced by choline. Curve B shows the potassium current by itself. The difference between curves A and B (bottom trace, A–B) shows the magnitude and time course of the inward directed sodium current when the axon is in seawater. The time course corresponds to an experiment at 8.5°C. (From B. Katz, *Nerve, Muscle, and Synapse*, McGraw-Hill, Book Company, New York, 1966.)

When a step depolarization was applied to a voltage-clamped membrane, the potassium conductance increased gradually, with an inflection, to its final value (see Fig. 1-12). In response to a step repolarization, or hyperpolarization, the potassium conductance decreased without an inflection toward its final, lower value. Both the rate at which the potassium conductance rose and the final magnitude that it attained increased monotonically as the amplitude of the step depolarization was increased. In order to describe the dynamics of this conductance, Hodgkin and Huxley sought a mathematical function that would have an inflected rise and a noninflected fall. They selected

$$g(t) = (a + be^{-ct})^d \tag{1-30}$$

The initial value of this function is $(a + b)^d$ and the final value is a^d. If the final value is greater than the initial value, b must be negative and the transition from $a + b$ to a will be inflected. On the other hand, if the final value is less than the initial value, b must be positive and the transition will be noninflected. Thus this function did describe the inflected rise and noninflected fall of the potassium conductance. This function fixed the form of their model.

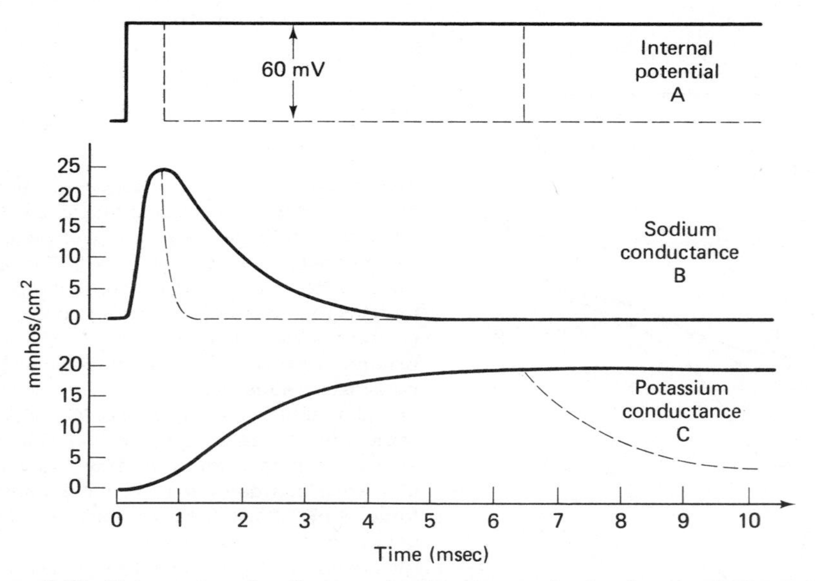

Figure 1-12 Time course of sodium conductance g_{Na} and potassium conductance g_K associated with a depolarization of 60 mV. The continuous curves are for a maintained voltage step. The dashed curves show the effect of repolarizing the membrane after two intervals indicated by dashed vertical lines in the trace at the top. [From A. L. Hodgkin, "The Croonian Lecture, Ionic Measurements and Electrical Activity in Giant Nerve Fibers," *Proceedings of the Royal Society London* (*B*) 148 (1958), 1–37.

The next stage in the model development was to select values of the parameters a, b, c and d so that they fit the experimental data. For example, $d = 4$, $a = 1$, $b = -1$, and $c = -1$ would yield a curve of the same shape as that of Fig. 1-11C, which reflects the behavior of the potassium conductance in response to a depolarization. But these values would not yet allow the equation to match the physiological data quantitatively. For this quantitative fit d was chosen to be four, and the parameters a, b and c were remapped into new variables n_{ss} and τ_n, so that the equations for potassium conductance became

$$g_K = \bar{g}_K n^4 \tag{1-31}$$

and

$$\frac{dn}{dt} = \frac{n_{ss} - n}{\tau_n} \tag{1-32}$$

where $\bar{g}_K$ was a constant and n_{ss} and τ_n depended on membrane potential. Thus the potassium conductance was a function of both voltage and time. The voltage dependence was incorporated into the two parameters n_{ss} and τ_n, and the time dependence was described by the differential equation. The dependence of n_{ss} and τ_n on the membrane potential is derived from voltage-clamp experiments and is shown in Fig. 1-13.

Similar equations were derived to describe the sodium conductance. When a step depolarization was applied to a voltage-clamped membrane, the sodium conductance rose with an inflection, passed through a peak, then fell again to a low magnitude. In response to a hyperpolarizing step,

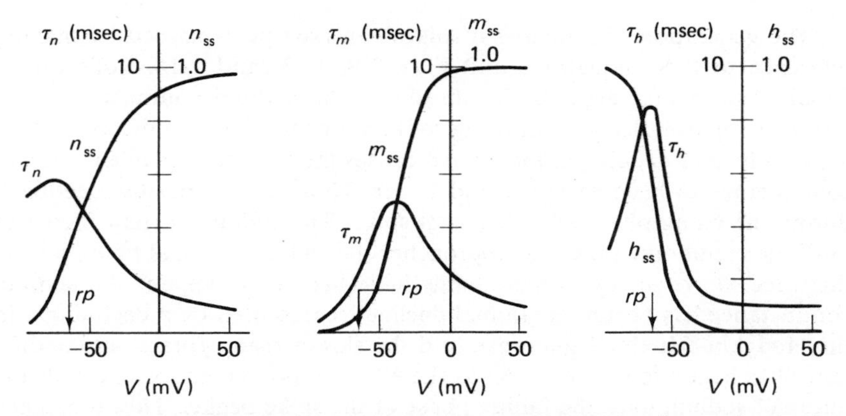

Figure 1-13 Voltage dependences of the six key parameters of sodium and potassium conductances in the Hodgkin–Huxley model. (From K. S. Cole, *Membranes, Ions, and Impulses*, University of California Press, Berkeley, 1968, 1972.)

the sodium conductance fell rapidly and without inflection. To describe the inflected rise and the rapid, noninflected fall of the sodium conductance, Hodgkin and Huxley again used the function of Eq. (1-30) with $d = 3$. To describe the slower fall that took place when the depolarization was maintained, they used the same function, with $d = 1$. Again, the parameters (τ_m, m_{ss}, τ_h, and h_{ss}) were matched to the data (see Fig. 1-13), and the conductance dynamics were embodied in linear differential equations.

$$g_{Na} = \bar{g}_{Na} m^3 h \tag{1-33}$$

$$\frac{dm}{dt} = \frac{m_{ss} - m}{\tau_m} \tag{1-34}$$

$$\frac{dh}{dt} = \frac{h_{ss} - h}{\tau_h} \tag{1-35}$$

We can use these equations and Fig 1-9, neglecting the sodium–potassium pumps, to express the total membrane current, i, as a function of time and voltage.

$$i = C_m \frac{dv}{dt} + g_K(v - E_K) - g_{Na}(v - E_{Na}) + \bar{g}_L(v - E_L) \tag{1-36}$$

These equations, (1.31) to (1.36), now known as the Hodgkin–Huxley model, were derived for the voltage- and space-clamped squid axon. However, it was later shown that with minor changes they could be applied to nonclamped, vertebrate dendrites and myelinated axons. The phenomena described by these equations are almost universal in spike-generating tissue.

The generation of a neuronal spikes can now be explained by making reference to these equations and Figs. 1-9, 1-13, and 1-14. Following a depolarization that exceeds the threshold, the sodium conductance rises rapidly with an inflection, allowing sodium ions to flow into the axon. This leads to further depolarization, which causes the voltage-dependent sodium conductance to become larger and larger. Thus there is positive feedback during the rising phase of a neuronal spike. The sodium current increases until the membrane potential approaches the sodium reversal potential. By this time, however, two other things have begun to happen: the sodium conductance has begun its gradual decline (represented by a declining h in the Hodgkin–Huxley equations), and the slower-rising potassium conductance has begun to rise. As soon as the efflux of potassium ions exceeds the influx of sodium ions, the falling phase of the spike begins. This, too, has a positive feedback aspect, brought about by the rapid decline in sodium conductance in response to repolarization. The potassium reversal poten-

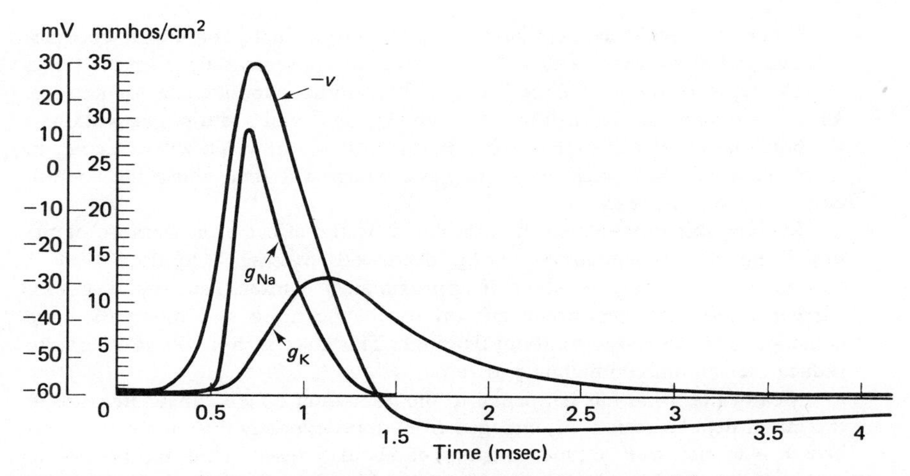

Figure 1-14 Numerical simulation of the Hodgkin–Huxley model showing a propagated action potential (curve-v) and sodium and potassium conductances, using experimental constants appropriate to 18.5°C. (After Hogkin and Huxley 1952d.)

tial is below the resting membrane potential in the squid giant axon. When the membrane potential reaches its original resting value, the potassium conductance is still large, because of its slow fall in response to repolarization. Therefore, the membrane potential approaches the potassium reversal potential. Gradually, as the potassium conductance falls toward its resting value, the membrane potential rises back to its resting level.

EXAMPLE

Use the Hodgkin–Huxley equations and Fig. 1-13 to explain the generation of a voltage spike.

In steady state,

$$\frac{dm}{dt} = \frac{dh}{dt} = \frac{dn}{dt} = 0$$

and at the resting membrane potential,

$$n_{ss} = 0$$
$$m_{ss} = 0$$
$$h_{ss} = 1$$

Therefore, from Eqs. (1-31) to (1-35),

$$n = 0$$
$$m = 0$$
$$h = 1$$

When the membrane depolarizes, m_{ss} increases. Therefore, dm/dt becomes positive and m increases with a time constant of approximately 0.4 msec (for a 10-mV depolarization). This will cause the sodium conductance to increase, allowing sodium ions to rush into the cell. The influx of sodium ions will drive the potential in the cell's interior in a positive direction, which will cause m_{ss} to increase, which will cause m to increase, which will then cause the sodium conductance to increase.

In time, this process will be terminated by the effect of h. Because of the initial membrane depolarization, h_{ss} decreased, causing h to drop from 1 toward 0, with a time constant of approximately 6 msec. But as the sodium current caused the membrane current to change more and more, the time constant for h decreased to about 0.6 msec. This change in h will now serve to reduce the sodium conductance to zero.

Meanwhile, what has happened to the potassium conductance? Because of the initial depolarization, n_{ss} changed from approximately 0 to about 0.4. This causes n to rise with a time constant of about 5 msec. Thus the potassium conductance should start to become appreciable when the sodium conductance is decaying.

This increase in potassium conductance will cause potassium ions to flow out of the cell, leaving behind a net negative charge which drives the potential in a negative direction. When the net flow of potassium ions is equal to the net flow of sodium ions, the voltage spike will peak and the potential will start to go in a negative direction. As the potential decreases, n_{ss} will decrease, causing n to decrease, which will cause the potassium conductance to finally shut off. However, this turnoff of the potassium conductance will have a time constant of 5 or 6 msec in its end stages, whereas the change in membrane potential will have a time constant of a fraction of a millisecond. Therefore, the potassium channels will still be on when the voltage reaches its resting level. This will cause the potential to overshoot, or hyperpolarize, the membrane. It will also produce a relative refractory period after the spike, raising the threshold voltage.

1-10 THE IRON-WIRE MODEL

The Hodgkin–Huxley model is currently accepted as the best model of a neuron; however, there were many older models which it had to supplant. One of the most interesting is the iron-wire model proposed by Lillie and explained by Lewis (1968) as follows.

If an iron wire is placed in a strong nitric acid solution, the surface of the iron wire becomes passive and will no longer precipitate copper sulfate or nitrate solutions. If the passive iron is scratched, that area becomes active, exhibiting a temporary change of color. This active region propagates down the length of the iron wire with a velocity similar to nerve conductive velocity.

Between 1909 and 1925, Lillie discovered numerous similarities between transmission in a nerve and transmission in an iron wire. A few of these are listed below.

1. Localized chemical, mechanical, or electrical changes may initiate a propagated disturbance in either system.
2. Both systems show stimulus thresholds for initiation of the propagated disturbance.
3. Once initiated, the propagated disturbance in either system has a form and velocity determined by the system itself and is independent of the characteristics of the stimulus.
4. Both systems have similar propagation velocities, which have strong temperature dependencies.
5. The conducting velocity may be increased in either system by creating saltatory conduction. This is done by wrapping the nerve in a myelin sheath and by sliding 1-mm sleeves of glass tubing over the iron wire, leaving short lengths of exposed wire between the segments to simulate the nodes of Ranvier.
6. Following the passage of a propagated disturbance, both systems exhibit temporary refractory periods when their thresholds are higher than normal.
7. In either system, propagation may be blocked by certain chemical or electrical changes.
8. In either system, an electric stimulus may either excite or inhibit, depending on its polarity.
9. Inhibitory electric stimuli increase mechanical and chemical thresholds in both systems, and may block passage of a disturbance initiated elsewhere.

This physical homolog model enjoyed great popularity for many years, until finally a new model replaced it. These simulations, as well as those reported in over 100 papers on the iron-wire model, are discussed by Lewis (1968).

1-11 SUMMARY

We have presented two neuronal models in this chapter: Lillie's iron-wire model and the Hodgkin–Huxley model. The iron-wire model is a homolog model. Its physical structure is similar to that of a neuron. It was a popular model before people really understood how neurons worked. The Hodgkin–Huxley model explained how neurons worked and then became the most popular neuronal model. With this model we gained an understanding of how the axon of a single neuron worked. Subsequent research explained how complete neurons worked and how individual neurons communicated with each other. Ultimately, we wish to understand the large network of neurons called the human brain. However, this neural network is much too complex for present techniques. (See the September 1979 issue of *Scientific American*.) So our best bet for understanding how individual neurons function in complete neural networks is to study small neural networks such as those mediating the tail flip of the crayfish, the gill withdrawal of the aplisia, or swimming movements of the leech. The latter has been nicely explained by Friesen and Stent (1977) and Stent et al. (1978). Many small neural networks have been analyzed on a cell-by-cell basis; their functioning is now understood. Quantitative models for many small neural networks are discussed by MacGregor and Lewis (1977). One of the most interesting studies of neural networks is "What the Frog's Eye Tells the Frog's Brain" by Lettvin, Maturana, McCulloch, and Pitts (1959). I encourage you to read about some of these neural networks.

In the rest of this book we do not dwell upon the individual behavior of single neurons. Rather, we study large groups of nerve or muscle cells and assume that the overall output is just the average of many cells. A muscle, for example, is composed of many cells, and the overall response is quite adequately modeled as the average response of many cells.

REFERENCES

BENNETT, M. V. L., "Electric Organs," and "Electroreception," in *Fish Physiology*, Vol. 5, *Sensory Systems and Electric Organs*, ed. W. S. Hoar and D. J. Randall. New York: Academic Press, Inc., 1971, Chaps. 10 and 11, pp. 347–574.

BROWN, A. M., J. L. WALKER, and R. B. SUTTON, "Increased Chloride Conductance as the Proximate Cause of Hydrogen Ion Concentration Effects in *Aplysia* Neurons," *Journal of General Physiology*, 56 (1970), 559–82.

COLE, K. S., "Dynamic Electrical Characteristics of the Squid Axon Membrane," *Archives des Sciences Physiologiques*, 3 (1949), 253–58.

EINSTEIN, A. (1908), *Investigations on the Theory of Brownian Movement*, ed. R. Furth, trans. A. D. Cowper. New York, Dover Publications, Inc., 1956, p. 75.

Erlij, D., and S. Grinstein, "The Number of Sodium Ion Pumping Sites in the Skeletal Muscle and Its Modification by Insulin," *Journal of Physiology*, 256 (1976), 13–45.

Friesen, W. O., and G. S. Stent, "Generation of a Locomotory Rhythm by a Neural Network with Recurrent Cyclic Inhibition," *Biological Cybernetics*, 28 (1977), 27–40.

Gibbons, J. F., *Semiconductor Electronics*, New York: McGraw-Hill Book Company, 1966.

Hodgkin, A. L., and A. F. Huxley, "Currents Carried by Sodium and Potassium Ions through the Membrane of the Giant Axon of *Loligo*," *Journal of Physiology*, 116 (1952a), 449–72.

Hodgkin, A. L., and A. F. Huxley, "The Components of Membrane Conductance in the Giant Axon of *Loligo*," *Journal of Physiology*, 116 (1952b), 473–96.

Hodgkin, A. L., and A. F. Huxley, "The Dual Effect of Membrane Potential on Sodium Conductance in the Giant Axon of *Loligo*," *Journal of Physiology*, 116 (1952c), 497–506.

Hodgkin, A. L., and A. F. Huxley, "A Quantitative Description of Membrane Current and Its Application to Conduction and Excitation in Nerve," *Journal of Physiology*, 117 (1952d), 500–44.

Hodgkin, A. L., A. F. Huxley, and B. Katz, "Measurement of Current-Voltage Relations in the Membrane of the Giant Axon of *Loligo*," *Journal of Physiology*, 116 (1952), 424–48.

Junge, D., *Nerve and Muscle Excitation*. Sunderland, Mass.: Sinauer Associates, Inc., 1976.

Katz, B., *Nerve, Muscle, and Synapse*. New York: McGraw-Hill Book Company, 1966.

Keynes, R. D., "Ion Channels in the Nerve-Cell Membrane," *Scientific American* 240(3) (1979), 126–35.

Kittel, C., *Elementary Statistical Physics*. New York: John Wiley & Sons, Inc., 1958.

Kuffler, S. W., and J. G. Nicholls, *From Neuron to Brain*. Sunderland, Mass.: Sinauer Associates, Inc., 1976.

Lettvin, J. Y., H. R. Maturana, W. S. McCulloch, and W. H. Pitts, "What the Frog's Eye Tells the Frog's Brain," *Proceedings of the IRE*, 47(11) (1959), 1940–51.

LEWIS, E. R., "The Iron-Wire Model of a Nerve, a Review," in *Cybernetics Problems in Bionics*, ed. H. L. Oestreicher. New York: Gordon and Breach, Science Publishers, Inc., 1968.

MACGREGOR, R. J., and E. R. LEWIS, *Neural Modeling, Electrical Signal Processing in the Nervous System*. New York: Plenum Press, 1977.

MARMONT, G., "Electrode Clamp for Squid Axon: Studies on the Axon Membrane," *Journal of Cellular Comparative Physiology*, 34 (1949), 351–82.

MOORE, J. W., T. NARAHASHI, and T. I. SHAW, "An Upper Limit to the Number of Sodium Channels in Nerve Membrane?" *Journal of Physiology*, 188 (1967), 99–105.

PLONSEY, R., *Bioelectric Phenomena*. New York: McGraw-Hill Book Company, 1969.

SELKURT, E. E., *Physiology*. Boston: Little, Brown and Company, 1976.

STENT, G. S., W. B. KRISTAN, JR., W. O. FRIESEN, C. A. ORT, M. POON, and R. L. CALABRESE, "Neuronal Generation of the Leech Swimming Movement," *Science*, 200 (1978), 1348–57.

PROBLEMS

1-1 A beaker similar to that of Fig. 1-3 containing calcium chloride is divided by a membrane. Initially, the concentrations in the right and left sides are 1 and 2 millimolar (mM), respectively.

(a) Use Ohm's law, Fick's law, and the Einstein relationship to write an expression for the net flow of calcium through the membrane. If the membrane is permeable to calcium but not to chlorine, what will the equilibrium voltage across the membrane be? How much will the concentrations change by the time equilibrium is reached?

(b) State the conditions for space-charge neutrality. What does this tell you about the relationship between J_{Ca} and J_{Cl}?

(c) If the membrane is permeable to both ions, how large will the voltage be at equilibrium? Assuming that the membrane is permeable to both ions, find the relationship between the voltage across the membrane and the concentrations of $CaCl_2$ before equilibrium is reached.

(d) Calculate the concentration ratios for Donnan equilibrium if the membrane is permeable to calcium and chlorine but not to some large organic radical added to the solution on the right side.

(e) What is the direction of the resulting osmotic pressure?

1-2 A tank similar to that of Fig. 1-5a has a 500 m*M* solution of sodium–chloride on the left side and a 300 m*M* sodium–chloride/200 m*M* sodium–aspartate solution on the right side. The membrane has channels large enough to allow sodium to pass through but too small to allow aspartate to pass through. Calculate the resulting membrane potential. Show the direction of all currents, electric fields, and pressure gradients.

1-3 The following concentration values have been calculated for the giant cell of the sea snail *Aplysia* (Brown, Walker, and Sutton, 1970):

$$[\mathrm{Na}]_o = 337\ \mathrm{m}M$$
$$[\mathrm{Na}]_i = 50\ \mathrm{m}M$$
$$[\mathrm{K}]_o = 6\ \mathrm{m}M$$
$$[\mathrm{K}]_i = 168\ \mathrm{m}M$$
$$[\mathrm{Cl}]_o = 340\ \mathrm{m}M$$
$$[\mathrm{Cl}]_i = 41\ \mathrm{m}M$$

For the resting cell membrane,

$$P_{\mathrm{K}}:P_{\mathrm{Na}}:P_{\mathrm{Cl}} = 1.0:0.019:0.381$$

What is the resting potential predicted by the Goldman equation? What would be the effect of a tenfold increase in the external potassium concentration on the resting membrane potential? Calculate the Nernst potential and draw a model similar to that of Fig. 1-9 for this cell.

The following conductances have been measured in this cell:

$$g_{\mathrm{K}} = 0.57\ \mu\mathrm{mho}$$
$$g_{\mathrm{Na}} = 0.11\ \mu\mathrm{mho}$$
$$g_{\mathrm{Cl}} = 0.32\ \mu\mathrm{mho}$$

What is the predicted membrane resting potential? (Based on Junge 1976.)

1-4 The membrane shown in Fig. P1-4 is permeable to all three ions. The pump is

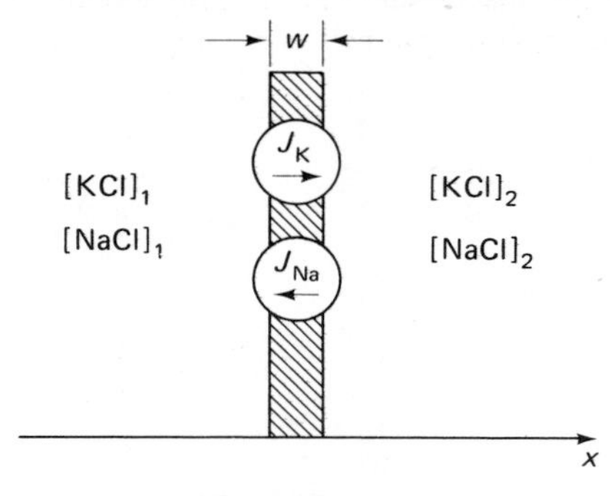

Figure P1-4

noneletrogenic, that is, $J_p(K^+) = J_p(Na^+)$. Find the equilibrium chloride concentration gradient across the membrane as a function of the magnitude of J_p.

1-5 To totally block the sodium channels in lobster nerves, it requires about 1.6×10^{-11} mole of tetrodotoxin, a virulent poison from the puffer fish, per gram of nerve (Moore, Narahashi, and Shaw 1967). How many tetrodotoxin molecules per gram of nerve are required to block these channels? Light- and electron-microscope studies have shown that these lobster nerves have a total membrane area of about 0.7×10^4 cm^2/g. How many tetrodotoxin molecules per square micron of membrane are required to block the sodium channels? If one tetrodotoxin molecule is required to block each sodium channel, we have an estimate of the number of sodium channels per square micrometer of membrane. (Based on Junge 1976.)

Studies with radioactively labeled α-bungarotoxin, a snake venom, allowed acetylcholine synaptic receptor sites to be counted in a photograph. Their density was $10^4/\mu m^2$. Is the density of sodium channels greater or less than the density of acetylcholine sites? How would this be reflected in a model such as that of Fig. 1-9?

1-6 In the Hodgkin–Huxley model, m responds faster to a step depolarization than do n and h (see Fig. 1-13). Therefore, in a voltage response the sodium gates open, with consequent regenerative effects. However, if the depolarization is done very slowly, voltage spikes will not occur. This is called accommodation. Explain this phenomenon in terms of the behavior of n, m, and h.

1-7 Provide a qualitative answer to Problem 4-14.

1-8 The resting membrane potential is the same as the chlorine Nernst potential for a motoneuron. Why? This is not true for a smooth muscle cell. The resting potential is about -55 mV and the chlorine Nernst potential is about -20 mV. What additional physiological process is necessary to make these potentials different? In what direction does it work?

1-9 Acetylcholine is applied to a motoneuron membrane at $t = 1$ ms. Sketch the resulting voltage in millivolts versus time in milliseconds. At $t = 5$ msec a transmitter that opens up potassium channels is applied. Sketch the resulting voltage.

2

BIOINSTRUMENTATION

Biomedical instruments are used to extend the human senses so that we can look inside the skull, hear a fetal heartbeat, or record the trajectory of a rapidly moving eyeball. Furthermore, they provide a means of producing a permanent quantitative record of these events. This chapter introduces the reader to some of the considerations that go into the design of biomedical instruments.

Most biomedical instruments have a transducer that changes energy from one form into another. Motors, generators, and batteries are transducers: an electric motor changes electrical energy into mechanical energy, a battery changes electrical energy into chemical energy and vice versa.

Most instrumental transducers have electrical energy as the output. For example, photodiodes and phototransistors change light energy into electrical energy. Many transducers convert displacement information into electrical signals: strain gauges and piezoelectric devices are ideal for very small displacements, and potentiometers are simple and inexpensive, moving one plate of a capacitor or moving an iron slug in an inductor also works. Thermocouples and thermistors are common transducers for converting temperature into an electrical signal. Chemical properties such as pH, oxygen content, and oxygen saturation can also be measured with transducers that have electrical outputs. These transducers and many more are discussed in detail in biomedical instrumentation textbooks (Cobbold 1974; Geddes and Baker 1975; Webster 1978).

2-1 ELECTRODES

We might think that many biomedical measurements could be made without using transducers, because nerve and muscle potentials are already electrical. All that would be needed is an electrode attached to the biological source. But if it were so simple, it would indeed be surprising to find chapters (Cobbold 1974, Geddes and Baker 1975, Webster 1978), and even entire books (Geddes 1972) dedicated to the topic of electrodes.

The major problem encountered by biological electrodes is caused by the fact that electric current in the body is carried by ions, whereas electric current in a wire is carried by electrons. Hence the electrode must act as a transducer, changing ionic current into electronic current. This transduction takes place in the electrolyte near the surface of the electrode. The electrolyte is the salt solution that dissociates into ions. It may be an electrolytic solution or paste used with surface electrodes, or it may be tissue fluids, or it may be perspiration accumulated under the electrode.

2-1-1 The Electrode–Electrolyte Model

The electrode–electrolyte interface is schematically illustrated in Fig. 2-1. The net current crossing the interface is composed of electrons in the electrode and cations (C^+) and anions (A^-) in the electrolyte. There are no free electrons in the electrolyte and no free cations or anions in the electrode. Therefore, chemical reactions at the interface must change the electronic current of the metal into the ionic current of the electrolyte. One such reaction is the oxidation of the atoms of the metal electrode:

$$C \rightleftharpoons C^{z+} + ze^- \qquad (2\text{-}1)$$

where z is the valence of the atom. Subsequently, the electric field pushes the electron to the left in the wire and propels the ion to the right into the electrolyte. Another reaction is the oxidation of the anion in the electrode–electrolyte interface.

$$A^{z-} \rightleftharpoons A + ze^- \qquad (2\text{-}2)$$

The anion is oxidized into a neutral atom, giving off one or more free electrons that are swept away in the wire. The neutral atom moves away

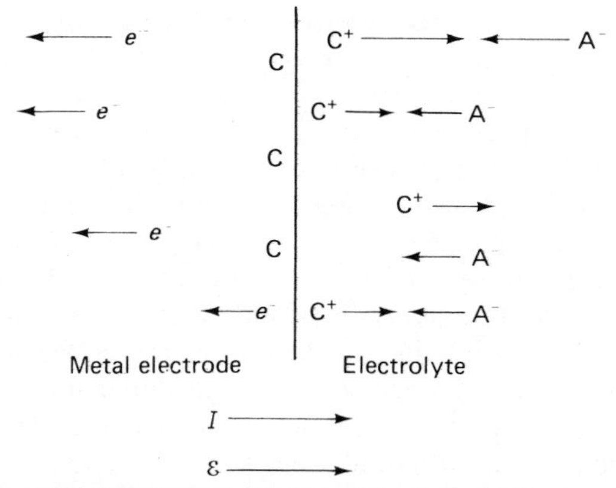

Figure 2-1 Model of the electrode–electrolyte interface with current crossing it from left to right. The electrode consists of metallic atoms C. The electrolyte is an aqueous solution containing cations C^+ and anions A^-.

from the interface by diffusion. Both reactions are reversible; reduction reactions (from right to left in the equations) also occur.

When the current flows from electrode to electrolyte, as indicated in Fig. 2-1, the oxidation reactions dominate; when the current is in the opposite direction, the reduction reactions dominate. When no current is crossing the electrode–electrolyte interface, these reactions still occur. But the rate of oxidation reactions equals the rate of reduction reactions, so that there is no net transfer of charge across the interface.

2-1-2 The Half-Cell Potential

Most metals are good conductors and have a large quantity of loosely held valence electrons. When an electrode made of such a metal is immersed in an electrolyte, a salt solution, some of these electrons will enter the solution, leaving behind positively charged ions on the surface of the metal. Thus the electrode becomes positively charged with respect to the electrolyte. This potential difference is called the *half-cell potential*. For a silver electrode, the half-cell potential is 0.80 V (from Fig. 2-2). If a copper electrode were placed in the same solution, it too would lose electrons; its half-cell potential is 0.34 V. The potential between the silver electrode and the copper electrode would be 0.46 V. Such half-cell potentials are responsible for common devices, such as batteries, but the actual chemical reactions and electronic behaviors are very complex. They are discussed in detail by Geddes (1972) and Cobbold (1974).

It is not possible to measure the half-cell potential of a single electrode because we cannot provide a connection between the electrolyte and the voltmeter unless we use a second electrode. Because this second electrode also has a half-cell potential, we can only measure the difference between the half-cell potentials of the metal and the second electrode. Because of this difficulty, one specific electrode, called the *hydrogen electrode*, has been chosen as a standard. The measurement of half-cell potentials of all other electrodes are made with respect to the hydrogen electrode.

The standard hydrogen electrode system consists of a specially prepared platinum surface, immersed in an acid solution with hydrogen gas bubbled through the solution. The half-cell potential of this system is defined as zero. The electrode whose half-cell potential is to be measured is placed in the same solution and is connected by a high-impedance voltmeter to the hydrogen electrode. The resulting half-cell potential is then recorded. The half-cell potentials of several popular electrode metals are given in Fig. 2-2.

If two identical electrodes were used for a physiological measurement, the effect of the two half-cell potentials should cancel out. But in practice they do not. As a further complication, half-cell potentials vary with time, temperature, and ionic concentration. These variations are not all black magic—some of them are partially understood. Geddes and Baker (1975) discuss several ways of solving these problems.

Process			Potential (V)
Al	⟶	$Al^{3+} + 3e^-$	−1.66
Zn	⟶	$Zn^{2+} + 2e^-$	−0.76
Cr	⟶	$Cr^{3+} + 3e^-$	−0.74
Fe	⟶	$Fe^{2+} + 2e$	−0.44
Ni	⟶	$Ni^{2+} + 2e$	−0.25
Sn	⟶	$Sn^{2+} + 2e^-$	−0.14
Pb	⟶	$Pb^{2+} + 2e$	−0.13
H_2	⟶	$2H^+ + 2e^-$	0.00 (by definition)
$Ag + Br^-$	⟶	$AgBr + e$	+0.10
Sn^{2+}	⟶	$Sn^{4+} + 2e^-$	+0.15
$Ag + Cl^-$	⟶	$AgCl + e^-$	+0.22
Cu	⟶	$Cu^{2+} + 2e^-$	+0.34
Cu	⟶	$Cu^+ + e^-$	+0.52
Ag	⟶	$Ag^+ + e^-$	+0.80
Au	⟶	$Au^{3+} + 3e^-$	+1.50
Au	⟶	$Au^+ + e^-$	+1.68

Figure 2-2 Half-cell potentials for common electrode materials at 25°C. The metal undergoing the reaction shown has the sign and potential, *E*, in volts, when referenced to the hydrogen electrode.

2-1-3 Silver–Silver Chloride Electrodes

One of the most popular physiological electrodes is the *silver–silver chloride* (Ag–AgCl) *electrode*. It has an insulated wire attached to a piece of silver which is coated with a layer of AgCl. This electrode should be used with an electrolyte having Cl^- as its principal anion.

Current is carried by electrons in the wire and chlorine ions in the solution. The silver and the silver chloride are involved in the reaction occurring at the electrode–electrolyte interface. First silver is oxidized:

$$Ag \rightleftharpoons Ag^+ + e^- \tag{2-3}$$

Then the newly formed silver ions combine with Cl^- ions in the solution to form the ionic compound AgCl. The AgCl precipitates out of solution and is deposited on the silver electrode.

$$Ag^+ + Cl^- \rightleftharpoons AgCl\downarrow \tag{2-4}$$

There are two common techniques for manufacturing Ag–AgCl electrodes. The electrolytic method is shown in Fig. 2-3. The electrode to be plated and another larger silver electrode are placed in a NaCl solution. The electrode to be plated, the anode, is connected to the positive terminal of a 1.5-V battery, and the other electrode, the cathode, is connected to the negative terminal of the battery. An ampmeter and a resistance of about 500 Ω, to limit the peak current, are inserted in this series circuit. When the switch is closed, the reactions of Eqs. (2-3) and (2-4) occur, and the current rises to its maximum value. As the thickness of the deposited AgCl layer

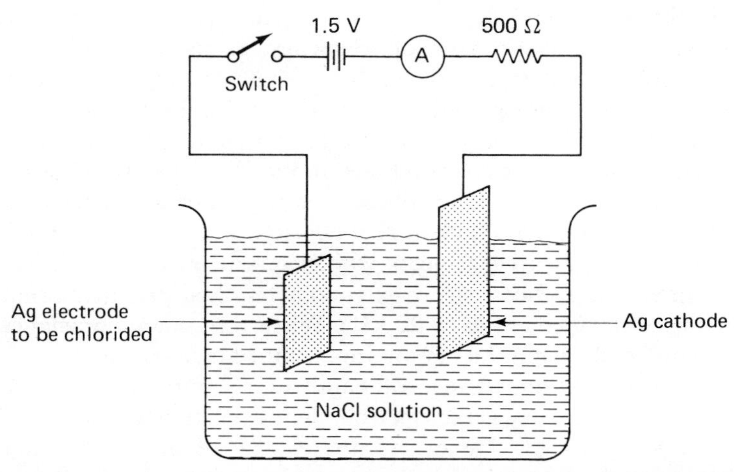

Figure 2-3 Simple circuit for making or replating silver–silver chloride electrodes.

increases, the reaction slows down and the current drops. A few milliamperes for a few minutes is sufficient to plate or replate most electrodes.

A sintering process is also used for making Ag–AgCl electrodes. A wire is placed in a die filled with a mixture of powdered Ag and AgCl. The die is compressed to form a pellet which is then baked at 400°C for several hours. These electrodes are more durable than the electrolytically deposited AgCl electrodes.

Silver–silver chloride electrodes are not very expensive, they are stable, and they are relatively noise-free. Moderate care must be taken to avoid abrading the AgCl surface during use and storage. The electrolyte must be washed off the electrode after each use. Once a month or so, they may be gently cleaned with cotton and replated. For very critical applications, the electrodes may be stored in a solution of NaCl in distilled water with the leads attached to the external projection of a carbon rod. This technique and other useful suggestions for the care of silver chloride electrodes are discussed by Geddes and Baker (1975). However, most bioengineers usually take no special precautions with their electrodes.

2-1-4 Electrode Models

A simple model of two electrodes and the intervening biological tissue is shown in Fig. 2-4. The battery, E, represents the half-cell potential. The resistance, R_B, represents the bulk resistance of the electrolyte and the elements C_{eq} and R_{eq} model the behavior at the electrode–electrolyte junction. The capacitance is from the double layer of ionic charges at the interface. Surprisingly, these elements have values that are current- and frequency-dependent.

Both the resistance, R_{eq}, and the capacitance, C_{eq}, are inversely proportional to the square root of frequency, as shown in Fig. 2-5. The impedance therefore decreases with a constant phase angle as the frequency increases. This relationship holds at least into the kilohertz range.

When the current density increases, the resistance, R_{eq}, decreases but the capacitance, C_{eq}, increases, as shown in Fig. 2-6. The overall impedance decreases again with constant phase angle. Electrode impedance is a function of both frequency and current density.

The reason for the great concern with electrode impedance is that years ago the measuring devices, string galvanometers and primitive amplifiers, had low input impedances (5 to 20 kΩ). When the source impedance is of the same order of magnitude as the input impedance, artifacts result. For example, a low input impedance will cause a large current to flow through the electrode. This may bias the electrode in the region where its impedance depends upon current. Hence, if the signal changes, the electrode impedance also changes, and therefore the measured signal will be distorted.

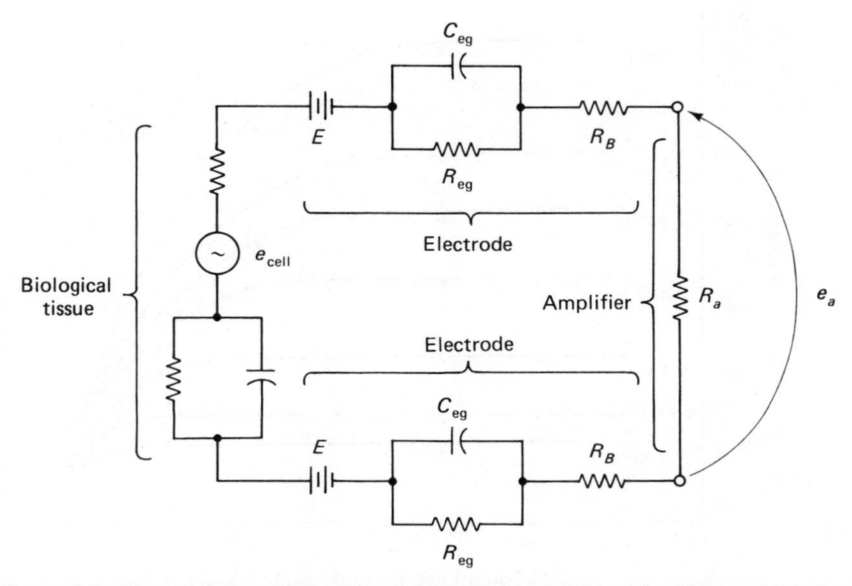

Figure 2-4 Model of two electrodes and the biological tissue in between. The batteries, E, represent the half-cell potentials of the electrodes. R_a is the amplifier input impedance, R_B is the bulk resistance of the electrode, and C_{eq} and R_{eq} are the frequency- and current-dependent impedances of the electrodes.

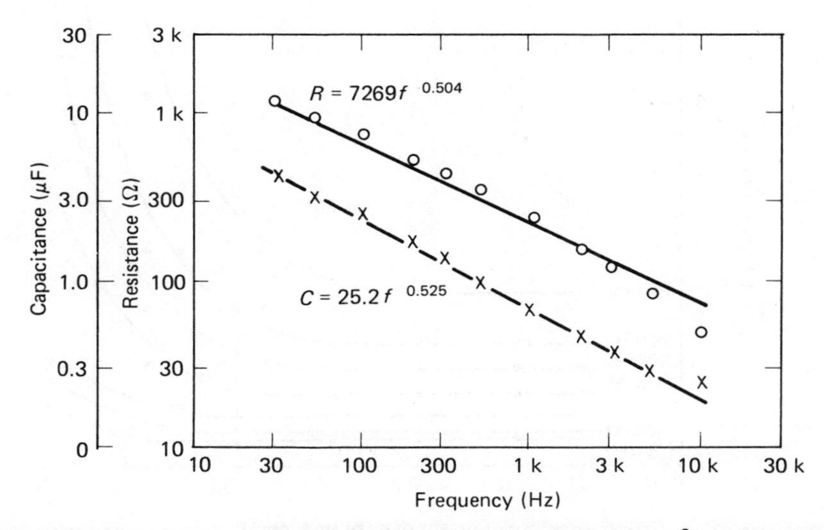

Figure 2-5 Capacitance and resistance of a pair of parallel 0.157-cm^2 stainless-steel electrodes immersed in 0.9% saline with current density of 0.025 mA/cm^2. [From L. A. Geddes, C. P. da Costa, and G. Wise, *Medical and Biological Engineering and Computing*, 9 (1971), 511–21.]

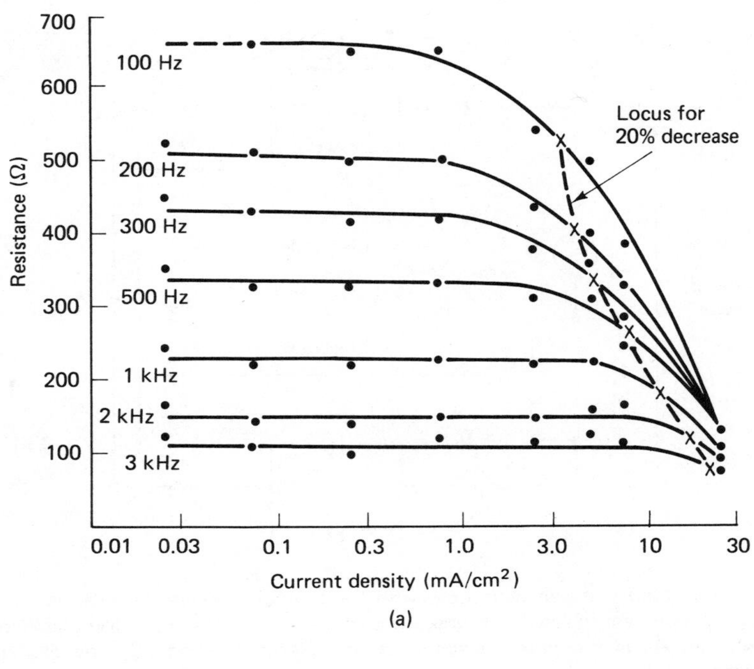

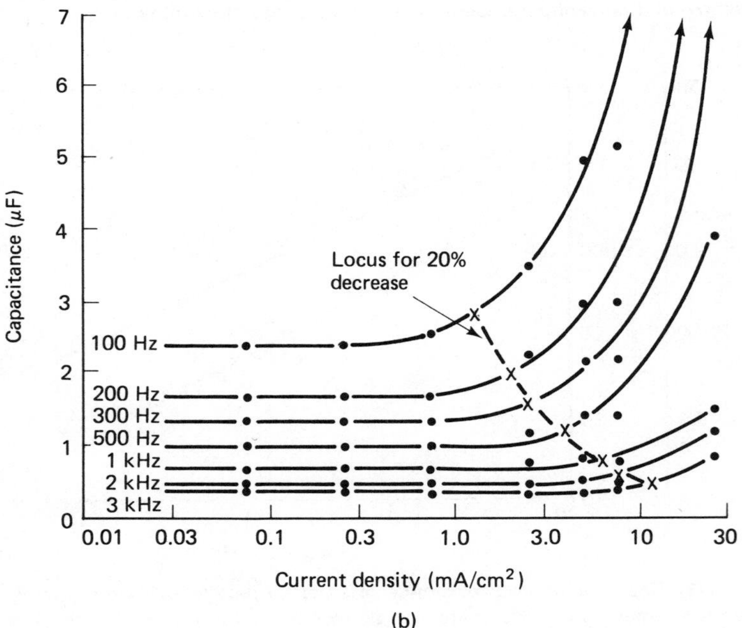

Figure 2-6 Dependence of the series-equivalent (a) resistance and (b) capacitance of a stainless steel–0.9% saline solution interface on current density and frequency. [From L. A. Geddes, C. P. da Costa, and G. Wise, *Medical and Biological Engineering and Computing* 9 (1971), 511–21.]

If the amplifier input impedance, R_a, is high, very little current will flow in the circuit and the amplifier input voltage, E_a, will approximately equal that of the biological signal (E_{cell} in Fig. 2-4). However, if the amplifier impedance is low, the input voltage, calculated by voltage division, will depend upon the electrode impedance. In fact, the input voltage function will have a frequency term in the numerator, which means high-frequency signals will be attenuated less than will low-frequency signals: this is high-pass filtering, or differentiation.

To ensure that the electrode impedance was small, the skin was often scraped, sanded, or abraded and strong chemicals were used to cleanse the skin. Today, high-impedance amplifiers are available, so the concern about electrode impedance has lessened. Current practice is merely to wash the skin with soap, perhaps an acne soap with mild abrasives, and to allow the electrolytes to settle for 15 or 20 minutes before taking measurements. Many engineers are putting great efforts into designing high-impedance amplifiers (100 MΩ or higher) that can be used without electrode creams. These instruments are not only more convenient but they also reduce artifacts produced by changes in electrode impedance.

2-1-5 Microelectrodes

Microelectrodes are used to study the electrical behavior of single cells. These electrodes are usually made of metal or glass tubing and have small-diameter tips (0.05 to 10 μm). They can penetrate a cell membrane without destroying the cell. Because of their small size, microelectrodes have impedances of 1 to 100 MΩ.

A metal microelectrode is typically made of a strong metal and is insulated except for its tip. A simplified model of a metal microelectrode in a cell is given in Fig. 2-7.

The activity of the living cell is represented by the voltage source, E_{cell}. The metal–electrolyte interface is modeled by R_1 and C_1, which decrease in value as the frequency increases. The difference in half-cell potentials between the microelectrode and the reference electrode in the bathing fluid is given by E. For the portion of the electrode immersed in the bathing fluid there are two conductors, the metal and the bathing fluid, separated by an insulator. This shunting capacitance is modeled by C_2. The input impedance of the amplifier is R_a, and the voltage measured by the amplifier is E_a. The transfer function between the cell voltage and the amplifier input voltage is

$$\frac{E_a}{E_{cell}} = \frac{R_1 C_1 s + 1}{R_1(C_1 + C_2)s + 1 + R_1/R_a} \tag{2-5}$$

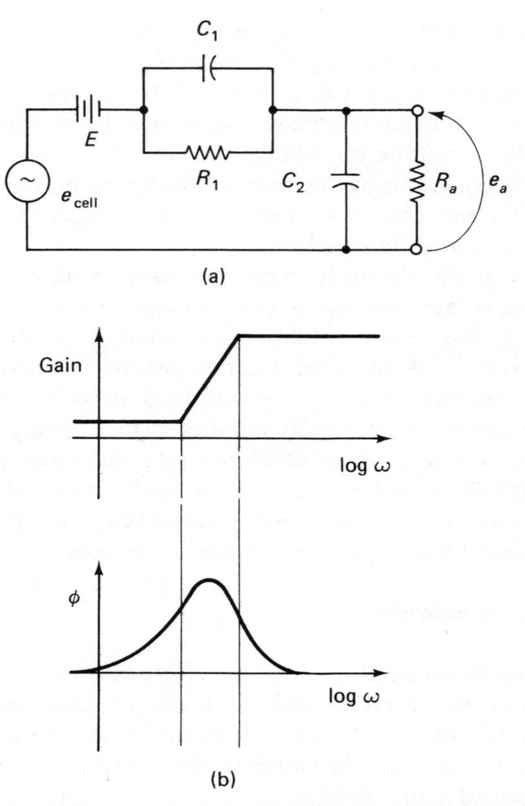

Figure 2-7 (a) Model of a metal microelectrode and (b) its associated Bode diagrams. R_a is the amplifier input impedance and E_a is the voltage measured by the amplifier. Over a certain frequency range this type of microelectrode acts as a high-pass filter.

If the amplifier input impedance is of the same order of magnitude as R_1, and if C_1 is greater than C_2, as is the case for most metal microelectrodes, the circuit behaves as a lead network. (This is essentially the same as a high-pass filter over a limited range of frequencies.) The Bode diagram for such a circuit is shown in Fig. 2-7b. It has been said that metal microelectrodes act as high-pass filters and fluid-filled glass microelectrodes act as low-pass filters (Gesteland, Howland, Lettvin, and Pitts 1959).

Microelectrodes can be made from glass tubing by heating the glass tube and pulling on the ends either manually or with an automatic puller. The tube is generally filled with a 3 *M* KCl solution. The circuit model of a glass microelectrode is similar to that of the metal microelectrode shown in Fig. 2-7, except that the capacitance C_1 is very small and the circuit behaves as a low-pass filter.

2-2 AMPLIFIERS

After a physiological signal is detected and transformed by a transducer, it must be amplified. Amplifiers for physiological measurements are usually very special. The physiological systems, the medical environment, and the government place constraints on the design of instruments for physiological measurements. In this section we discuss some of these constraints as well as some general instrumentation techniques.

In an amplifier circuit a small amount of input power controls a large amount of output power. Figure 2-8 shows a simple example of a common-emitter transistor amplifier. Its operation is briefly summarized as follows. If v_{in} increases, the voltage at the base will increase. This causes the base emitter voltage to increase (giving the junction more forward bias). The base current will therefore increase, resulting in an increase in collector current (i_c is equal to βi_B because the device is biased in the linear region). When collector current increases, this will give a larger voltage drop across R_L and thus v_{out} will decrease. The output waveform v_{out} is similar to the input waveform, v_{in}, but it is larger and is inverted, or phase-shifted by 180°.

2-2-1 Differential Amplifiers

The input voltage is usually transmitted from the transducer to the amplifier in a conductor. During this transmission, noise, for example 60-Hz noise induced from the power lines, becomes added to the signal. Often, physiological measurements are made with a pair of transducers

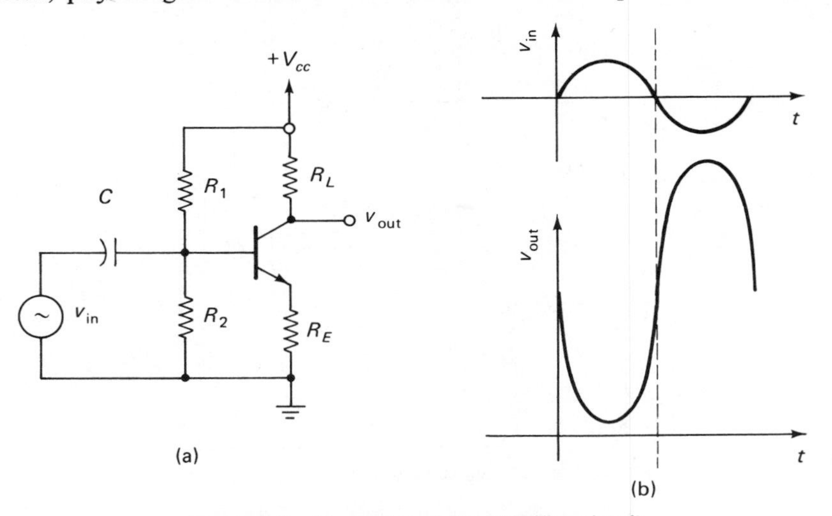

Figure 2-8 Simple transistor amplifier circuit.

arranged so that the signal travels up one wire and back down the other. The noise induced into each lead wire is almost identical. Therefore, if the difference of the two signals is taken, the noise from the two leads will cancel while the signals are reinforced. A single circuit for such a *differential amplifier* is shown in Fig. 2-9. If the source voltage, e_s, causes an increase in voltage at the base of Q_1, it will cause a decrease in the voltage at the base of Q_2. The base and collector currents of Q_1 will increase while the base and collector currents of Q_2 will decrease. Thus the collector voltage of Q_1 will decrease while the collector voltage of Q_2 increases. The difference of these two collector voltages is defined as the output voltage: it will be similar in form to the input voltage. However, the two lead wires will have the same noise induced upon them. So if this noise signal increases, the collector voltage of Q_1 will increase, and the collector voltage of Q_2 will also increase. The output voltage, the difference between these two voltages, will not change.

The signal common to both inputs is called the *common-mode signal*. It is produced primarily by differences in ground potentials between the transducer and the amplifier and by coupling to the ambient magnetic and electric fields. In real amplifiers the cancellation of the common-mode voltage is not perfect. The ratio of the difference-mode gain to the common-mode gain, called the *common-mode rejection ratio* (CMRR), is a measure of how well the amplifier rejects these common-mode voltages.

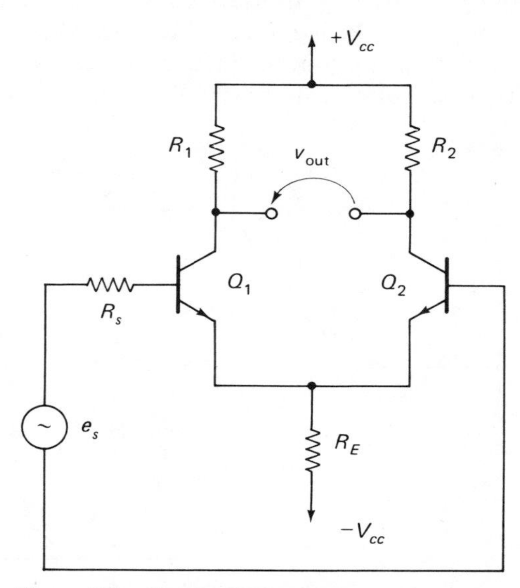

Figure 2-9 Simple differential amplifier circuit.

The resistor R_E and the negative voltage supply act as a constant-current source. With integrated-circuit technology, the resistor would be replaced with a transistor circuit.

2-2-2 Operational Amplifiers

Since the advent of integrated circuits, very few amplifiers have been designed using individual transistors. Instead, large circuits are now manufactured on small chips of silicon. The addition of an extra transistor or two adds relatively little to the cost or complexity of the circuit. A 741 operational amplifier, a typical integrated circuit that performs the task of the differential amplifier shown in Fig. 2-9, contains 22 transistors, 12 resistors, and one capacitor. When these amplifiers are used with negative feedback they are called *operational amplifiers* (Tobey, Graeme, and Huelsman 1971, p. xvii). The operational amplifier is a general-purpose device: operational-amplifier circuits can be used for amplifiers, integrators, filters, and for special waveshaping purposes.

To analyze operational-amplifier circuits, it is convenient to assume that we have an ideal operational amplifier which has infinite gain, infinite input impedance, zero output impedance, infinite bandwidth, and no offset voltage (if the two input voltages are identical, the output voltage is zero). Although this ideal amplifier may seem a bit remote from reality, the closed-loop gain relations which are derived in the following sections are applicable to real circuits and yield results that are correct within a small fraction of 1% for practical operational amplifiers.

Operational-amplifier circuits are general devices. They do not depend upon any particular technology. The first operational amplifiers were made with vacuum tubes. These were succeeded by discrete transistor circuits and then by integrated circuits. Several integrated-circuit technologies have been used for manufacturing operational amplifiers. Each improvement reduces the cost and makes the operational amplifiers more nearly ideal. However, the fundamental design considerations to be presented in this section remain the same.

2-2-2.1 THE INVERTING AMPLIFIER

The *inverting amplifier*, shown in Fig. 2-10, and its many variations form the bulk of commonly used operational amplifier circuitry. Negative feedback is supplied by R_o. The application of negative feedback around an ideal operational amplifier causes the voltage between the amplifier inputs (e_i in Fig. 2-10) to approach zero. The currents flowing into the inputs of the amplifier are zero (because we assumed infinite input impedance); therefore, by Kirchhoff's node law,

$$\frac{e_1 - e_i}{R_1} = \frac{e_i - e_0}{R_0}$$

By definition, e_0 equals $-Ae_i$ and e_i equals $-e_0/A$. Therefore, as A approaches infinity (the case of the ideal operational amplifier), e_i approaches zero and

$$\frac{e_1}{R_1} = -\frac{e_0}{R_0}$$

$$e_0 = -\frac{R_0}{R_1} e_1$$

Hence the output signal is independent of the amplifier gain and depends only on the ratio of R_0 to R_1. The input impedance for the *circuit* is $Z = e_1/i = R_1$. Note that this is different from the operational amplifier input impedance.

We now have two basic rules for analyzing operational-amplifier circuits:

1. No current flows into any input terminal of the ideal operational amplifier.
2. When negative feedback is applied around the ideal operational amplifier, the voltage between the two input terminals approaches zero.

These two rules will be used repeatedly in the analysis of operational amplifier feedback circuits.

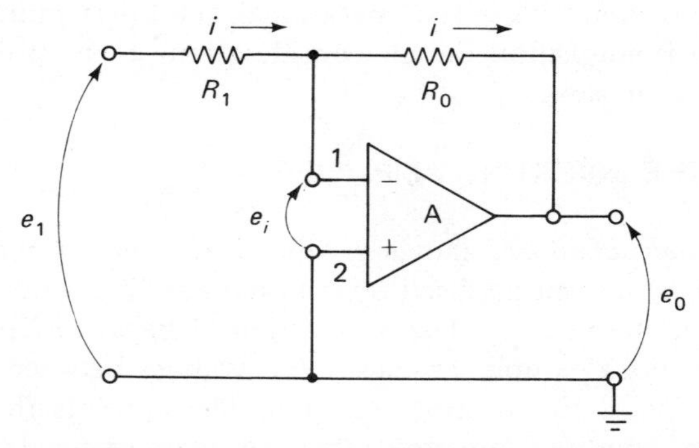

Figure 2-10 Inverting amplifier circuit using an operational amplifier, $e_0 = e_1(-R_0/R_1)$.

2-2-2.2 THE NONINVERTING AMPLIFIER

The circuit of Fig. 2-11 is of a *noninverting amplifier*. Rule 2 prescribes that the voltage of input 1 with respect to input 2 is zero; therefore, the voltage at input 1 is simply e_2. By rule 1 the current into input 1 is zero, so we can sum currents at the junction of the resistors to get

$$\frac{e_2}{R_1} = \frac{e_0 - e_2}{R_0}$$

$$e_2 R_0 = e_0 R_1 - e_2 R_1$$

$$e_0 = e_2 \frac{R_0 + R_1}{R_1}$$

The output is merely the input multiplied by the positive gain $(R_0 + R_1)/R_1$. The circuit input impedance is the voltage e_2 divided by the current flowing into input 2. But by rule 1 this current approaches zero. Therefore, the noninverting amplifier has a very high input impedance—10 MΩ and up for common operational amplifiers.

2-2-2.3 THE VOLTAGE SUMMER

For special applications where it is desirable to have e_0 be the sum or perhaps the weighted sum of several voltages, the circuit of Fig. 2-12 is used. Rule 2 (the assumption of infinite gain) puts input 1 at ground potential. Rule 1 (assuming infinite input impedance) means

$$-i_0 + i_1 + i_2 + i_3 + \cdots = 0$$

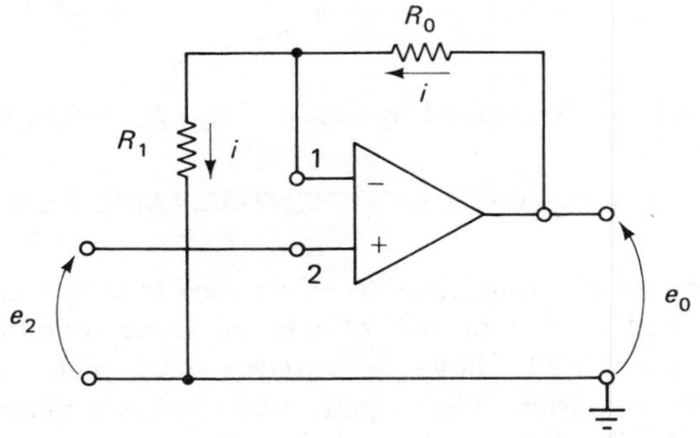

Figure 2-11 Noninverting amplifier, $e_0 = e_2(R_0 + R_1)/R_1$.

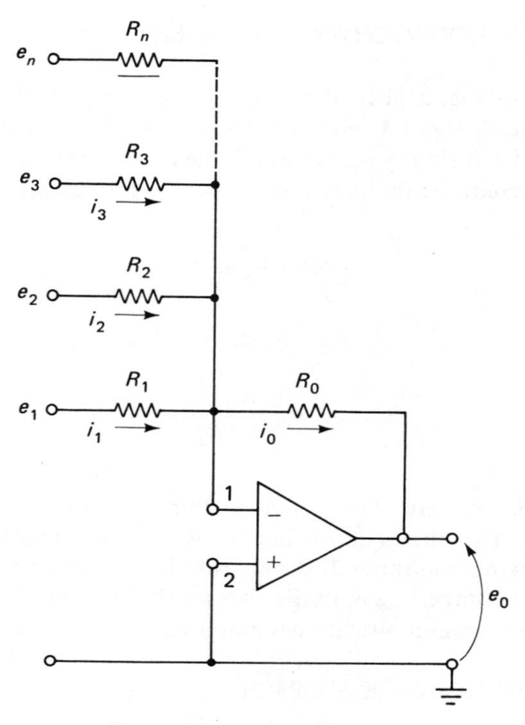

Figure 2-12 Voltage summer.

The current, i_0, is given by

$$i_0 = \frac{-e_0}{R_0} = \frac{e_1}{R_1} + \frac{e_2}{R_2} + \frac{e_3}{R_3} + \cdots$$

so that

$$e_0 = -R_0\left(\frac{e_1}{R_1} + \frac{e_2}{R_2} + \frac{e_3}{R_3} + \cdots + \frac{e_n}{R_n}\right)$$

Thus each input e_n is multiplied by factor $-R_0/R_n$ before summing.

2-2-2.4 VARIABLE-GAIN DIFFERENTIAL AMPLIFIER

Several operational amplifiers are often used in conjunction to optimize performance and minimize the effects of using nonideal operational amplifiers. Figure 2-13 shows a variable-gain, high-input impedance differential dc amplifier. The signal source is represented by the two difference-mode signals, e_1 and e_2, and by the common-mode component,

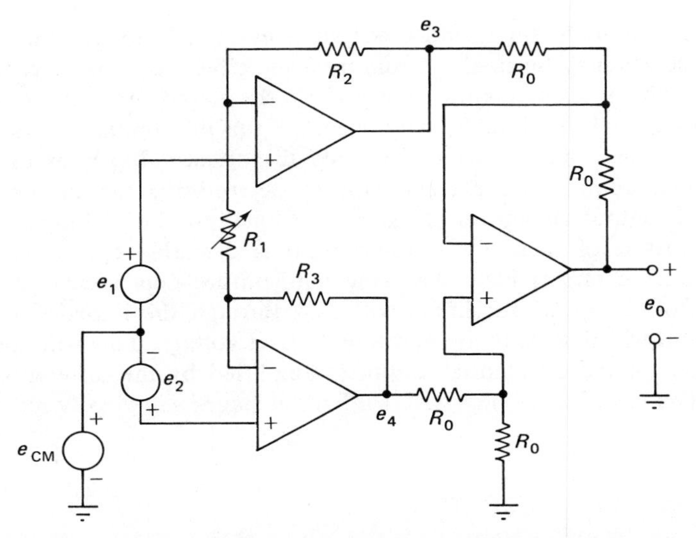

Figure 2-13 High-input-impedance adjustable-gain dc differential amplifier. (From G. E. Tobey, J. G. Graeme, and L. P. Huelsman, *Operational Amplifiers, Design and Application, Burr Brown*, McGraw-Hill Book Company, New York, 1971.)

e_{CM}. Application of our two basic operational-amplifier rules and Kirchhoff's voltage law yields

$$e_3 = \left(1 + \frac{R_2}{R_1}\right)e_1 - \frac{R_2}{R_1}e_2 + e_{CM}$$

$$e_4 = \left(1 + \frac{R_3}{R_1}\right)e_2 - \frac{R_3}{R_1}e_1 + e_{CM}$$

$$e_0 = e_4 - e_3$$

If R_2 equals R_3, the output voltage becomes

$$e_0 = \left(1 + \frac{2R_2}{R_1}\right)(e_2 - e_1)$$

This amplifier is a differential amplifier with a gain of $(1 + 2R_2/R_1)$ for differential signals, and zero gain for common-mode signals. The noninverting configuration of the input amplifiers ensures high input impedance. The gain is varied by resistor R_1. The effect of mismatching resistors R_2 and R_3 is simply to create a gain error without affecting the common-mode rejection of the circuit. The resistors R_0 of the output amplifier must be accurately matched, or trimmed, to ensure the rejection of common-mode signals at this point. This final amplifier acts as a differential-input to single-ended-output converter. Feedback impedances in both stages may be small, to minimize the effects of bias current, since these feedback elements do not affect the input impedance of the differential amplifier. Usually, all the gain of this differential amplifier is in the input stage, thus ensuring that only the offset voltages of these two operational amplifiers are significant in determining the output offset. Since the output-voltage offset is proportional to the difference of the voltage offsets of these two amplifiers, it is desirable to use amplifiers whose voltage offsets have the same temperature dependence. The bias currents of the input amplifiers will flow through the impedance of the source and will thus generate additional offset voltage. This will appear at the output of the differential amplifier amplified by the differential gain factor. The use of amplifiers with FET input stages will greatly reduce this effect.

2-2-2.5 THE INTEGRATOR

When a capacitor is used as the feedback element, as in Fig. 2-14, the circuit behaves as an *integrator*. The basis for this behavior is the law which states that current passing through a capacitor is proportional to the time derivative of the capacitor voltage:

$$i = C\frac{de}{dt}$$

Once again we assume that the currents flowing into the amplifier inputs are zero and that the voltage of input 1 relative to input 2 is zero. Now we

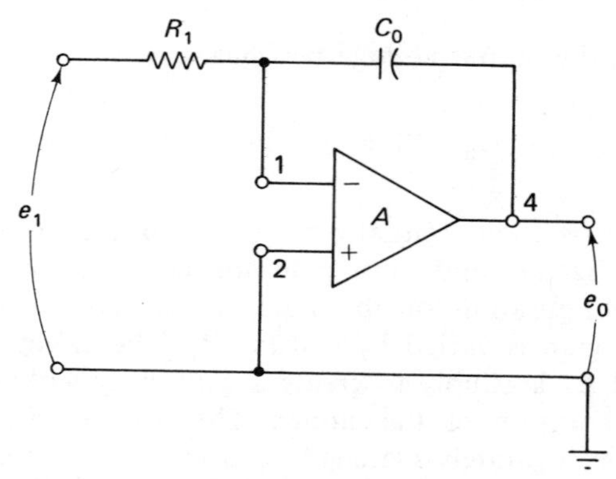

Figure 2-14 Integrating circuit using an operational amplifier, $e_0 = (-1/R_1C_0)\int e_1\, dt$.

can sum currents at the resistor–capacitor junction.

$$\frac{e_1}{R_1} + C_0 \frac{de_0}{dt} = 0$$

$$e_0 = \frac{-1}{R_1 C_0} \int e_1 \, dt$$

Thus we see that the output voltage is the integral of the input voltage.

2-2-2.6 ACTIVE FILTERS

The use of a reactive element in the amplifier circuit makes the circuit performance frequency dependent. For example, if the ideal capacitor of the previous integrator circuit is replaced with a resistor, R_0, in parallel with a capacitor, C_0, the output voltage becomes

$$e_0 = e_1 \left(\frac{1/C_0}{s + 1/R_0 C_0} \right)$$

This is a low-pass filter with a cutoff frequency at

$$\omega = \frac{1}{R_0 C_0}$$

Circuits such as this one, with denominators that are first order in s, are called *single-pole filters*.

Multiple-pole filters can be built using only one operational amplifier if multiple-feedback elements are used. Figure 2-15 shows the generalized form for a two-pole circuit. The correct choice of the circuit elements can produce low-pass, high-pass, bandpass, or band-reject filters. Such design details are given by Tobey, Graeme, and Huelsman (1971). As one such

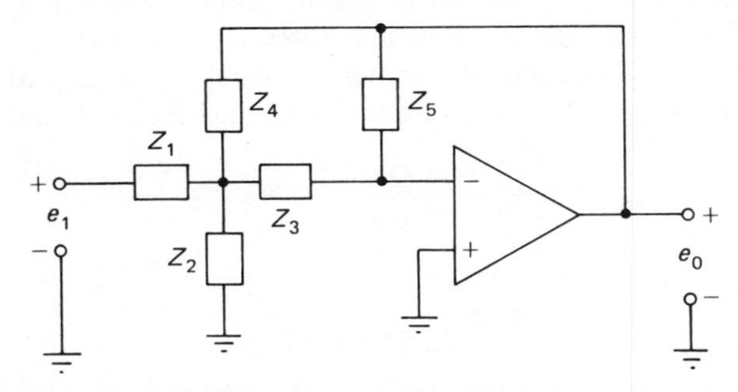

Figure 2-15 Generalized multiple-feedback filter circuit. Each Z_i represents a single resistor or capacitor. Correct choice of these elements will make the circuit either a high-pass, a low-pass, a bandpass, or a band-reject filter.

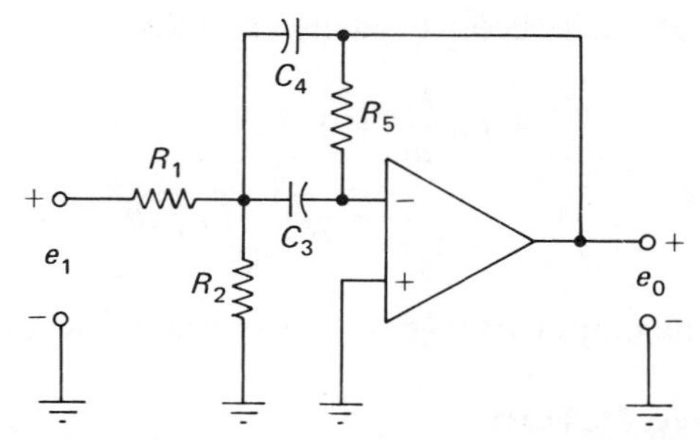

Figure 2-16 Multiple-feedback bandpass filter.

example, Fig. 2-16 shows a bandpass filter. The transfer function is

$$\frac{e_0}{e_1}(s) = \frac{-s(1/R_1C_4)}{s^2 + s(1/R_5)(1/C_3 + 1/C_4) + (1/R_5C_3C_4)(1/R_1 + 1/R_2)}$$

The frequency parameters are

$$\omega_0 = \left[\frac{1}{R_5C_3C_4}\left(\frac{1}{R_1} + \frac{1}{R_2}\right)\right]^{1/2}$$

$$\frac{1}{Q} = \left[\frac{1}{R_5(1/R_1 + 1/R_2)} \right]^{1/2} \left[\sqrt{\frac{C_3}{C_4}} + \sqrt{\frac{C_4}{C_3}} \right]$$

where

$$Q = \frac{\omega_0}{\omega_2 - \omega_1}$$

ω_2 and ω_1 are the frequencies where the magnitude response is 3 dB down from ω_0. The actual design of such a filter is easier than the equations might suggest. If Q, ω_0, and the center frequency gain, H_0, are specified, choose $C = C_3 = C_4$ to be a convenient value and then calculate

$$R_1 = \frac{Q}{H_0 \omega_0 C}$$

$$R_2 = \frac{Q}{(2Q^2 - H_0)\omega_0 C}$$

$$R_5 = \frac{2Q}{\omega_0 C}$$

2-2-2.7 BEHAVIOR OF ACTIVE FILTERS

The behavior of active filters can be illustrated in either the time domain or the frequency domain. Figure 2-17 shows Bode diagrams and time responses for two bandpass filters. The low-frequency cutoffs, 20 Hz and 100 Hz for these filters, are associated with the droop or slow drift back to the base line in the time response. The high-frequency cutoffs, 4 kHz and 20 kHz for these filters, are associated with the rise time of the voltage response. In fact, if the filters are single-pole filters, the high-frequency cutoff, f_h, will be related to the time constant, τ, as follows:

$$\tau = \frac{1}{\omega} = \frac{1}{2\pi f_h}$$

Most physiological amplifiers are bandpass amplifiers. They usually have front-panel dials to adjust the high- and low-frequency cutoff values. Many of the better amplifiers allow a setting of zero for the low-frequency cutoff. These are called *dc amplifiers*.

2-3 DIGITAL TECHNIQUES

A modern trend in instrumentation is to convert the data into digital form, to do the signal processing in digital form, and then to convert back into analog form. Integrated circuits, such as the Intel 2920, have been designed for this purpose. The economy of microcomputers and these special circuits has made it possible for most instruments to have a digital signal processing capability.

Digital signals are not as susceptible to corruption by noise as analog signals are. Analog amplifiers can drift; digital signals will not. Furthermore, the use of digital processing allows the computer to monitor the drift of the analog circuits and make compensating adjustments. Major modifications of the design can be done in digital machines merely by changing a software program. Analog circuits would require major redesign. Digital circuits can compute fast Fourier transforms, coherence functions, and correlations rapidly. Analog circuits to implement these functions are not common. However, analog circuits must still be used before the analog-to-digital conversion, and analog processing is usually faster than digital processing. Although real-time, 40-pole, bandpass filters with center frequencies as high as 5 kHz can be implemented with digital signal processing chips (Hoff and Townsend 1979; Jacobs, Landsburg, White, and Hodges 1979).

A classical data acquisition system has each measured parameter serviced by its own transducer and instrumentation amplifier. An

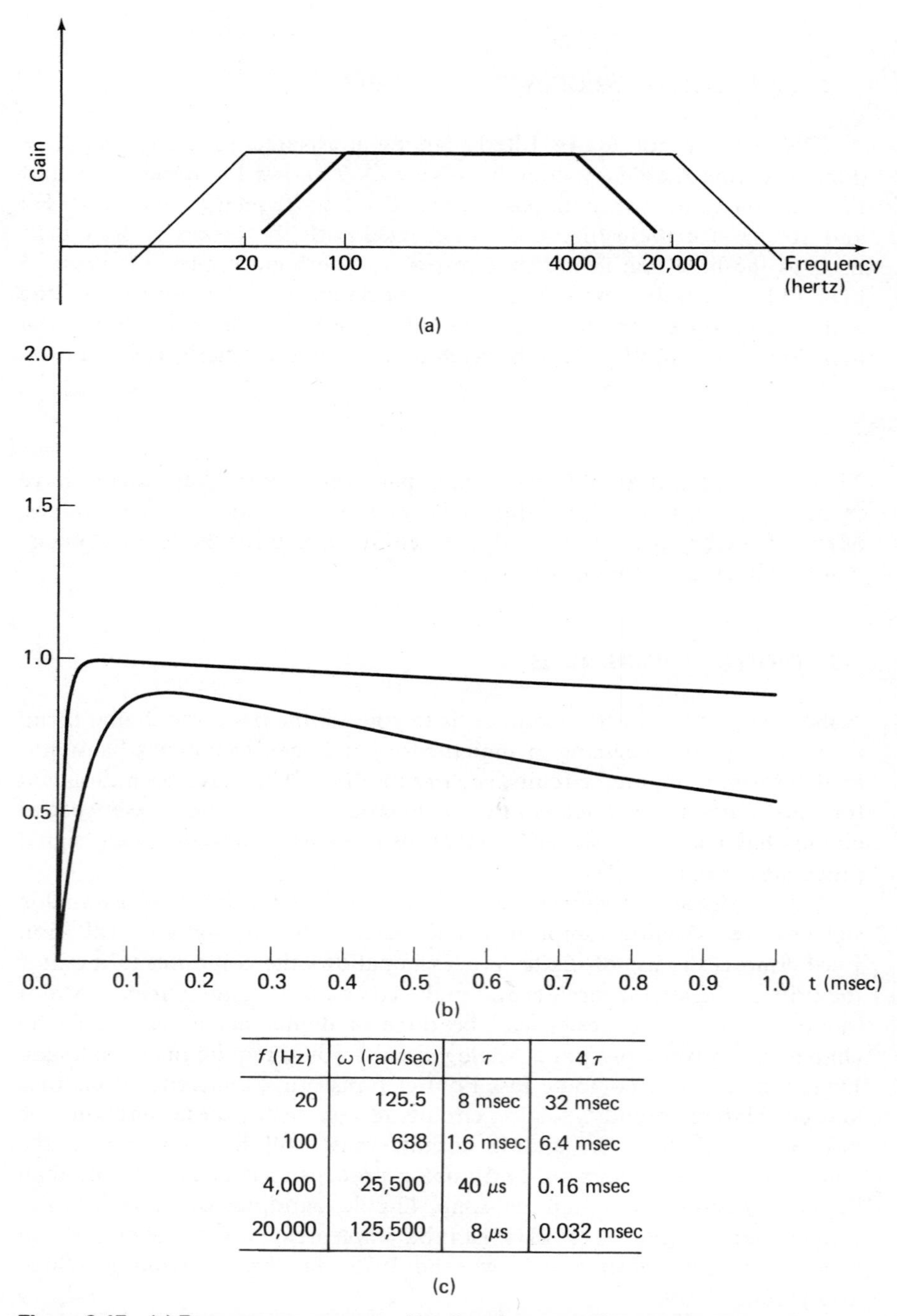

f (Hz)	ω (rad/sec)	τ	4τ
20	125.5	8 msec	32 msec
100	638	1.6 msec	6.4 msec
4,000	25,500	40 μs	0.16 msec
20,000	125,500	8 μs	0.032 msec

(c)

Figure 2-17 (a) Frequency responses of two bandpass amplifiers having low-frequency cutoffs of 20 and 100 Hz and high-frequency cutoffs of 4 and 20 kHz. (b) Time response for a unit-step input to the same two amplifiers. The high-frequency cutoffs influence the step-response rise time, and the low-frequency cutoffs are responsible for the gradual fall back to the baseline.

instrumentation amplifier is a high-input-impedance, dc, differential amplifier with modifiable gain and large common-mode rejection ratio, such as that of Fig. 2-13. From these instrumentation amplifiers the signals go to an analog multiplexor, then to a sample-and-hold circuit, and finally to an analog-to-digital converter. These digital signals are then to be processed by the computer.

Another type of data acquisition system converts the data into digital form at a much earlier stage (Morrison, 1978). In this system each parameter is still measured with its own transducer and instrumentation amplifier. However, the output of the amplifier is now fed to a voltage-to-frequency converter, whose period is directly proportional to the input voltage. This is now a digital signal. If the signal voltage is distorted by noise during transmission, the signal information is not contaminated. This frequency-coded signal can be changed into a binary number by hardwired logic or a microprocessor. If a multiplexor is used, it would be a digital multiplexor. This type of data acquisition system is not as fast or as accurate as the classical system, but it is less expensive and has a great deal of noise immunity for transmission.

A very important part of circuit design is the incorporation of test and diagnostic circuits into the instrument. To test or troubleshoot an analog instrument, a specified input voltage is applied and voltage waveforms of the circuit under test are compared to waveforms shown in a schematic diagram. The digital counterpart to this is *signature analysis* (Gordon and Nadig 1977; Pynn 1979). A specific train of pulses is applied to a special test point on the device under test and the signature analyzer's test probe is connected to a test point on the device. The serial bit stream from the test point on the instrument is fed to a linear-feedback shift register with tapped outputs driving, typically a four-digit alphanumeric display. The resulting four-digit display is then compared to the specified number printed on the schematic, the circuit board, or in a troubleshooting guide, or this four-digit number can be compared to the number displayed on a board known to be good. Circuit failures are isolated by moving the analyzer's probe from test point to test point until a signature is found which differs from the expected result. Machines with digital processors can and should be designed so that, when they are not busy taking or analyzing data, they will be running self-test diagnostic programs.

2-4 PATIENT LEAD DEVICES

There are certain restrictions put on the design of biomedical instruments because of the susceptibility of human beings to electric shock. As a result, most biomedical instruments have patient-protection devices. When possible, all measurements should be made without contacting the patient, but unfortunately, electrical contact usually cannot be avoided. In these cases, danger to the patient can be minimized if a device is used which limits the amount of current that can flow in the patient leads if faults occur. One such device is a fuse. Some patient protection can be gained by putting a fuse in each patient lead. This solution, however, is expensive, inconvenient, and the fuses may not blow before the patient is injured.

2-4-1 Diode Circuits

Diodes can be used to limit the voltage potential between the patient leads and thus control the fault current (Fig. 2-18). If voltages are present, one of these diodes will always be conducting. The largest voltage that can appear across a conducting diode is about 0.7 V. Therefore, no voltage larger than this could be applied to the patient. However, it is current, not voltage, which should be regulated. Therefore, we will install a resistor R_1, as shown in Fig. 2-19. The value of R_1 will be chosen so that only 10 μA of current can flow through the patient.

$$R_1 = \frac{0.7\text{ V}}{10\ \mu\text{A}} = 70\text{ k}\Omega$$

We have protected the patient with protective devices, but we must now

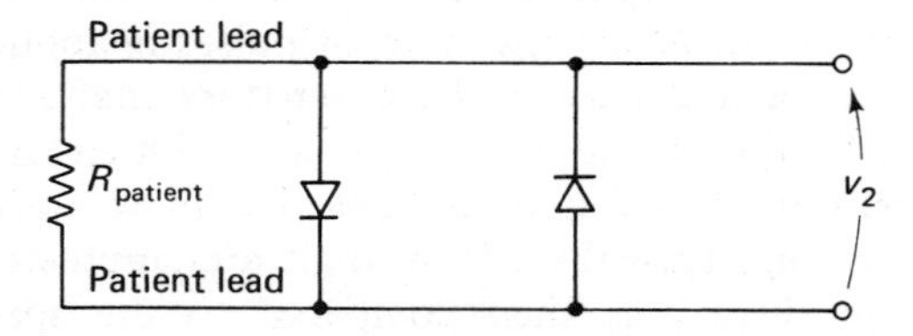

Figure 2-18 Two diodes used to protect the patient from voltages above 0.7 V.

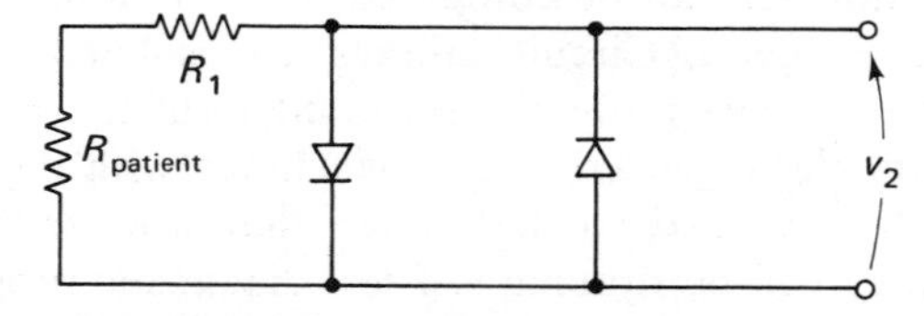

Figure 2-19 Resistor R_1 is added to limit the patient current.

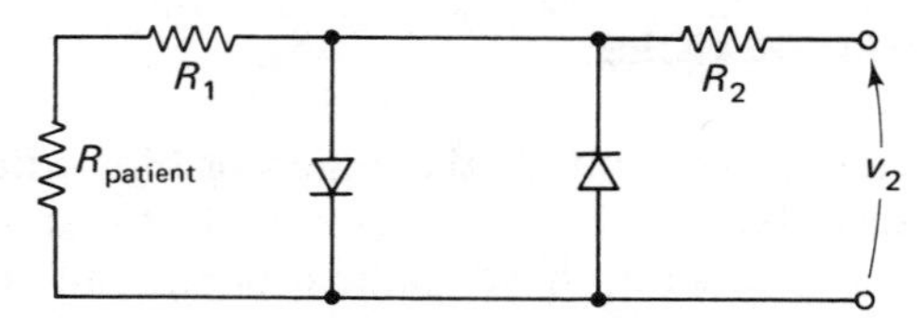

Figure 2-20 Resistor R_2 is added to protect the patient-protection diodes.

protect the protective devices. Specifically, if we apply 115 V at v_2, the diodes will blow out. So insert another resistor, R_2, as shown in Fig. 2-20. If the maximum allowable diode current is 23 mA rms, then

$$R_2 = \frac{115 \text{ V}}{23 \text{ mA}} = 5 \text{ k}\Omega$$

and the power dissipated is

$$P = I^2R = (23 \text{ mA})^2(5 \text{ k}\Omega) = 2.64 \text{ W}$$

Therefore, a 3-W 5-kΩ resistor should be used. Several diodes can be connected in series if the desired signal voltages are larger than 0.7 V. If a differential amplifier is used, it may be desirable to split R_1 into two resistors and put one of these in each lead. The same could be done with R_2. This would present a balanced load to the amplifier. Patient lead circuits such as these alter the physiological signals being measured. They attenuate the signals, introduce noise, and sometimes produce unwanted filtering of the data.

2-4-1.2 DIODE-BRIDGE CURRENT LIMITER

It is often desirable to ground a patient to reduce noise pickup. But grounding a patient is not recommended because of the dangers of providing a current path to ground. The circuit of Fig. 2-21 solves this dilemma in a simple fashion. It provides a bidirectional path for a maximum of 1 μA from the patient to ground. The 4-V sources and 4-MΩ resistors act as 1-μA current sources. If the voltage on the patient is such that it tries to push more than 1 μA through the diode bridge, one of the diodes will become reverse-biased and current will be limited. For example, if the patient ground were at the same potential as earth ground, the 1-μA current from the current sources would divide evenly and all four diodes would conduct 0.5 μA of current. However, if a potential difference existed such that 0.5 μA of current would flow from the patient ground to earth ground, the diode currents would become $D_1 = 0.25$ μA, $D_2 = 0.75$ μA, $D_3 = 0.75$ μA, and $D_4 = 0.25$ μA, assuming ideal diodes with no voltage drop across them. If the potential difference tried to push more than 1 μA through the bridge, diodes D_1 and D_4 would become nonconducting, thus limiting the current to 1 μA.

This same circuit can also be used to deglitch the output of a D to A converter. The left side of the bridge, V_o, is connected to the D to A converter, and the right side of the bridge, V_o', is connected to the load and to ground through a capacitor. Normally $V_o = V_o'$, so the circuit output

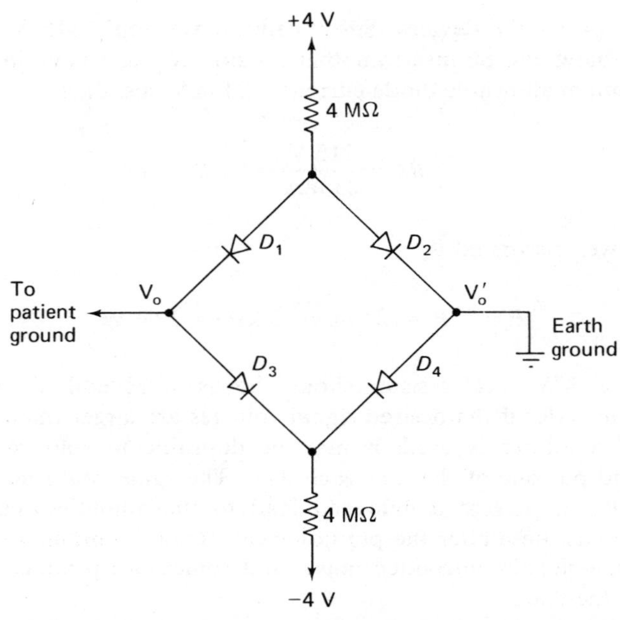

Figure 2-21 Diode bridge circuit that restricts current flow between patient ground and earth ground to 1 μA or less.

follows the output of the *D* to *A* converter. However, large, fast voltage spikes do not pass through the circuit; thus the glitches produced by the *D* to *A* converter are removed.

2-4-2 JFET Limiters

The cross section of a simplified *junction-field-effect transistor* (JFET) is shown in Fig. 2-22a. A positive voltage is applied to one side, of the *n* region arbitrarily termed the *drain*, and the other side, the *source*, is grounded. Since there is a continuous *n*-type semiconductor region between the two terminals, the *channel*, a current will flow. The resistance of the channel is determined by its dimensions and the resistivity of the *n*-type region.

If a negative voltage is applied to the *gate*, a depletion region will be created in the *n* material, reducing the effective cross section of the channel and thus increasing its resistance. The depletion region is deeper near the drain because the voltage between the gate and the *n* region increases in this direction, owing to the voltage drop along the channel.

For simplicity, assume that the gate is tied to the source terminal: the shape of the depletion region will be roughly the same as if a voltage were applied. When a voltage V_{DS} is applied between the drain and source, a

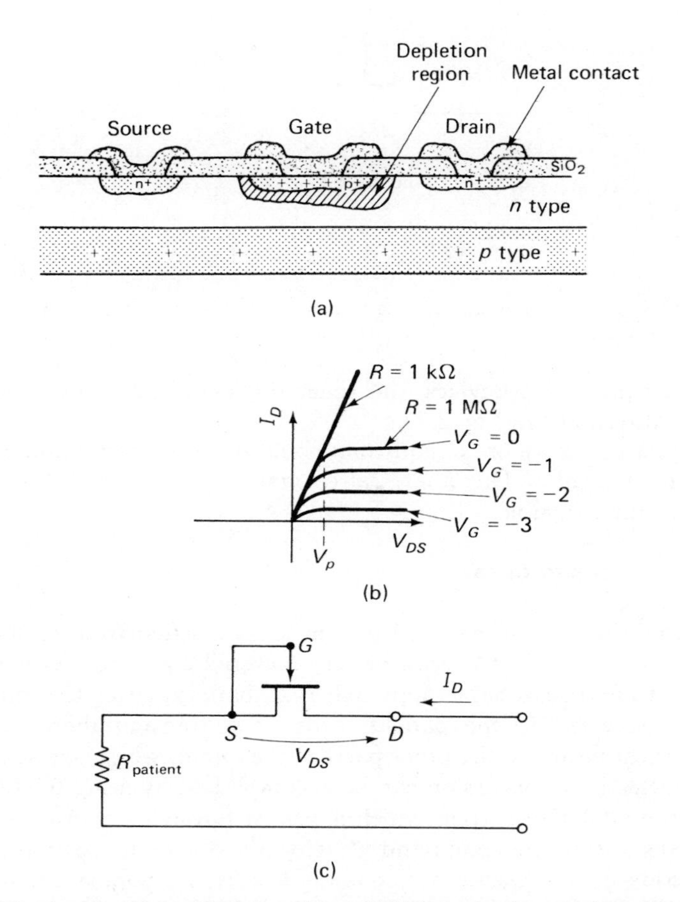

Figure 2-22 Junction-field-effect transistor (JFET) protection device: (a) physical structure, (b) V–I characteristic, and (c) circuit. The gate terminal (G) may be connected to the source terminal (S) as shown, or to a negative potential.

current will flow. This current will increase until the depletion region grows large enough to reach the bottom of the channel. This point is called *pinch-off*. The current does not drop to zero once the channel is pinched off, because if current stopped, the shape of the depletion region would change, allowing more current to flow. Therefore, just enough current flows through the channel to support the depletion region. The pinch-off voltage is a function of both the gate and the drain-to-source voltage. It is usually between 2 and 10 V. Therefore, if a JFET is placed in a patient lead, it will limit the current that can flow in that lead. Of course, this protective device should be protected from excessive voltages. The JFET can be protected with zener diodes, as shown in Fig. 2-23. Furthermore,

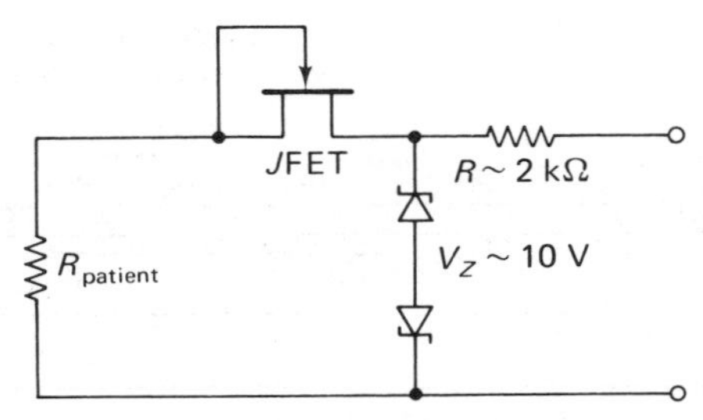

Figure 2-23 Zener diodes are added to protect the JFET and a resistor is added to protect the diodes.

these new protective devices, the zener diodes, can be protected with a resistor, also shown in Fig. 2-23.

There are many more circuits that could be designed to limit current in the patient leads. There are also many patient-protection devices commercially available.

2-4-3 Isolated Leads

It would be nice to have all patient leads isolated from earth ground. One way to do this is by using battery-powered equipment exclusively. A somewhat more reasonable approach is to battery-power the initial stage that is connected to the patient leads, then transmit these data by a nonelectrical means to the other parts of the circuit which are ac-powered. This nonelectric transmission can be accomplished by using the bioelectric signal to modulate a carrier, which is passed through a transformer. Then after transmission the compound waveform is demodulated to retrieve the bioelectric signal. Transformer coupling has been a popular technique for constructing isolation amplifiers. New variations of transformer coupling are being exploited in new isolation amplifier designs (Olschewski, 1978).

Most recently designed isolation amplifiers use optical coupling. Figure 2-24 shows this concept. Two matched photodiodes (PD) are arranged so that they receive equal amounts of light from a light-emitting diode (LED). This enables the effects of nonlinearities and temperature dependencies to be canceled. The operational amplifier A1 (assumed to be ideal) is connected in a negative feedback configuration so that

$$i_1 = i_{\text{in}} = \frac{v_{\text{in}}}{R_G}$$

where R_G is the user-supplied gain-setting resistor. PD1 and PD2 are closely matched and receive the same amount of light, so

$$i_2 = i_1 = i_{\text{in}}$$

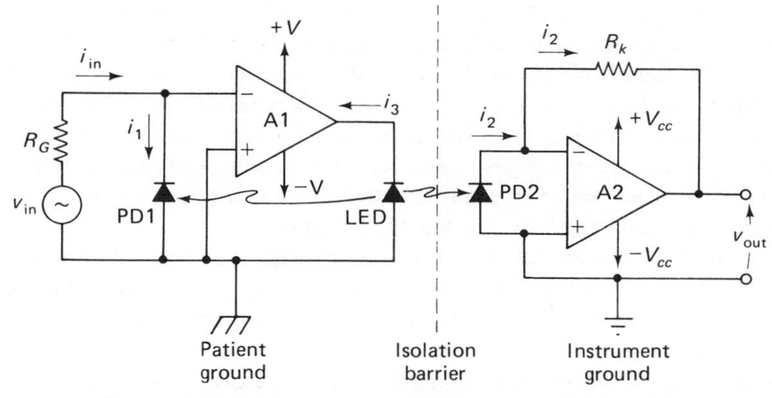

Figure 2-24 Simplified equivalent circuit for Burr–Brown 3650 series, optically coupled, isolation amplifiers, showing principles of operation for positive input voltages.

Amplifier A2 is connected as a current-to-voltage converter, so

$$v_{out} = i_2 R_K = v_{in} \frac{R_K}{R_G}$$

This output voltage is a linear function of the input voltage. However, more important, the patient is also isolated from ground. If a 60-Hz, 5000-V peak-to-peak sine wave were applied between the patient ground and the instrument ground, less than 3.7 μA rms of current would flow.

Isolation amplifiers are also used to protect computers and other control systems from inductive spikes produced by the motors and generators they are controlling.

2-5 SUMMARY

As a means of summarizing this chapter, let us investigate some techniques for making a particular physiological measurement: the measurement of eye position. To make a physiological measurement, one should examine the system to be studied and weigh the alternative methods for making the measurement. For example, eye movements can be measured either photographically or with television cameras. If, however, it is permissible to make physical contact with the eyeball, a contact lens can be placed on the eye. This contact lens can carry a coil of wire, a mirror, one plate of a capacitor, or accelerometers. These devices can then be used to sense eye position. Sometimes, as in the case of eye movements, the movements can be based on some unrelated physiological properties.

The technique called *electro-oculography* (EOG) uses the electronic charge on the eyeball to facilitate the measurement of eye movements. As a part of the metabolic process, the pigment epithelium, which lies behind the retina, pumps ions across a membrane. As a result, the retina carries a negative charge. The eyeball behaves as a sphere that is three-fourths covered with negative charges. These charges produce an electric field that can be measured with external electrodes. As the sphere rotates, the electrical field measured by the electrodes changes.

To measure horizontal eye movements with electro-oculography, the skin is cleansed with alcohol or an acne soap and three *silver–silver chloride electrodes* are taped to the skin: one on the nose, one on the temple, and a ground electrode on the ear lobe. These electrodes are connected to a *differential amplifier*. The signals obtained are *low-pass-filtered* and then analyzed. The technique is old, inexpensive, common, and easy to use. However, it is noisy, the dc level drifts, and it requires electrical contact with the subject. Electrical contact could be, but seldom is, eliminated by using *optically coupled isolation amplifiers*. Newer techniques avoid electrical contact by using a different measurement principle.

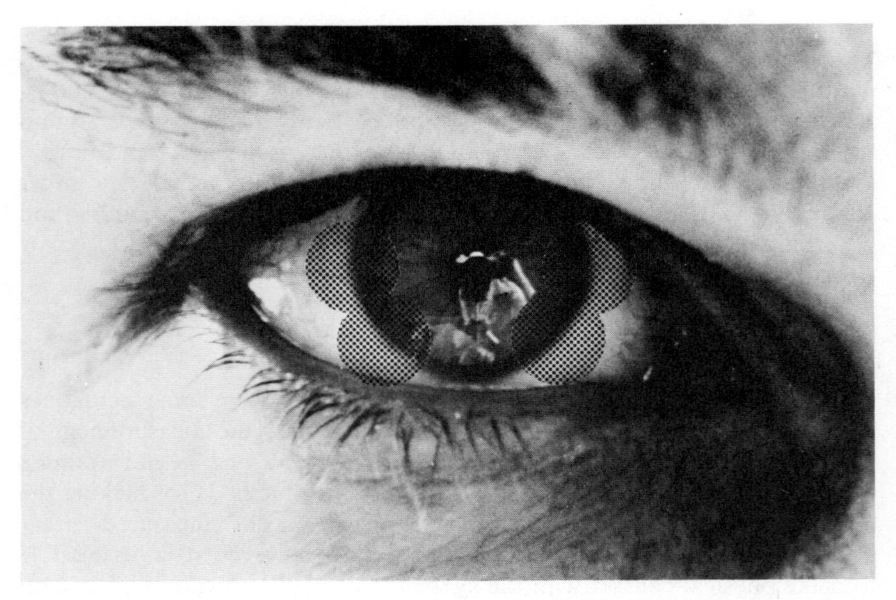

Figure 2-25 To measure human eye movements with the photoelectric technique, photodiodes are aimed so that they receive light from the stippled areas. The difference of the photocurrents from the upper two diodes is used to indicate horizontal eye position, and the sum of the photocurrents from the lower two diodes is used as a measure of vertical eye position. The reflection in the pupil is me and my camera.

The photoelectric technique for measuring eye movements employs four photodiodes which are mounted on a spectacle frame and aimed at the iris–sclera border, as shown in Fig. 2-25. The photodiodes are aimed so that they receive light from the stippled areas. The two upper areas are for horizontal eye movements. As the eye moves toward the nose, the temporal photodiode (the one nearest the temple) will see more of the white sclera, and its photocurrent will increase; meanwhile, the nasal photodiode will see more of the colored iris and its photocurrent will decrease. The difference of these two photocurrents is a measure of eye position. Any noise that is common to both sensors will be eliminated when the difference of the two signals is taken.

The photocurrents are changed into voltages by connecting the photodiodes to the input of inverting operational amplifiers similar to A2 in Fig. 2-24. The resulting voltages are then *differentially amplified*, *low-pass-filtered*, and passed through an *analog-to-digital converter* for subsequent digital signal processing.

These two instrumentation examples illustrate most of the principles discussed in this chapter.

REFERENCES

COBBOLD, R. S. C., *Transducers for Biomedical Measurements: Principles and Applications*. New York: John Wiley & Sons, Inc., 1974.

GEDDES, L. A., *Electrodes and the Measurements of Bioelectric Events*. New York: John Wiley & Sons, Inc., 1972.

GEDDES, L. A., and L. E. BAKER, *Principles of Applied Biomedical Instrumentation*. New York: John Wiley & Sons, Inc., 1975.

GEDDES, L. A., C. P. DA COSTA, and G. WISE, "The Impedance of Stainless-Steel Electrodes," *Medical and Biological Engineering*, 9 (1971), 511–21.

GESTELAND, R. C., B. HOWLAND, J. Y. LETTVIN, and W. H. PITTS, "Comments on microelectrodes," *Proceedings of the IRE*, 47 (1959), 1856–62.

GORDON, G., and H. NADIG, "Hexadecimal Signatures Identify Troublespots in Microprocessor Systems," *Electronics*, 52 (March 3, 1977), 89–96.

HOFF, M. E., and M. TOWNSEND, "Single-Chip *n*-MOS Microcomputer Processes Signals in Real Time," *Electronics*, 52 (March 1, 1979), 105–10.

JACOBS, G. G., G. F. LANDSBURG, B. J. WHITE, and D. A. HODGES, "Touch-Tone Decoder Chip Mates Analog Filters with Digital Logic," *Electronics*, 52 (February 15, 1979), 105–12.

MORRISON, R. L., "Microcomputers Invade the Linear World," *IEEE Spectrum*, *July* 1978, 38–41.

OLSCHEWSKI, W., "Unique Transformer Design Shrinks Hybrid Isolation Amplifier's Size and Cost," *Electronics*, 51 (July 20, 1978), 105–12.

PYNN, C., "In-Circuit Tester Using Signature Analysis Adds Digital LSI to Its Range," *Electronics*, 52 (May 24, 1979), 153–57.

TOBEY, G. E., J. G. GRAEME, and L. P. HUELSMAN, *Operational Amplifiers, Design, and Application*. New York: McGraw-Hill Book Company, 1971.

WEBSTER, J. G., *Medical Instrumentation, Application, and Design*. Boston: Houghton Mifflin Company, 1978.

PROBLEMS

2-1 A Zn electrode and an Ag–AgCl electrode are placed in a beaker of ZnCl at 25°C. Describe the chemical reactions occurring at these electrodes. If the free ends of the wires are connected to a high-impedance voltmeter, what voltage will be measured? What will be the direction of current flow?

2-2 If a copper wire and an aluminum wire are connected to a high-impedance voltmeter and the free ends are lowered into a jar of salt water, what voltage will be registered? Would the type of salt (e.g., NaCl) be important? If a gold electrode and an aluminum electrode were used to record an electrocardiograph, what artifacts would appear?

2-3 An exotic new animal, recently hypothesized by M. R. Neuman in Webster (1978), has an unusual electrolyte makeup with the principal anion being Br^- rather than Cl^-. Attempts to make physiological measurements with normal Ag–AgCl electrodes yielded noisy signals. Why? Postulate and explain a better electrode system.

2-4 Assume that you are using electrodes with the behavior described by Figs. 2-5 and 2-6 to measure a physiological event. Current density is 0.025 mA/cm^2. The physiological signal has uniform power in all bands from zero to 10 kHz. Sketch the amplitude-versus-frequency response of the original signal, and of the signal after it passes through the electrode and a 100-kΩ input impedance amplifier. How do these curves change if current density increases to 10 mA/cm^2?

2-5 Derive the input/output ratio for the circuit of Fig. P2-5. Sketch the Bode diagram. What name would you give to this circuit?

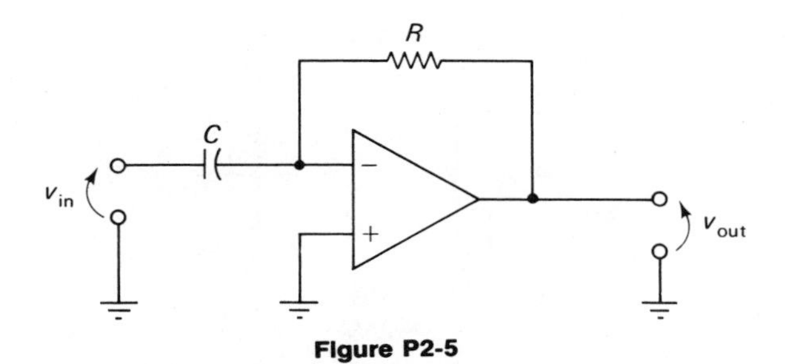

Figure P2-5

2-6 Derive the input/output ratio for the circuit of Fig. P2-6. Sketch the Bode diagram. What name would you give to this circuit? This circuit has improved noise and stability over that of Fig. P2-5.

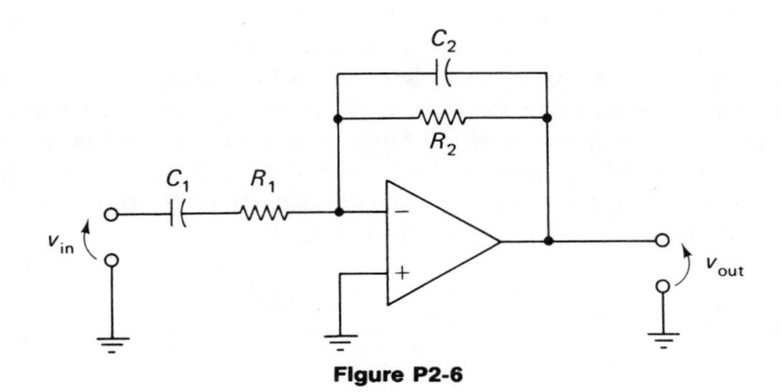

Figure P2-6

2-7 Find expressions for the differential-mode and common-mode gain for the circuit of Fig. 2-9. Express the common-mode rejection ratio, CMRR, as the ratio of these two gains. How does this ratio depend upon the matching of R_1 and R_2?

2-8 Design an integrator with an input resistance of 1 MΩ that will drive the output from 0 to −5 V in 1 sec if an input of +5 V is applied. If the amplifier is not ideal and 1 μA of bias current flows into the negative terminal, how long will it take the output to drift from 0 to ±5 V if the input is grounded? Suggest a method of improving this circuit.

2-9 A photodiode acts as a light-sensitive current source when it is reverse-biased. Sketch an operational amplifier circuit that will have a voltage output that is proportional to light intensity.

2-10 Compute the transfer function v_o/v_{in} for the circuit of Fig. P2-10. Propose a model of this circuit that uses only two passive components. What are the values of these components? Calculate the input impedance for the circuit.

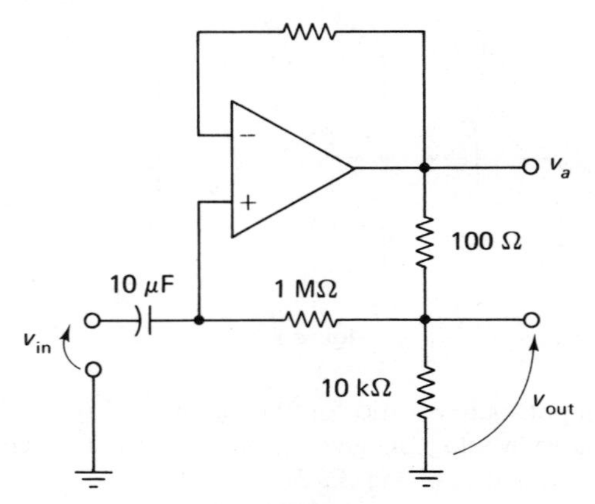

Figure P2-10

2-11 The most obvious way to build a filter circuit is with inductors, capacitors, and resistors. But it is difficult to build integrated circuits containing inductors, so they are simulated with operational amplifiers and capacitors. In some processes capacitors are easier to make than resistors; so why not simulate resistors with operational amplifiers and capacitors? Show that the circuit of Fig. P2-11 (from Jacobs, Landsburg, White, and Hodges 1979) does just that. The switch must operate at a much higher frequency than the signals of interest. What type of circuit is this? Derive an input/output equation for it.

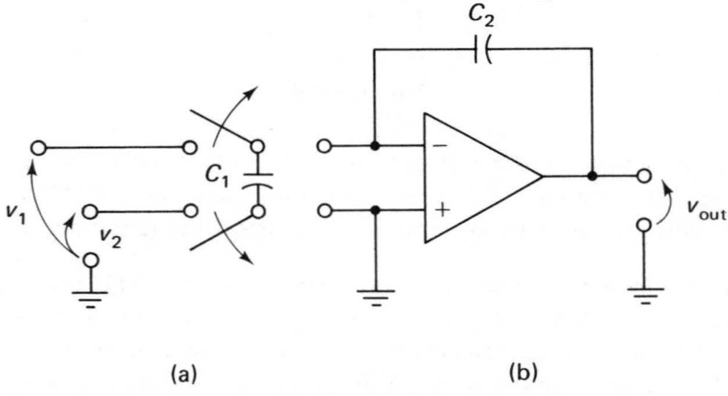

Figure P2-11

2-12 The thin coating on Ag–AgCl electrodes is the layer of Ag–AgCl. How would cleaning these electrodes with steel wool affect biological measurements? Is there a way to salvage electrodes that have been cleaned in such a manner?

3

OPEN-LOOP SYSTEMS

3-1 WHY USE LAPLACE TRANSFORMS?

There are many mathematical techniques for analyzing systems. Differential equations are often convenient. Electrical circuits with sinusoidal sources are often analyzed with phasor techniques, and bioengineering models are often analyzed using Laplace transforms. To demonstrate clearly how the differential equation technique (often called the time-domain technique because time is the independent variable in the equations) is parallel to the Laplace transform technique (often called the frequency-domain technique because the complex frequency, s, is the independent variable in the equations), we analyze the electrical circuit of Fig. 3-1 with each technique.

Find v_{out} when v_{in} is a unit step at time $t = 0$. Assume that all initial conditions are zero. The fundamental defining relationships for the circuit elements are

$$v_R(t) = Ri(t) \tag{3-1}$$

$$v_C(t) = \frac{1}{C}\int_0^t i(t)\,dt + v_C(0) \tag{3-2}$$

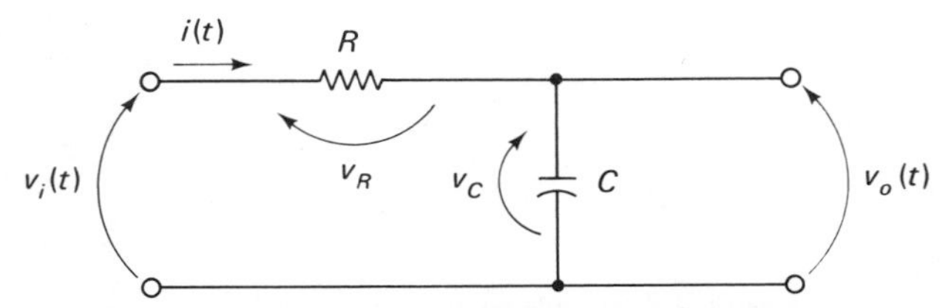

Figure 3-1 *RC* circuit with time-domain notation.

For simplicity, let $v_C(0) = 0$. By Kirchhoff's voltage law

$$v_{\text{in}}(t) = v_R(t) + v_C(t) \tag{3-3}$$

Substituting the defining relations of Eqs. (3-1) and (3-2) yields

$$v_{\text{in}}(t) = Ri(t) + \frac{1}{C}\int_0^t i(t)\,dt \tag{3-4}$$

Solving differential equations is easier than solving integral equations. When $t > 0$, v_{in} is a constant, so differentiating both sides of the equation with respect to time yields

$$0 = R\frac{di}{dt} + \frac{1}{C}i(t)$$

Separation of variables yields the form

$$\frac{di}{i(t)} = \frac{-1}{RC}\,dt$$

This equation can now be integrated:

$$\ln i(t) = \frac{-t}{RC} + K_i$$

where K_i is the constant of integration. This constant is most conveniently written in logarithmic form:

$$\ln i(t) = \frac{-t}{RC} + \ln K$$

This equation can be rearranged as

$$\frac{-t}{RC} = \ln i(t) - \ln K$$

$$\frac{-t}{RC} = \ln\frac{i(t)}{K}$$

By the definition of a logarithm, $e^a = \ln b$, we get

$$\frac{i(t)}{K} = e^{-t/RC} \tag{3-5}$$

The constant K can be evaluated by noting that the voltage across a capacitor cannot change instantaneously. Therefore, at $t = 0$,

$$v_{\text{out}}(0) = 0 = v_{\text{in}} - Ri(t)$$

and from Eq. (3-5) at $t = 0$, $i = K$. Therefore, $v_{\text{out}}(0) = 0 = 1 - KR$ and

$$K = \frac{1}{R}$$

Now the complete solution for the current is

$$i(t) = \frac{e^{-t/RC}}{R} \tag{3-6}$$

Equation (3-6) can be substituted into Eq. (3-2), and by noting that $v_C(t) = v_{\text{out}}(t)$, we get

$$v_{\text{out}}(t) = \frac{1}{C}\int_0^t i(t)\,dt = \frac{1}{RC}\int_0^t e^{-t/RC}\,dt$$

Carrying out the indicated integration yields

$$v_{\text{out}}(t) = \frac{1}{RC}(-RC)e^{-t/RC}\Big|_0^t$$

$$v_{\text{out}}(t) = -(e^{-t/RC} - 1)$$

and

$$v_{\text{out}}(t) = 1 - e^{-t/RC} \tag{3-7}$$

This equation describes the output voltage as a function of time when a unit step is applied to the input.

Now let us find V_{out} using frequency-domain techniques. The circuit of Fig. 3-2 is described by Kirchhoff's voltage law as

$$V_{\text{in}} = RI + \frac{1}{sC}I \tag{3-8}$$

where V_{in} means $V_{\text{in}}(s)$, which is equal to the Laplace transform of the

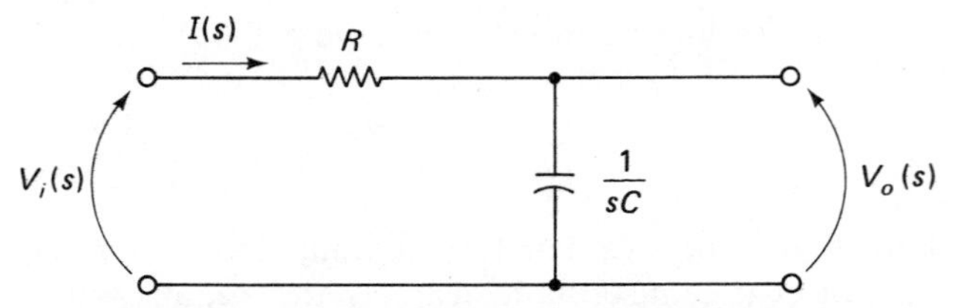

Figure 3-2 *RC* circuit with frequency-domain notation.

input, $\mathcal{L}[v_{\text{in}}(t)]$. Likewise, $I = I(s) = \mathcal{L}[i(t)]$. Equation (3-8) can be rearranged as

$$I = \frac{V_{\text{in}}}{R + 1/sC} \tag{3-9}$$

and

$$V_{\text{out}} = \frac{1}{sC} I \tag{3-10}$$

$$V_{\text{out}} = \frac{1}{sC} \frac{V_{\text{in}}}{R + 1/sC} \tag{3-11}$$

$$V_{\text{out}} = \frac{V_{\text{in}}/RC}{s + 1/RC} \tag{3-12}$$

If V_{in} is a unit step, then

$$V_{\text{out}} = \frac{1/sRC}{s + 1/RC} \tag{3-13}$$

From a table of Laplace transforms such as Table 3-1, we find the resulting time response to be

$$v_{\text{out}}(t) = 1 - e^{-t/RC} \tag{3-14}$$

By using Laplace transforms, we were able to solve the problem using algebraic equations instead of differential equations; for this reason, the frequency-domain solution is usually considered to be easier.

Often, it is desirable to characterize a system for any general input. A common way of doing this is to compute the ratio of $V_{\text{out}}(s)$ to $V_{\text{in}}(s)$. This is called a *transfer function*. All initial conditions are assumed to be zero. The transfer function of the circuit described above is

$$M(s) = \frac{V_{\text{out}}(s)}{V_{\text{in}}(s)} = \frac{1/RC}{s + 1/RC} \tag{3-15}$$

TABLE 3-1 A short table of Laplace transforms.

1	Unit impulse, $\mu_0(t)$
$1/s$	Unit step, $\mu_{-1}(t)$
$1/s^2$	Unit ramp, t
$\frac{1}{s+a}$	e^{-at}
$\frac{a}{s(s+a)}$	$1-e^{-at}$
$\frac{1}{(s+a)^2}$	te^{-at}
$\frac{b-a}{(s+a)(s+b)}$	$e^{-at}-e^{-bt}$
$\frac{\omega}{s^2+\omega^2}$	$\sin \omega t$
$\frac{s}{s^2+\omega^2}$	$\cos \omega t$
$\frac{\omega^2}{s(s^2+\omega^2)}$	$1-\cos \omega t$
$\frac{\omega}{(s+a)^2+\omega^2}$	$e^{-at}\sin \omega t$
$\frac{s+a}{(s+a)^2+\omega^2}$	$e^{-at}\cos \omega t$
$\frac{\omega_n^2}{s^2+2\zeta s\omega_n+\omega_n^2}$	$\frac{\omega_n}{\sqrt{1-\zeta^2}}[e^{-\zeta\omega_n t}\sin(\omega_n\sqrt{1-\zeta^2}\,t)]$
$\frac{\omega_n^2}{s(s^2+2\zeta s\omega_n+\omega_n^2)}$	$1-\frac{e^{-\zeta\omega_n t}}{\sqrt{1-\zeta^2}}\sin(\omega_n\sqrt{1-\zeta^2}\,t+\Phi)$ where $\Phi = \text{Arc tan}\,\frac{\sqrt{1-\zeta^2}}{\zeta}$
$e^{\pm as}$	$\mu_0(t \pm a)$

Now for any particular applied voltage, we can compute $V_{out}(s) = M(s)V_{in}(s)$ and then find the time response by inverse Laplace transforms.

3-2 THE IMPULSE RESPONSE

A unit impulse is defined, in the limit as $1/A$ goes to zero, as a function with height A and width $1/A$. It is convenient because the Laplace transform of a unit impulse is unity.

$$\mathcal{L}[\mu_0(t)] = 1$$

Instead of describing a system by its transfer function, we often describe it by its response to a unit impulse. For the previous example

$$V_{out}(s) = \frac{1/RC}{s+1/RC}$$

Thus $V_{out}(s) = M(s)$ and

$$m(t) = \mathcal{L}^{-1}[M(s)]$$

This is called the *impulse response*, or the *weighting function*. Thus if a system with transfer function $M(s)$ is excited by an impulse $\mu_0(t)$, then

$$v_{out}(t) = m(t)$$

Unit impulses are not readily available for studying physical systems. However, the mathematics is so neat that impulse responses are often used to describe systems. In many cases, unit steps, sinusoids, or white-noise inputs are used to determine the system transfer function, and then the results are presented as an impulse response. The form of the impulse response is the same as the unforced response of a system to its initial conditions.

3-3 THE IDENTIFICATION PROBLEM

The basic nature of bioengineering problems is different from other engineering problems. In each, the engineer has to make a model for a physical system: this means that the engineer must design a control system subject to certain constaints. However, for bioengineering problems, the constraints are unusual. They are not specified in terms of rise time, overshoot and sensivity to parameter variations: they are specified by nature. The bioengineer has to figure out what performance criteria would describe the system best. Then he has to design such a system and see how closely it matches the physiological system. The bioengineer must be familiar with more performance criteria and more types of control.

A typical problem in an electrical engineering course would be to construct a Bode diagram of a given circuit. We will now perform the inverse task. You are given a black box containing resistors, inductors, or capacitors, and are asked to identify what is inside.

Types of inputs that may be helpful in this analysis are:

1. Transients
   a. Impulse
   b. Step
   c. Ramp
2. Sinusoids
3. Noise
   a. White, Gaussian
   b. Pseudo-random binary sequences

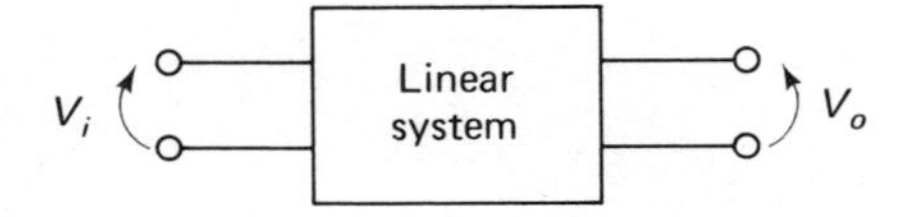

Figure 3-3 Unknown linear network.

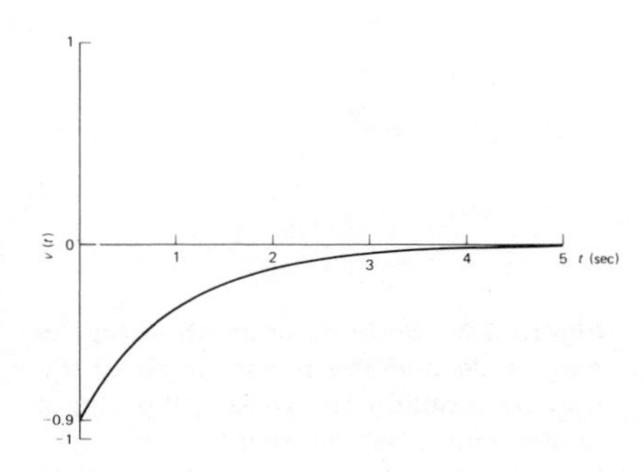

Figure 3-4 Output of the system under study, as a function of time, in response to a unit impulse.

When an impulse of voltage is applied to the circuit of Fig. 3-3, the output waveform of Fig. 3-4 results. Therefore, we can conclude that the box does not contain a simple *RC* parallel or series circuit because the peak magnitude is $\frac{9}{10}$, not 1. We should be able to use this waveform to answer the question of what is in the box, but it looks difficult, so let us try another input waveform.

Applying a unit step yields the output waveform of Fig. 3-5. Once again this waveform contains all the information needed to answer the question, but the answer is not yet obvious. So let us try another technique.

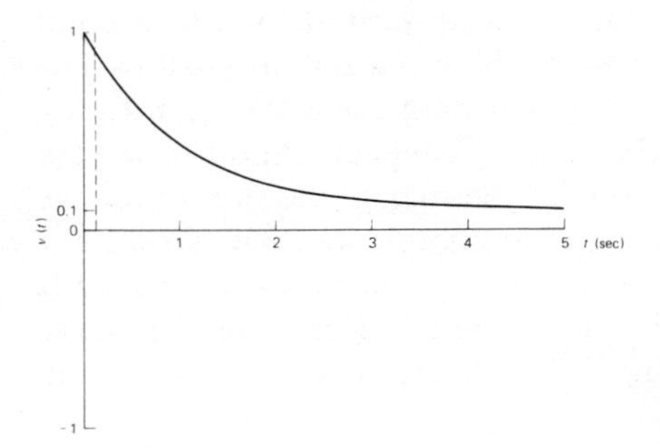

Figure 3-5 System output, as a function of time, in response to a unit-step input.

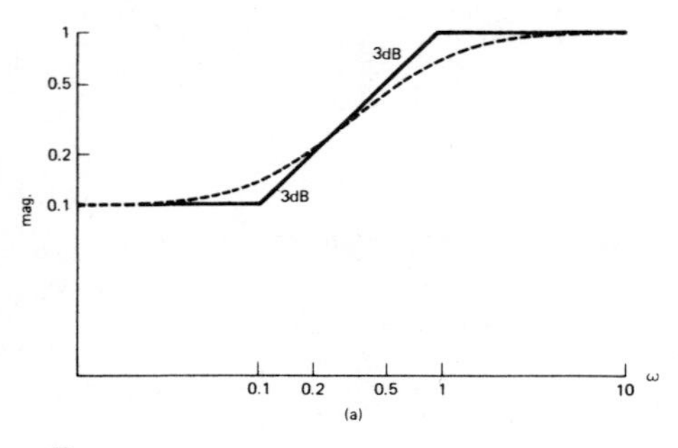

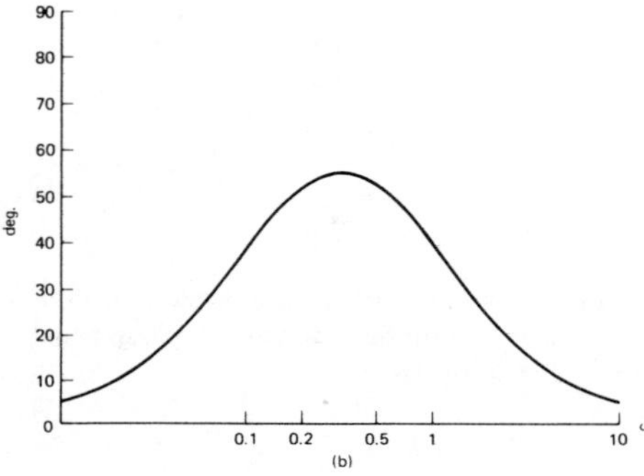

Figure 3-6 Bode diagram showing the magnitude and the phase angle of the transfer function, the ratio of the output to the input, as the function of log ω. Dashed lines are actual data, straight lines are asymptotic approximations.

Applying sinusoids yields the Bode diagrams of Fig. 3-6. From these Bode diagrams we can infer the following transfer function.

$$M(s) = \frac{V_0(s)}{V_{in}(s)} = \frac{1}{10} \frac{(10s+1)^n}{1} \frac{1}{(s+1)^n} \tag{3-16}$$

$$M(s) = \frac{1}{10} \frac{10s+1}{s+1} \tag{3-17}$$

The fraction one-tenth is present because one-tenth is the gain when s approaches zero. The term $(10s + 1)$ is in the numerator because the magnitude portion of the Bode diagram bends upward at the break point $s = 0.1$ and the phase angle is increasing (i.e., there is a zero at $s = 0.1$). The exponent n equals 1 because the experimental data are 3 dB, or 0.15 log unit, away from the asymptotes at the break frequency. Another way to determine this is by noting that the slope of the magnitude portion of the Bode diagram is 1 when the plot is on log-log coordinates. The term $s + 1$ is in the denominator because there is a break point at $s = 1$ where the magnitude portion turns down and the phase angle is decreasing (i.e., there is a pole at $s = 1$). The circuit of Fig. 3-7 with the following values will

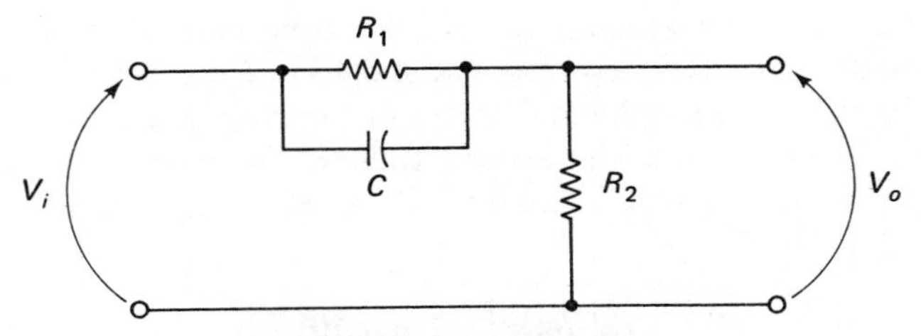

Figure 3-7 Unknown circuit.

yield this transfer function: $R_2 = 1.1\ \text{M}\Omega$, $R_1 = 10\ \text{M}\Omega$ and $C = 1\mu\text{F}$.

To show clearly that this is different from normal engineering problems, let us now use frequency-domain techniques and the normal engineering approach to study the system, whose transfer function is given in Eq. (3-17) and repeated here.

$$\frac{V_0(s)}{V_{\text{in}}(s)} = \frac{1}{10}\,\frac{10s+1}{s+1}$$

Find the impulse response.

$$V_0(s) = \frac{1}{10}\,\frac{10s+1}{s+1} = \frac{s}{s+1} + \frac{1/10}{s+1}$$

From a table of Laplace transforms, we find that

$$v(t) = \frac{d}{dt}e^{-t} + \frac{1}{10}e^{-t}$$

so the impulse response is

$$v(t) = \frac{-9}{10}e^{-t} \tag{3-18}$$

Find the step response of the system.

$$V_0(s) = \frac{1}{10s}\,\frac{10s+1}{s+1} = \frac{1}{s+1} + \frac{1/10}{s(s+1)}$$

From a table of Laplace transforms, we find that

$$v_0(t) = e^{-t} + \tfrac{1}{10}(1 - e^{-t}) = \tfrac{1}{10} + \tfrac{9}{10}e^{-t} \tag{3-19}$$

Note that these two equations fit the experimental graphs for the impulse response and the step response shown in Figs. 3-4 and 3-5.

Bode diagrams are graphical displays of the magnitude and phase, as a function of frequency, for the system transfer function, the ratio of the output to the input. To sketch the Bode diagram for this system, substitute $s = j\omega$ into Eq. (3-17):

$$\frac{V_0(j\omega)}{V_{in}(j\omega)} = \frac{1 + j\omega 10}{10(1 + j\omega)}$$

Usually, the magnitude ratio and the frequency axes are logarithmic while the phase axis is linear. To find the magnitude plot, take the logarithm of this expression:

$$\log\left[\frac{V_0(j\omega)}{V_{in}(j\omega)}\right] = \log[1 + j\omega 10] - \log[1 + j\omega] - \log 10$$

To find the low-frequency approximation for this function, let ω be very small,

$$\log\left[\frac{V_0(j\omega)}{V_{in}(j\omega)}\right] \approx -\log 10 = -1$$

This is a horizontal line. It is the low-frequency asymptote. The phase angle in this region is zero. To find the high-frequency asymptote, let ω be very large. Equation (3-17) becomes

$$\frac{V_0(s)}{V_{in}(s)} \approx \frac{10(j\omega)}{10(j\omega)} = 1$$

The logarithm of 1 is zero, so this asymptote is also a horizontal line and the phase angle is also zero. To find the midfrequency asymptote, let ω be near 0.1; then

$$\log\left[\frac{(V_0)}{(V_{in})}\right] = \log[1 + j\omega 10] - \log[10(1 + j\omega)]$$

which is approximately equal to

$$\log 10\omega - \log 10$$

The first term is a linear function of ω and the second term is a constant. Therefore, the portion of the curve connecting the high- and low-frequency asymptotes is a straight line with slope of +1. The phase angle will vary as ω varies between 0.1 and 1. It can be determined by direct numerical calculation or by using the sketching techniques presented in systems and control theory textbooks. These Bode diagrams are the same as those given in Fig. 3-6.

This circuit is called a *lead-compensation network*. The metal microelectode model of Chap. 2 was such a network. This type of circuit can be used to give preemphasis. If a lead compensation network is inserted before a lagging load, the response will speed up. Systems like this with phase lead are unusual; most physical and biological systems have phase lag.

3-4 LAPLACE TRANSFORM OF A TIME DELAY

Biological systems usually have time delays that are important in modeling the system. The time delay that is required for the action potential to be transmitted along the axon has already been mentioned. The idealized pure time delay is defined as an element that delays, but transmits without distortion any input signal presented to it. No such idealized element can exist, of course, but the axon approximates it for the action potentials.

The time delay also arises as the appropriate model for transportation delay. For example, a time delay occurs when the blood transports nutrients and hormones which are exchanged at various sites after traveling some distance in the blood stream.

Traditional engineering problems seldom consider such time delays, so we must derive the Laplace transform for the pure delay element. Consider the function shown in Fig. 3-8.

$$g(t) = f(t-a)\mu_{-1}(t-a)$$

where a is a constant and μ_{-1} is the step function. By definition of the

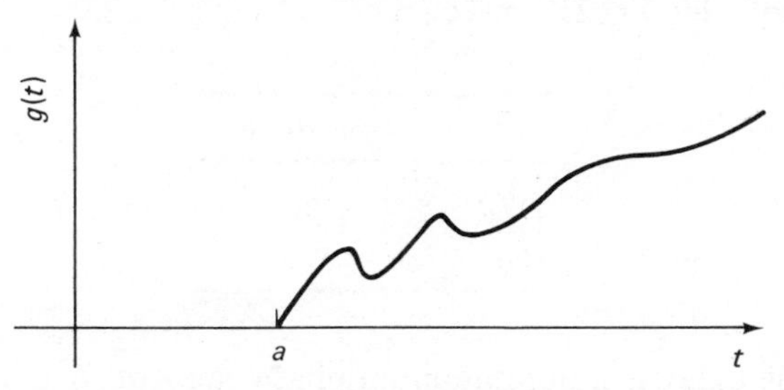

Figure 3-8 Function whose onset is delayed a seconds after time $t = 0$.

Laplace transform,

$$G(s) = \int_0^\infty f(t-a)e^{-st}\,dt$$

The unit step ensures that $g(t) = 0$ for $t < a$; therefore,

$$G(s) = \int_a^\infty f(t-a)e^{-st}\,dt$$

Now substitute $x = t - a$ and $dx = dt$, and note that $t = a$ when $x = 0$ (for the lower limit of integration). Then

$$G(s) = \int_0^\infty f(x)e^{-(a+x)s}\,dx = e^{-as}\int_0^\infty f(x)e^{-xs}\,dx$$

x is simply a dummy variable of integration. Therefore, we can replace it with t.

$$G(s) = e^{-as}\int_0^\infty f(t)e^{-ts}\,dt$$

this integral is by definition the Laplace transform of $f(t)$. Therefore,

$$G(s) = e^{-as}F(s) \tag{3-20}$$

meaning that a shift to the right of a seconds in the time-domain corresponds to multiplying by e^{-as} in the frequency domain. Bode diagrams for a time delay are shown in Fig. 3-9.

$$H(s) = e^{-sa}$$
$$H(j\omega) = e^{-j\omega a}$$

The magnitude is always 1. The phase equals $-\omega a$ and is always increasing. For example, let $a = 250$ msec; then

f	*Phase Angle*
1 Hz	90°
2 Hz	180°
4 Hz	360°

A pure time delay is a nonminimum phase element. A minimum phase element has the smallest phase lag possible for a realizable system with any

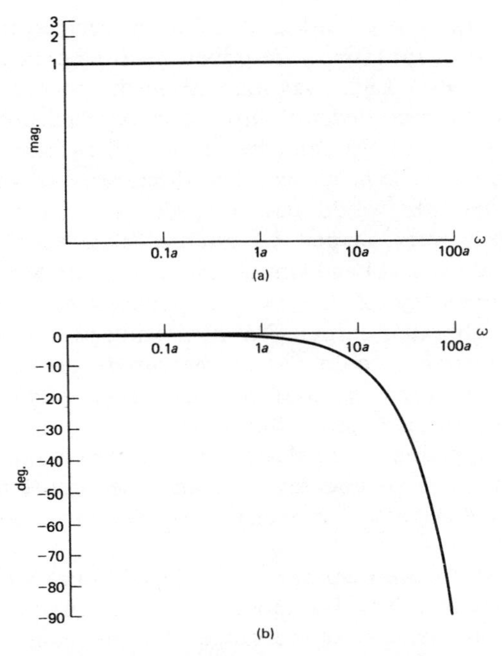

Figure 3-9 Bode diagram for a pure time delay, $Y(j\omega) = ke^{-j\omega a}$.

given magnitude characteristic. The time delay has a constant magnitude but increasing phase angle, so it is nonminimum phase. Another definition of a minimum phase system is that if the magnitude ratio (from the Bode diagram) is fixed, the phase is also. If the phase is fixed, the magnitude ratio is also. If the magnitude ratio is fixed over an interval and the phase is fixed over the rest of the frequency range, the system is specified. Examples of nonminimum phase are pure time-delay systems and unstable systems. This brief review of frequency-domain analysis techniques will allow us to examine a few examples of engineering analysis of biological systems.

3-5 TRANSFER FUNCTION OF A CRAYFISH PHOTORECEPTOR GANGLION

The terminal abdominal ganglion of the crayfish acts as a photoreceptor even though it has no specialized optical organization. When light is shone on this sensory receptor, the crayfish will begin walking randomly into a region of subdued illumination. This is a very primitive, low-level reflex and only functions if no other, more active behavior is needed.

In the 1950's, Lawrence Stark and his group decided that this crayfish behavior could be quantitatively described by frequency-domain-analysis techniques (Stark 1968). Light was focused on the photoreceptor ganglion and the response was recorded with intra- and extracellular electrodes. The extracellular gross recordings contained axon spikes from several types of cells. An electronic pulse-height-window discriminator was used so that spikes of only one size would pass through. The midrange pulses, the pulses of interest, were shaped to produce constant-size pulses, then integrated to produce a voltage whose level corresponded to the instantaneous output frequency of the neuron population. The stimulus was recorded as the calibrated output of a monitoring photocell.

An impulse response could not be measured because an impulse of light is difficult to produce, and because range nonlinearities of the biological system can invalidate a linear systems analysis.

The result of applying a step change in light level is shown in Fig. 3-10. The fit between the experimental data and the model function (to be derived later) is not perfect. The error is attributed to noise and nonlinearities.

Almost all systems are nonlinear if the input values are chosen from a large enough range of values. But most engineering analysis techniques are intended for linear systems. If the range of input values is small, most systems can be modeled adequately as linear systems. A small-signal analysis was the next technique used for this crayfish preparation. The

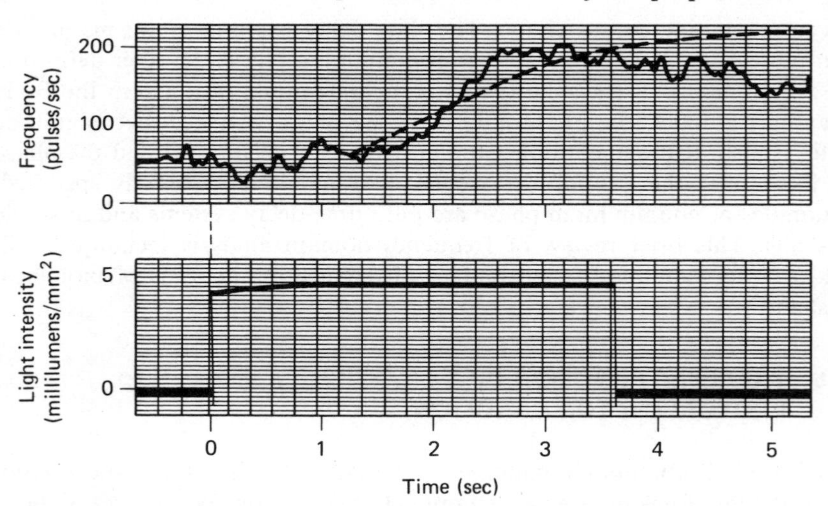

Figure 3-10 Response of the crayfish photosensitive ganglion to a step change of light. The dashed line represents the response of the linear model. (From L. Stark, *Neurological Control Systems Studies in Bioengineerings*, Plenum Press, New York, 1968.)

light focused on the photoreceptor was sinusoidally modulated around a dc level (Fig. 3-11). The modulation frequency was varied, and the magnitude and phase angle of the neuronal response were plotted in Fig. 3-12. Gain is plotted on a log-log scale, phase on a linear-log scale. Points are experimental. Dashed lines of the gain plot are the asymptotes on either side of the break frequency. Response is down 6 dB at a break frequency

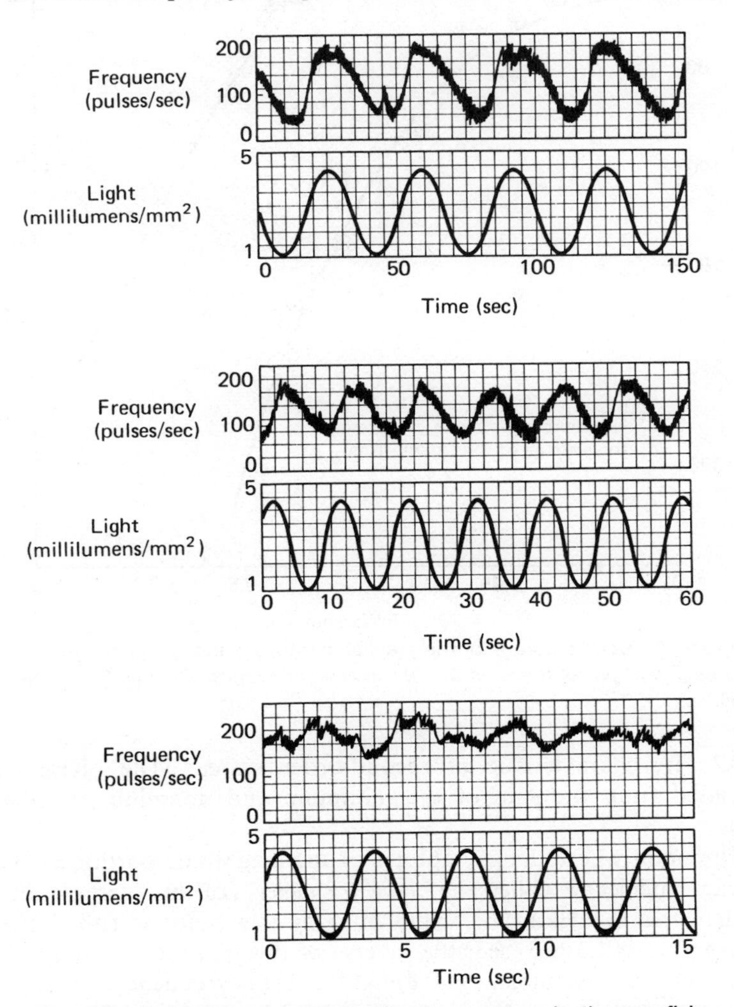

Figure 3-11 Frequency of firing for the neurons in the crayfish photoreceptor ganglion and intensity of the light shone upon the ganglion are both plotted as functions of time. (From L. Stark, *Neurological Control Systems Studies in Bioengineering*, Plenum Press, New York, 1968.)

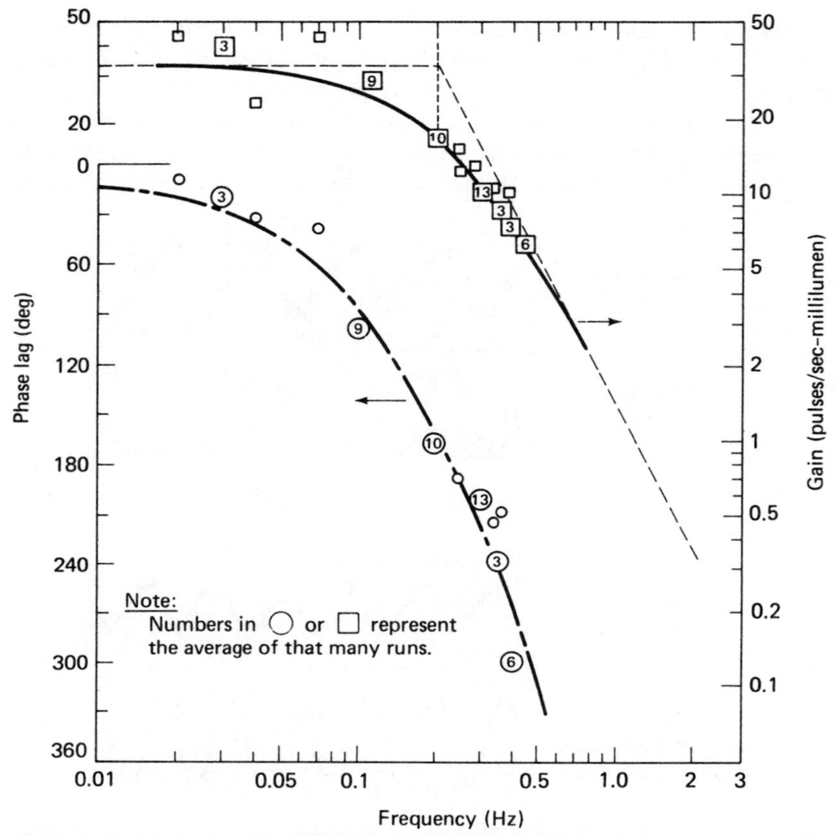

Figure 3-12 Bode diagram for the crayfish photosensitive ganglion. (From L. Stark, *Neurological Control Systems Studies in Bioengineering*, Plenum Press, New York, 1968.)

of 0.2 Hz, characteristic of second-order system. The phase curve is computed from the sum of the minimum and nonminimum phase elements.

The point where the asymptotes of the magnitude portion of the Bode diagram intersect is called the *break frequency*. (In this example, the break frequency is 0.2 Hz.) The value of ω at this point is called the *cutoff frequency*, $\omega_c = 1.3$ rad/sec; the inverse of this quantity is the system time constant (in this example, $\tau = 0.8$). At this break frequency, the experimental data are 6 dB below the intersection of the asymptotes. The slope of the asymptote after the breakpoint is -2. Both of these facts suggest a double pole at this break frequency. However, a double pole would only contribute 90° of phase lag at this frequency, whereas the actual phase lag is 162°. The extra 72° of phase shift must be due to a pure time delay. At 0.2 Hz,

72° of phase shift would result from a time delay of 1.0 sec. Combining all these data yields the transfer function:

$$M(s) = \frac{32e^{-1.0s}}{(1 + 0.8s)^2} \frac{\text{pulses}}{\text{sec-millilumen}} \tag{3-21}$$

The standard form of writing transfer functions forces the s coefficient of the highest power of s in the denominator to be unity. Therefore, we can rearrange this equation to obtain

$$M(s) = \frac{50e^{-1.0s}}{(s + 1.3)^2}$$

This is the transfer function that was fitted to the data of Fig. 3-10; it was derived from a small-signal analysis. The experimental data in that figure were from a large step response. The discrepancies show that the large step drove the system into its nonlinear region.

The data of Figs. 3-10 to 3-12 were recorded with extracellular electrodes from the entire cell population. Stark also used intracellular electrodes and computed transfer functions for individual nerve fibers. Different cells had slightly different gains, time constants, and time delays, but the general forms of the transfer functions were the same.

For these studies, the transfer functions were fit to the data by hand analysis. The fact that the data could be fit so well with a critically damped second-order system was merely a coincidence. There is no a priori reason to suspect that any biological system can be best fit with such a simple, well-studied engineering model. Computer programs are now available that will fit arbitrary ordered transfer functions to the data contained in Bode diagrams (Seidel 1975; Peterka 1980).

This application of engineering systems theory allowed a quantification of the behavior of this biological system. The previous studies of this crayfish behavior had been physiological, anatomical, or behavioral; none of them were as successful in quantifying the transduction of light energy into nerve-axon firing frequency. The low-pass characteristics of the transfer function show that this system will not respond to high-frequency changes in illumination: it is a tonic system. The maintenance of gain to very low frequencies (0.01 Hz) demonstrates that the system does not show adaptation.

The range of linearity of the system and the nonlinear effects caused by large signals were illustrated by the contrasting results of the large step input and the small-signal analysis. An analysis of the noise in the system and the interrelation of the firing patterns of individual neurons were also a part of Stark et al.'s study of the crayfish photoreceptor. These studies enabled them to conclude that the average neuronal firing rate carried the information about the light intensity, and that neither nerve pulse train patterns nor firing relationships between fibers carried any information. However, these conclusions are not necessarily true for nerve fibers in general.

3-6 MATHEMATICAL ANALYSIS OF LINEAR SECOND-ORDER SYSTEMS

3-6-1 The Transfer Function

The linear second-order system may be the most common and most intuitive model of physical systems. The crayfish transfer function is one example of such a second-order system. In this section we investigate many different properties of linear second-order systems.

The Newtonian equation for the spring–mass–dashpot system of Fig. 3-13 is

$$f(t) = M\ddot{x} + B\dot{x} + Kx \tag{3-22}$$

where M represents the mass of the object, B the viscosity and K the elasticity. The transfer function is

$$\frac{X(s)}{F(s)} = \frac{1}{Ms^2 + Bs + K} \tag{3-23}$$

We now define two new parameters, ζ and ω because they have physical

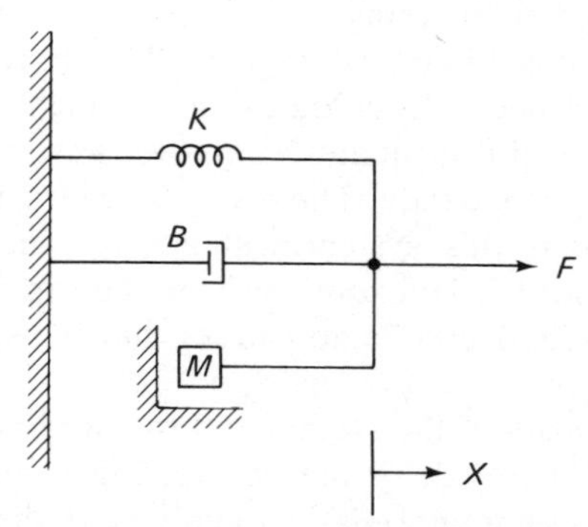

Figure 3-13 Simple spring–mass–dashpot system.

significance and create mathematical simplicity.

$$\zeta = \frac{B}{2\sqrt{KM}} \tag{3-24}$$

$$\omega_n = \sqrt{\frac{K}{M}} \tag{3-25}$$

The undamped natural frequency, ω_n, is the frequency at which the system would oscillate if the damping, B, were zero. When we substitute the new

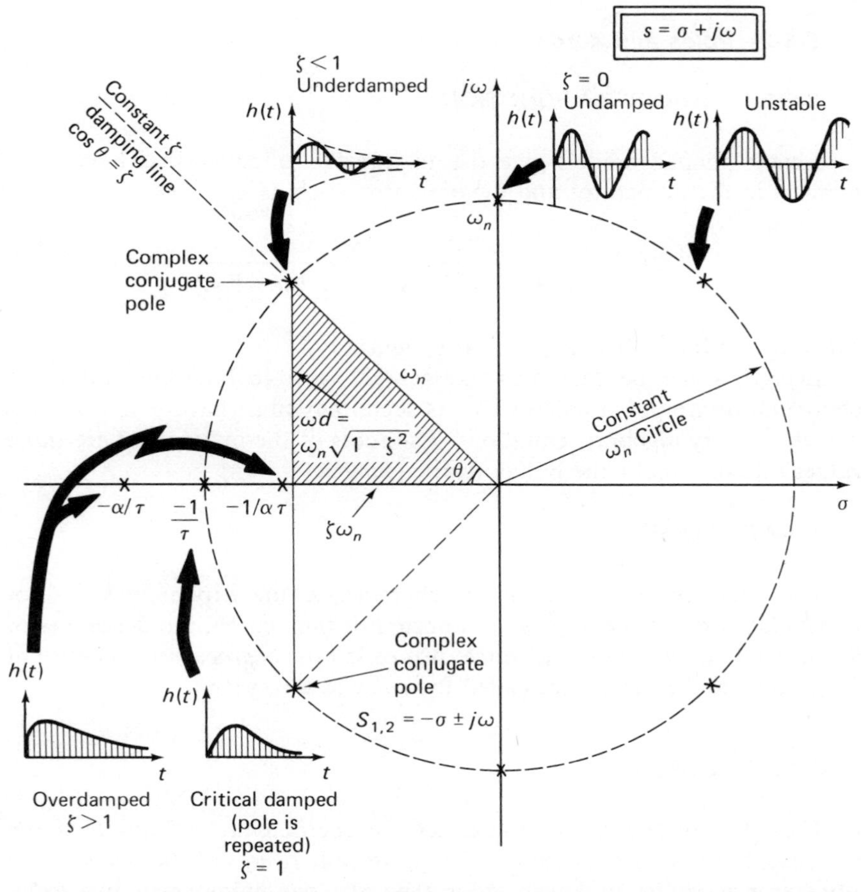

Figure 3-14 Pole–zero diagrams and impulse responses for second-order systems with various pole locations. (From J. H. Milsum, *Biological Control Systems Analysis*, McGraw-Hill Book Company, New York, 1966.)

parameters, the transfer function becomes

$$\frac{X(s)}{F(s)} = \frac{1}{K} \frac{\omega_n^2}{s^2 + 2\zeta\omega_n s + \omega_n^2} \tag{3-26}$$

This system exhibits four different types of behavior; they are defined uniquely by the value of the damping ratio, ζ. Figure 3-14 summarizes these responses. These mathematical results will not be derived in this text. The interested reader is referred to a systems theory textbook such as Milsum (1966), Melsa and Schultz (1969), or Takahashi, Rabins, and Auslander (1970).

3-6-2 Poles and Zeros

3-6-2.1 TRANSFER FUNCTION

Linear lumped models give rise to transfer functions that have polynomials in the numerator and denominator, such as

$$M(s) = \frac{b_m s^m + ,\cdots, + b_1 s + b_0}{s^n + a_{n-1} s^{n-1} + ,\cdots, + a_1 s + a_0} \tag{3-27}$$

where $m < n$ for realizable physical systems.

Because s may be treated as an algebraic variable, both numerator and denominator are polynomials of s. As such, they must have m and n roots like elementary algebraic equations. The roots of the polynomial are those values of s that make the polynomial equal zero.

3-6-2.2 POLES

The roots of the denominator characterize the exponential and/or sinusoidal terms of the impulse response. For this reason, the denominator polynomial has special significance; it is called the *characteristic function* of the system, and its roots are called the *poles* of the system.

3-6-2.3 ZEROS

The roots of the numerator affect the coefficients and phase of the response, but not its basic nature. They are called *zeros* of the system. The numerator must be of lower order than the denominator in physically realizable systems. For example, the idealized first-order lead compensation network of Sec. 3-3 is nonrealizable at sufficiently high frequencies. In other words, such an electronic circuit could not be built to operate at a gigahertz.

3-6-2.4 FIRST-ORDER SYSTEM

The basic first-order system

$$M(s) = \frac{k}{\tau s + 1}$$

has a first-order denominator, which is equal to zero when

$$s = \frac{-1}{\tau}$$

This pole is the negative inverse of the time constant that characterizes the impulse response. In the frequency domain, it is the negative of the cutoff frequency $\omega_c = 1/\tau$. Figure 3-15 shows some of these relationships.

The first-order lead–lag system of Sec. 3-3 with its transfer function

$$M(s) = \frac{1}{10}\,\frac{10s + 1}{s + 1}$$

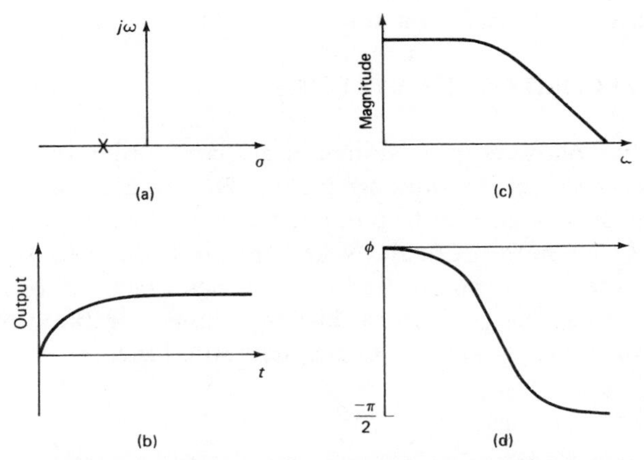

Figure 3-15 Pole–zero diagram, step response, $y(t) = 1 - e^{-t/\tau}$, and Bode diagram for the first-order system with the transfer function $G(s) = K/(\tau s + 1)$.

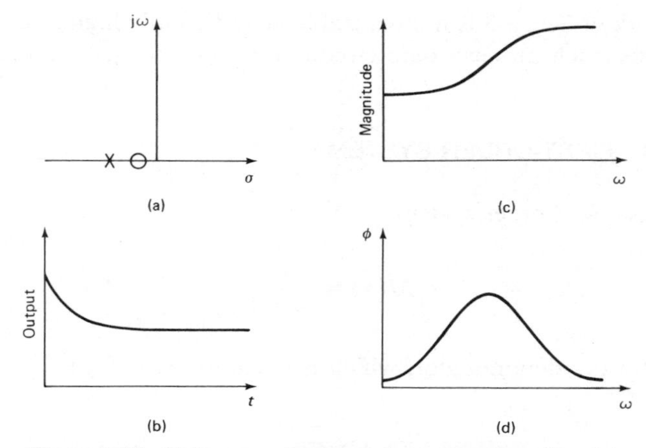

Figure 3-16 Pole–zero diagram, step response, $y(t) = 1 + e^{-t/\tau}$, and Bode diagram for the first-order system with the transfer function $G(s) = (10s + 1)/10(s + 1)$.

adds a numerator root, or zero, at

$$s = \frac{-1}{10}$$

This zero does not change the actual time constant of the exponential time response, but does affect the other constants. Its effect on the frequency response is more explicit, however, because a new break frequency is added, at $\omega = \frac{1}{10}$, where the magnitude ratio curve breaks upward, rather than downward as it would for a pole (Fig. 3-16).

3-6-2.5 SECOND-ORDER SYSTEM

The crayfish photoreceptor system is a second-order system (see Fig. 3-14) with two poles at the same frequency. However, in general, second-order systems do not have both poles at the same frequency, and finding the poles may be more difficult. When the underdamped second-order system is considered, we cannot find real s values which set the denominator equal to zero. This problem is like the classic algebraic problem of finding the two roots, x_1 and x_2, of the quadratic function

$$f(x) = ax^2 + bx + c$$

which is solved by setting $f(x) = 0$ and obtaining

$$x_{1,2} = \frac{-b \mp \sqrt{b^2 - 4ac}}{2a}$$

3-6-2.6 THE DISCRIMINANT AND COMPLEX ROOTS

The expression under the square root is called the *discriminant*. It has special significance because its square root is imaginary when $b^2 < 4ac$. If b does not equal 0, the root always has a real part, $-b/2a$. Thus, in general, the roots are complex numbers $x = \sigma + j\omega$. Applying this quadratic formula to the denominator of Eq. (3-26) shows that the two poles are

$$s_{1,2} = -\zeta\omega_n \pm \omega_n\sqrt{\zeta^2 - 1} \tag{3-28}$$

The ζ and ω_n parameters are most significant when $\zeta < 1$ and the discriminant is negative; therefore, it is preferable to rewrite Eq. (3-28) in a way that directly illustrates the real and imaginary components:

$$s_{1,2} = -\zeta\omega_n \pm j\omega_n\sqrt{1 - \zeta^2} = -\zeta\omega_n \pm j\omega_d \tag{3-29}$$

where ω_d is a new parameter called the *damped natural frequency*.

This important relation demonstrates that zeta determines whether the discriminant is positive or negative. In particular, the roots are:

1. Negative real $\zeta > 1$.
2. Repeated real when $\zeta = 1$.
3. Complex when $0 < \zeta < 1$.
4. Purely imaginary when $\zeta = 0$.

For the J, B, K-parametered model of Sec. 3-6-1, the roots are

$$s_{1,2} = -\frac{B}{2J} \pm \sqrt{\left(\frac{B}{2J}\right)^2 - \frac{K}{J}}$$

To summarize this section, the potentially oscillatory behavior of the second-order system is characterized mathematically by whether or not the poles are complex. First-order poles can only be real; therefore, a second-order denominator polynomial is the minimum order that allows complex poles to exist. However, the poles of high-order systems can be real or complex, depending on the numerical values of the system's parameters.

3-6-3 Pole–Zero Plots on the Complex Plane

When possible, transfer functions are expressed in factored form like this:

$$M(s) = \frac{(s+z_1)(s+z_2),\cdots,(s+z_m)}{(s+p_1)(s+p_2),\cdots,(s+p_n)} \tag{3-30}$$

where p_n represents the nth pole and z_m represents the mth zero. It is of conceptual value to plot the poles and zeros of a system on the complex plane because the characteristic patterns of dynamic response in different regions are readily remembered. This plotting also provides the basis for the root-locus technique, by which the movement of system poles and zeros can be followed as the system parameters are modified.

The Laplace transform variable s is exhibiting the properties of a complex variable. It is a complex frequency variable defined by

$$s = \sigma + j\omega$$

where σ and ω are, respectively, the real and imaginary parts. The real part σ characterizes the transient response of the system, while the imaginary part ω characterizes the steady-state, or sinusoidal, response.

The four pole configurations of the basic second-order system as functions of ζ are shown in Fig. 3-14, along with sketched impulse responses. The right-triangle relationship between the real and imaginary parts and the undamped natural frequency of the system ω_n should be noted. In particular, the damped natural frequency of the system ω_d is specified by the imaginary part,

$$\omega_d = \omega_n\sqrt{1-\zeta^2}$$

Similarly, the real part $-\zeta\omega_n$ corresponds to the inverse exponential decay time constant for the impulse response's envelope. A radial line from the origin is called a constant damping line because the angle from the negative real axis is given by

$$\zeta = \cos\theta$$

So if ζ is constant, so is the angle. A circle about the origin is at constant ω_n. For the overdamped case, the poles are arranged on either side of $-1/\tau$ (which also equals $-\zeta\omega_n$ for the $\zeta = 1$ condition).

The right half-plane is a region of instability, as sketched at one complex pole. The imaginary axis is clearly the dividing line between stable and unstable systems; that is, it is a region of zero damping.

3-6-4 Step Response of a Second-Order System

As a brief review of the mathematical techniques used in analyzing system responses, we will now derive the time response of a linear *JBK* system with critical damping. The reader is encouraged to perform derivations of the impulse and step responses for the over- and underdamped systems.

Find the step response for the *JBK* system of Fig. 3-13. Let $k = K$.

$$\frac{X(s)}{F(s)} = \frac{1}{k}\frac{\omega_n^2}{s^2 + 2\zeta\omega_n s + \omega_n^2}$$

for the particular case where zeta is unity (critically damped). If $f(t)$ is a unit step, then

$$X(s) = \frac{1}{s}\frac{1}{k}\frac{\omega_n^2}{s^2 + 2\omega_n s + \omega_n^2} = \frac{1}{k}\frac{\omega_n^2}{s(s+\omega_n)^2} \tag{3-31}$$

We will evaluate this by the method of partial fractions.

$$x(s) = \frac{1}{k}\frac{\omega_n^2}{s(s+\omega_n)^2} = \frac{A}{s} + \frac{B}{(s+\omega_n)} + \frac{C}{(s+\omega_n)^2} \tag{3-32}$$

To find A, multiply both sides of Eq. (3-32) by the denominator of the A term, s in this case, then let s take on a value that would make the denominator equal zero (in this instance, 0). We find that

$$A = \frac{1}{k}$$

To find C, multiply Eq. (3-32) by $(s+\omega_n)^2$ and then let $s = -\omega_n$.

$$C = \frac{-\omega_n}{k}$$

If we try the same trick for B, we will get

$$B = \frac{0}{0}$$

which is indeterminate. So we must apply L'Hospital's rule. Multiply Eq. (3-32) by $(s+\omega_n)^2$.

$$\frac{1}{k}\frac{\omega_n^2}{s} = (s+\omega_n)^2\frac{A}{s} + (s+\omega_n)B + C$$

Take the derivative with respect to s.

$$\frac{1}{k}\frac{-\omega_n^2}{s^2} = \frac{s^2-\omega_n^2}{s^2}A + B$$

Evaluate this at $s = -\omega_n$:

$$B = \frac{-1}{k}$$

Therefore, in response to a unit step of force, the position becomes

$$X(s) = \frac{1}{ks} + \frac{-1}{k(s+\omega_n)} + \frac{-\omega_n}{k(s+\omega_n)^2} \tag{3-33}$$

From a table of Laplace transforms, we find the resulting time response to

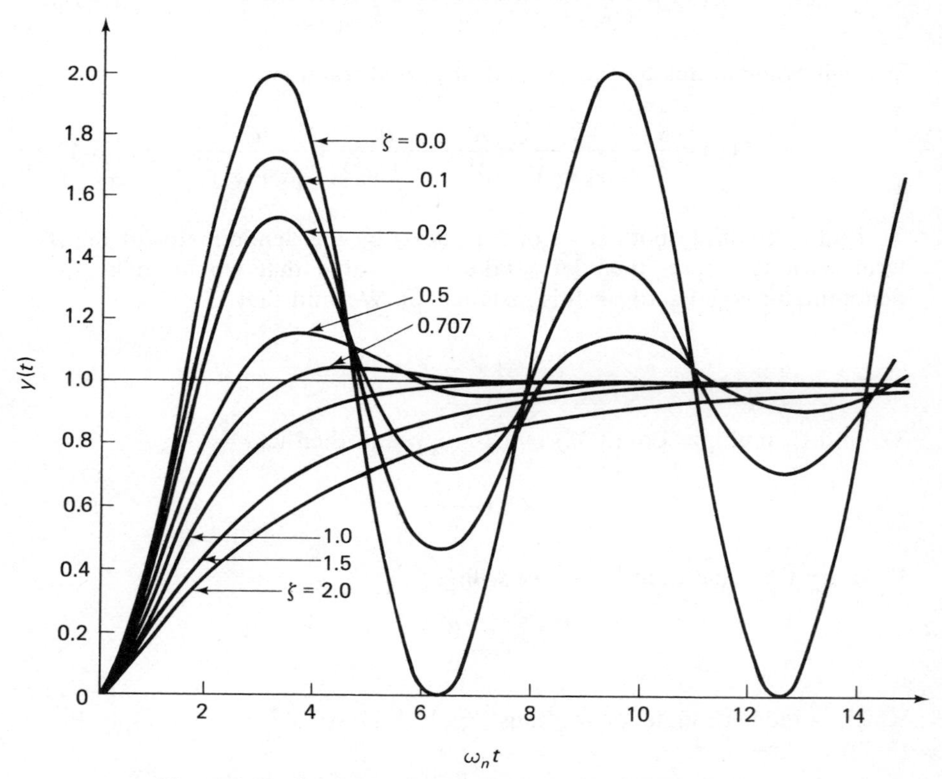

Figure 3-17 Step responses of linear second-order systems.

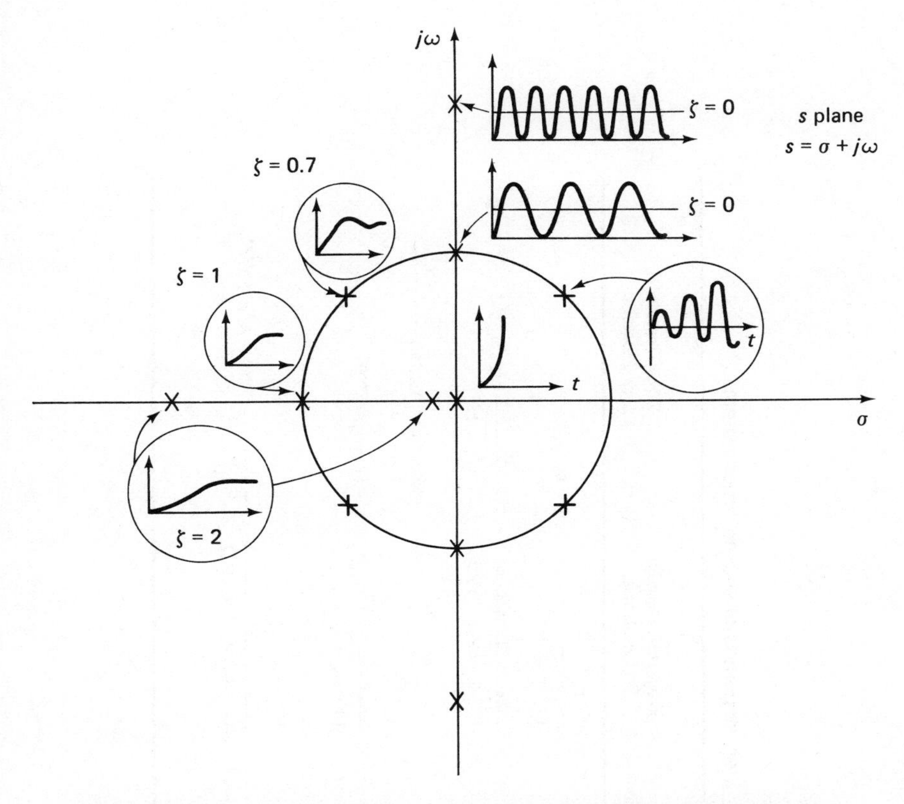

Figure 3-18 Step responses and pole locations of linear second-order systems.

be

$$x(t) = \frac{1}{k}\mu_{-1}(t) + \frac{-1}{k}e^{-\omega_n t} + \frac{-1}{k}\omega_n t e^{-\omega_n t}$$

For $t > 0$, this is

$$x(t) = \frac{1}{k}\left[1 - (1 + \omega_n t)e^{-\omega_n t}\right] \qquad (3\text{-}34)$$

Similar results for over- and underdamped systems are presented in Figs. 3-17 and 3-18 and in Table 3-2. The step response of an underdamped linear second-order system is derived in Sec. 3-7-3.1.

TABLE 3-2 Step responses of second-order systems.

Damping Ratio Range	*Transfer Function*	*Frequency-Domain Step Response*	*Time-Domain Step Response*
$\zeta = 0$ undamped	$\frac{\omega_n^2}{s^2 + \omega_n^2}$	$\frac{\omega_n^2}{s(s^2 + \omega_n^2)}$	$1 - \cos \omega_n t$
$0 < \zeta < 1$ underdamped	$\frac{\omega_n^2}{s^2 + 2\zeta\omega_n s + \omega_n^2}$	$\frac{\omega_n^2}{s(s^2 + 2\zeta\omega_n s + \omega_n^2)}$	$1 - \frac{e^{-\zeta\omega_n t}}{\sqrt{1-\zeta^2}} \sin(\omega_n \sqrt{1-\zeta^2}\, t + \varphi)$ where $\varphi = \text{Arc}\tan \frac{\sqrt{1-\zeta^2}}{\zeta}$
$\zeta = 1$ critically damped	$\frac{\omega_n^2}{(s + \omega_n)^2}$	$\frac{\omega_n^2}{s(s + \omega_n)^2}$	$1 - (1 + \omega_n t)e^{-\omega_n t}$
$\zeta > 1$ overdamped $\zeta \equiv \frac{1 + \alpha^2}{2\alpha}$	$\frac{\omega_n^2}{(s + \omega_n/\alpha)(s + \alpha\omega_n)}$	$\frac{\omega_n^2}{s(s + \omega_n/\alpha)(s + \alpha\omega_n)}$	$1 + \frac{1}{\alpha^2 - 1}(e^{-\alpha\omega_n t} - \alpha^2 e^{-\omega_n t/\alpha})$

3-6-5 *Frequency Response of a Second-Order System*

The basic transfer function of a linear second-order system is

$$M(s)=\frac{1}{k}\frac{\omega_n^2}{s^2+2\zeta\omega_n s+\omega_n^2}$$

Sometimes this will be written in the form

$$M(s)=\frac{1/k}{(s/\omega_n)^2+2\zeta(s/\omega_n)+1}$$

When Stark derived the Bode diagrams of Fig. 3-12, he applied sinusoids of various frequencies and recorded the responses. Although this is not the only way to obtain data for a Bode plot, it is conceptually the simplest. Therefore, our first step in constructing Bode plots for the basic second-order system will be to ignore transients and consider only the steady-state portion of the complex variable s. That is, we will replace s with $j\omega$ in the transfer function shown above.

$$M(j\omega)=\frac{1/k}{1-(\omega/\omega_n)^2+j(2\zeta\omega/\omega_n)} \tag{3-35}$$

Initially, we are only interested in the magnitude, or absolute value, of this transfer function. If $\omega < \omega_n$, we can approximate this with

$$|M(j\omega)|\approx\frac{1}{k}$$

Taking the logarithm of both sides yields

$$\log|M(j\omega)|\approx\log\left|\frac{1}{k}\right| \tag{3-36}$$

This is the low-frequency asymptote of the magnitude ratio of the Bode plot. For very high frequencies where $\omega > \omega_n$,

$$|M(j\omega)|\approx\frac{1/k}{(\omega/\omega_n)^2}$$

Taking the logarithm of both sides yields

$$\log|M(j\omega)| \approx \log\left|\frac{1}{k}\right| - 2\log\left|\frac{\omega}{\omega_n}\right| \tag{3-37}$$

This is of the form $y = b - mx$ and will plot as a straight line with a slope of -2. This is the high-frequency asymptote for the transfer function. These asymptotes clearly intersect at $\omega/\omega_n = 1$. In contrast to Bode plots of first-order systems, the actual data curves will not necessarily fall above or below the asymptotes, but rather the exact location of the true curves will depend upon ζ. In particular, as ζ approaches zero, the magnitude ratio approaches infinity at frequencies where ω/ω_n is close to unity. This is a *resonance* phenomenon. The exact value of the magnitude ratio may be calculated from Eq. (3-35).

The phase varies between 0° at zero frequency and $-180°$ as ω approaches infinity. Note that when the input frequency equals the undamped natural frequency $\omega = \omega_n$, the real component in the denominator of Eq. (3-35) disappears and the output lags the input by exactly 90°. We used this bit of information to compute the value of the time delay for the crayfish photoreceptor [Eq. (3-21)].

The phase angle θ is defined as

$$\theta = -\text{Arc}\tan\frac{2\zeta(\omega/\omega_n)}{1-(\omega/\omega_n)^2}$$

These relationships concerning the magnitude ratio and phase are summarized in the Bode plots of Fig. 3-19.

The following interesting facts are given without proofs. Some of them are illustrated in Fig. 3-20. The interested reader should refer to a systems theory textbook for the derivations.

1. For all values of damping ratio less than 0.707, the *magnitude ratio*, MR, exceeds unity in some frequency range for $k = 1$. In particular, the maximum magnitude ratio, or resonance, is reached at the resonant frequency ω_R. When $\zeta > 0.707$, there is no resonance.

2. The *resonant frequency* ω_R is less than the undamped natural frequency ω_n and less than the damped natural frequency ω_d:

$$\omega_R = \omega_n\sqrt{1-2\zeta^2}$$

$$\omega_d = \omega_n\sqrt{1-\zeta^2}$$

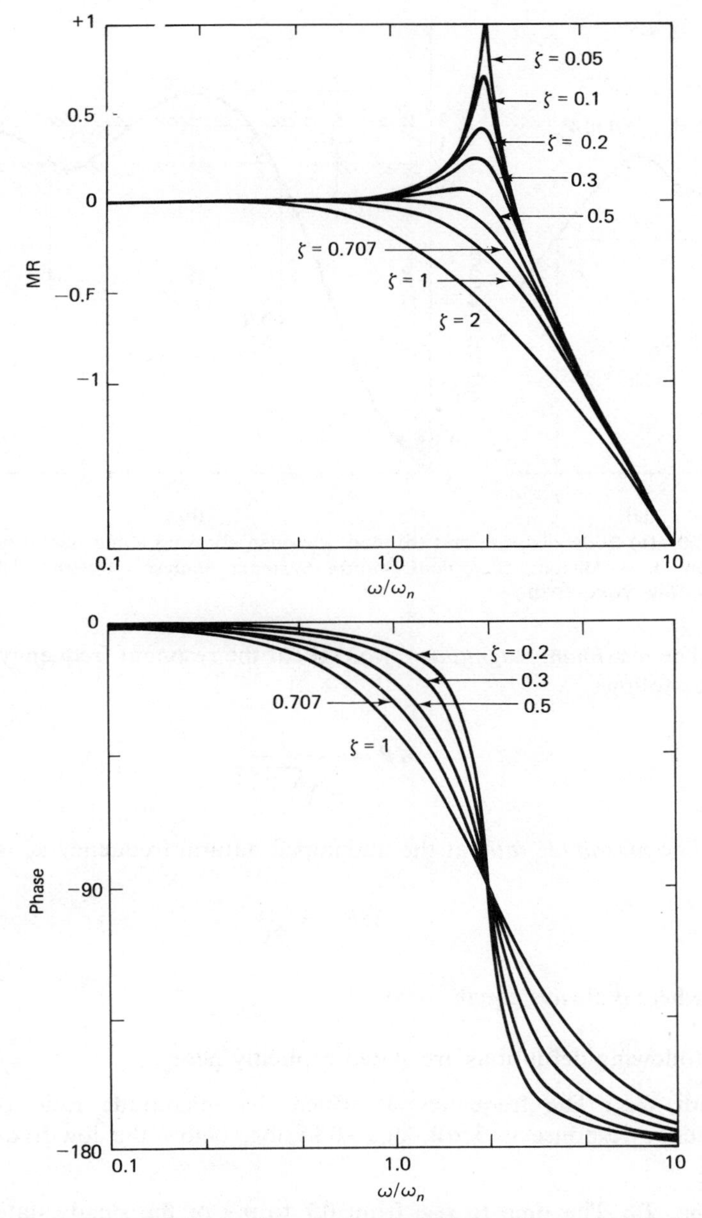

Figure 3-19 Bode diagrams for simple second-order systems.

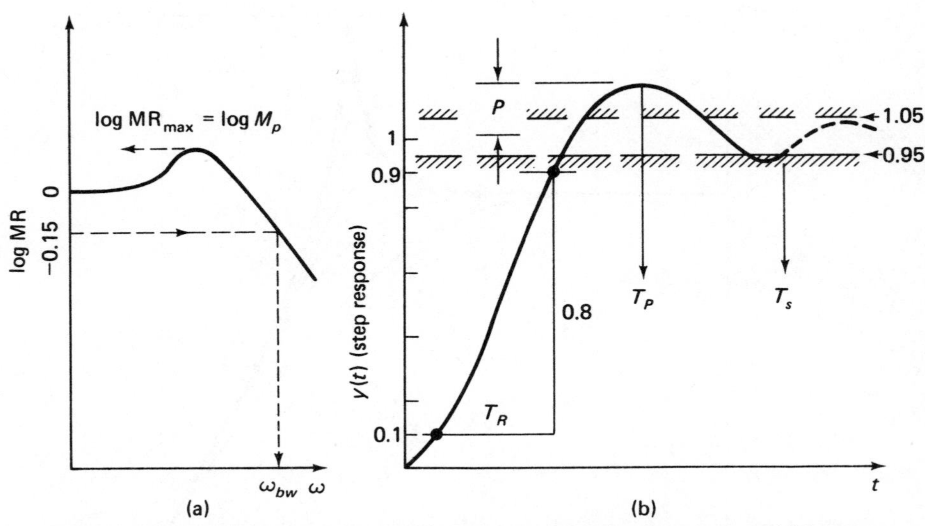

Figure 3-20 (a) Bode diagram and (b) step response showing some useful parameters. (From J. H. Milsum, *Biological Control Systems Analysis*, McGraw-Hill Book Company, New York, 1966.)

3. The *maximum magnitude ratio*, MP, at the resonant frequency ω_R is as follows:

$$MP = \frac{1}{2\zeta\sqrt{1-\zeta^2}}$$

4. The *magnitude ratio* at the undamped natural frequency ω_n is

$$MR_{\omega_n} = \frac{1}{2\zeta}$$

where θ always equals $-90°$.

The following definitions are stated explicitly here:

Bandwidth, ω_{bw}: The frequency at which the magnitude ratio of the frequency response is 3 dB, or -0.15 log, below the low-frequency value.

Rise Time, T_R: The time to rise from 0.1 to 0.9 of the steady-state step response.

Settling Time T_s: The time it takes to settle within $\pm 5\%$ of the steady-state step response.

Peak Overshoot Time, T_p: The time to the peak (first) overshoot in the step response (this is derived in Sec. 3-7-3.1).

Using these definitions and the ones shown in Fig. 3-20, the following interesting approximations may be given:

1. Peak overshoot and magnitude ratio: for $P = 0.15$, $MP \approx 1.3$.
2. Overshoot ζ and ω_n

$$T_p = \frac{\pi}{\omega_n\sqrt{1-\zeta^2}}$$

$$P = \exp\left(-\frac{\pi\zeta}{\sqrt{1-\zeta^2}}\right)$$

3. Rise time and bandwidth

$$T_R \approx \frac{1}{\omega_{b\omega}}$$

Compare this with $T_R = 2.2/\omega_c$ for a first-order system.

4. Settling time

$$\frac{3}{\omega_n} < T_s < \frac{5}{\omega_n}$$

or

$$T_s < \frac{4}{\zeta\omega_n}$$

Often ζ is chosen to be 0.707. There is nothing magic about this value; it is only a coincidence that many interesting properties are associated with the value. For example:

1. The magnitude-ratio portion of the Bode diagram has no resonance; that is, it is never above unity (for $k = 1$).
2. The overshoot is always less than 5%.
3. It is the value that will give minimum settling time (for a $\pm$ 5% band).
4. The roots are on a line with a 45° inclination to the negative real axis of the s plane.

5. All points on the Nyquist diagram are within the unit circle.
6. Equations can be solved by hand because most numbers turn out to be integers.
7. It is the value that yields the minimum integral of the absolute value of error between a step input and the system response, and also the minimum integral of time multiplied by the absolute error.

Armed with these tools, facts, and assertions we are now ready to analyze some models of some human movement control systems.

3-7 MODELS OF HUMAN MOVEMENT

A single textbook cannot present all the good models of the many human physiological systems that have been developed. For this reason, we have limited this section to a discussion of the neurological control of human movement. Because the locomotory-control systems and the hand and arm systems are both complicated and not well understood, we have further limited this discussion to a simple, but typical movement control system: the eye-movement control system.

3-7-1 The Eye-Movement Control System

The eye-movement systems are ideal for studying human control of movement: eye movements are easy to measure, and the control of saccadic eye movements is simpler than the control of other neuromuscular systems. It is simpler because the load presented by the eyeball and extraocular tissues is small and constant. Horizontal eye movements offer a further simplification because they primarily involve only two muscles of each eye. By scrutinizing the shapes of saccadic eye movements, we can infer the motoneuronal activity, deduce the central nervous system's control strategy, and observe changes in this control strategy caused by fatigue, alcohol, drugs, or pathology. These eye-movement control principles should generalize to other neuromuscular systems.

There are two additional reasons for choosing the eye-movement control system: (1) we can present the development of a model by nine authors over a span of more than 25 years (see Table 3-3); and (2) an in-depth evaluation of this final model can be presented. The eye-movement control models were not chosen because eye movements are thought to be so important, but because the physiological system is simple and typical, and the models are well studied.

TABLE 3-3 Evolution of the linear homeomorphic eye movement model.

Author	*Contribution*	*Improvement*
Descartes (1626)	First suggestion of reciprocal innervation	First explicit model
Westheimer (1954)	$G(s) = \dfrac{\omega_n^2}{s^2 + 2\zeta\omega_n s + \omega_n^2}$	Fit 10° saccades
Robinson (1964)	Pulse-step input	Fit various-size saccades
Cook and Stark (1968)	Implementation of reciprocal innervation	Realistic velocity shapes
Clark and Stark (1974)	Incorporated physiological parameters	Realistic acceleration shapes
Collins (1975)	Provided human physiological data	Clinical usage of model
Bahill, Clark, and Stark (1975a, b)	Effects of controller-signal variations	Simulated eye movements not used in design of model (e.g., glissades and dynamic overshoot)
Hsu, Bahill, and Stark (1976)	Sensitivity analysis	Method of validating model
Latimer and Bahill (1979)	Parameter estimation by function minimization, inclusion of length–tension diagram	Linearized the model

3-7-2 Four Eye-Movement Systems

As we look around a room or at pictures, read, walk or drive a car, we make a multitude of eye movements. Most of these eye movements are precise, staccato, *saccadic eye movements* like those shown in Fig. 3-21. They place the high-resolution fovea, the central $\frac{1}{2}°$ of the retina, on the important features of the scene by using information from the periphery of the retina to direct the movement.

The purpose of saccadic eye movements is to move the fovea of the eyes, to facilitate efficient information processing. However, we would not want our eyes to be constantly producing saccadic eye movements, because we do not see well during saccades. For example, if you move your eyes between two points in the front of the room, you do not see the world rushing past, although the image on your retina is this moving scene. We have a process that suppresses our visual acuity during saccadic eye movements. This process, called saccadic suppression, is specific for saccadic size and direction (Stark, Kong, Schwartz, Hendry, and Bridgeman 1976; Campbell and Wurtz 1978). Its effect can also be easily demonstrated by a simple experiment with a mirror. Look at the image of one of your eyes, then look at the image of the other eye. The rapid movement of

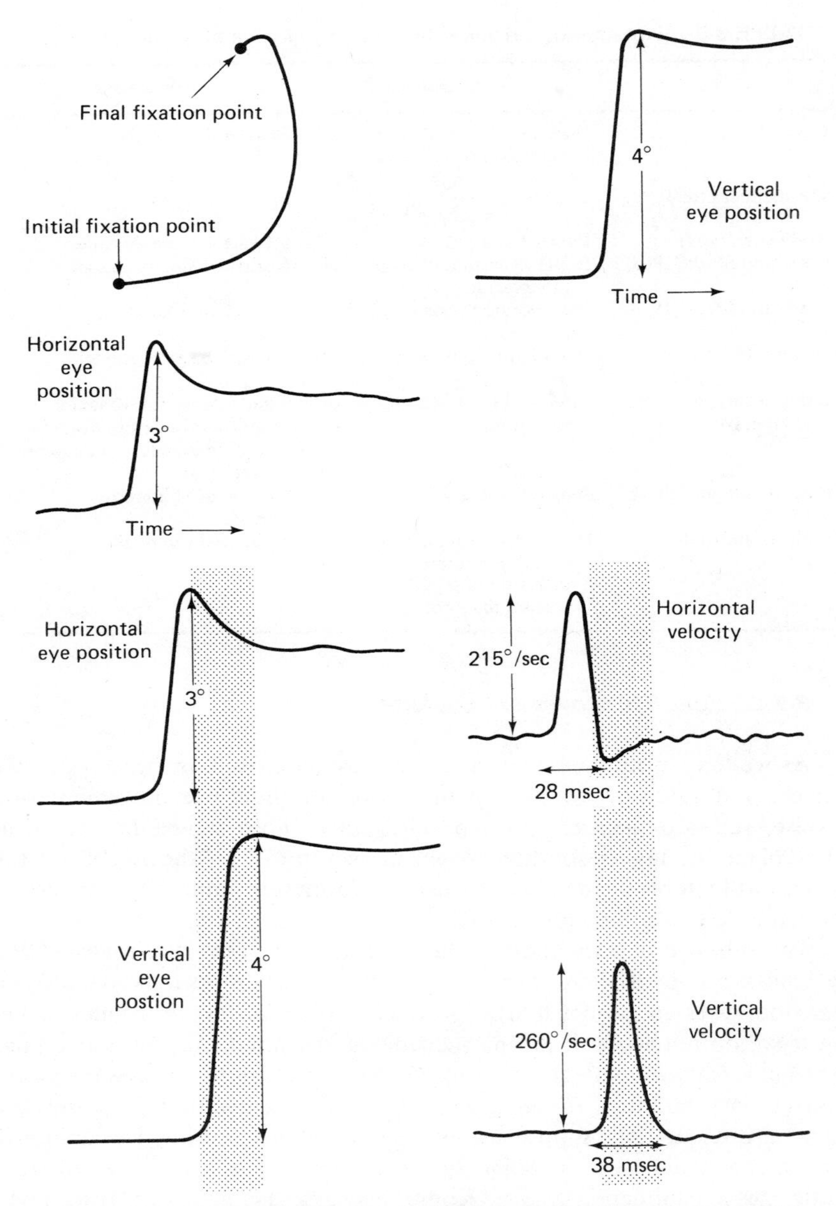

Figure 3-21 Eye movements are often in oblique directions. The horizontal and vertical saccades are dynamically and temporally independent. Sometimes one component (the horizontal, in this case) can be nearly finished by the time the other component (stippled band) begins. Horizontal saccades are generally faster than vertical saccades of the same size. (From A. T. Bahill and L. Stark, "The Trajectories of Saccadic Eye Movements." Copyright 1979 by Scientific American, Inc. All rights reserved.)

the image during the saccade and the movements of the eyes were not seen because vision was partially suppressed during the saccade. Saccadic suppression is specific to the saccadic-eye-movement system.

Three other neurological control systems also produce eye movements: vestibular ocular movements, which are used to maintain fixation during head movements; vergence eye movements, which are used when looking between near and distant objects; and smooth pursuit eye movements, which are employed when tracking moving objects, such as a flying bird.

These four eye-movement control systems are independent. Their dynamic properties, such as latency, speed and high-frequency cutoff values, are different. They are produced by different areas of the brain, and they are affected differently by fatigue, drugs, and disease.

The specific action of each system can be illustrated by considering a duck hunter sitting in a rowboat on a lake. (Hunting with a camera, of course). He scans the sky using saccadic eye movements. When he spots a duck, he tracks its movement using smooth pursuit eye movements. When the duck comes closer to him, his eyes must move toward each other using vergence eye movements. While he is doing all of this, the boat is rocking, which requires compensatory vestibular ocular eye movements. Thus all four control systems are continually used to move the eyes. For further information on these systems, consult Davson (1972) or Carpenter (1977). The rest of this chapter treats primarily saccadic eye movements.

3-7-3 Quantitative Eye-Movement Models

The first control systems model for the human saccadic-eye-movement system was proposed by Westheimer (1954). He recorded 20° saccades and suggested that they looked like the step response of a linear second-order system. He suggested the following equation as his model:

$$a\ddot{\theta} + b\dot{\theta} + c\theta = T(t) \tag{3-38}$$

where a, b, and c are constants and $\ddot{\theta}$, $\dot{\theta}$, and θ represent, respectively, eye acceleration, eye velocity, and eye position. The force applied to the globe by the extraocular muscles is represented by $T(t)$, the tension in the muscle. This can be expressed in our standard form as

$$\frac{\theta(s)}{T(s)} = \frac{\omega_n^2/k}{s^2 + 2\zeta\omega_n s + \omega_n^2}$$

Note that we have changed from linear movement, represented by the variable x, to rotary motion, represented by the variable θ. Accordingly, we should also change from a mass, M, to a rotational inertia, J. Westheimer suggested that the values of zeta and omega were $\zeta = 0.7$ and

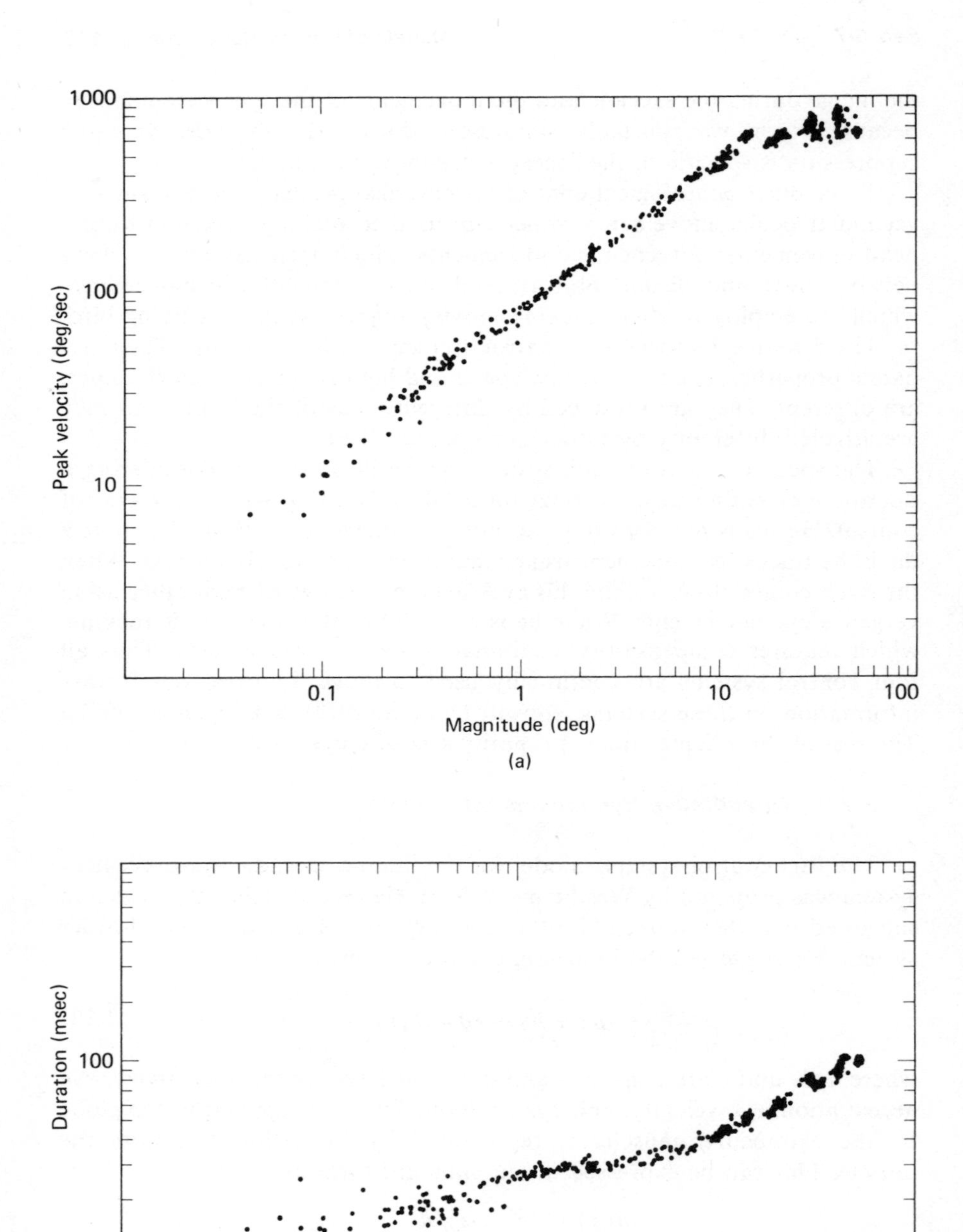

Figure 3-22 Main-sequence diagrams showing (a) peak velocity and (b) duration as functions of saccadic magnitude for normal human saccades. [From A. T. Bahill, M. R. Clark, and L. Stark, "The Main Sequence, A Tool for Studying Human Eye Movements," *Mathematical Biosciences*, 24 (1975), 191–204. Copyright Elsevier North-Holland, Inc.]

$\omega_n = 120$. The two major reservations that Westheimer had about his model were:

1. That the real system was not linear because the peak velocity-versus-magnitude plot was nonlinear (see Fig. 3-22a)
2. That the force input was probably an exponential, not a step.

After a model of a system has been developed, it is usually worthwhile to perform experiments on it. The model should be run in a variety of new modes. If any of these new situations yield interesting outputs, similar experiments should be designed and performed on the physiological system. If the results of the model simulation and the physiological experiments are similar, the validity of the model has been enhanced. As an example of this technique, let us rederive and simulate Westheimer's model.

3-7-3.1 MATHEMATICAL ANALYSIS OF AN UNDERDAMPED LINEAR SECOND-ORDER SYSTEM

The basic linear second-order spring–mass–dashpot system studied in Secs. 3-6-4 and 3-6-5 and suggested as an eye movement model by Westheimer has the following transfer function:

$$\frac{\theta(s)}{T(s)} = \frac{\omega_n^2/k}{s^2 + 2\zeta\omega_n s + \omega_n^2} \tag{3-39}$$

where ω_n and ζ are defined as

$$\omega_n = \sqrt{\frac{K}{J}}$$

and

$$\zeta = \frac{B}{2\sqrt{KJ}}$$

Therefore, to simulate eye movements on this model, let us find the step response of such a system when it is underdamped. That is, if $\zeta < 1$ and $T(s) = 1/s$. Then

$$\theta(s) = \frac{\omega_n^2/k}{s\left(s^2 + 2\zeta\omega_n s + \omega_n^2\right)}$$

This can be written as the sum of two fractions:

$$\theta(s) = \frac{A}{s} + \frac{Bs + C}{s^2 + 2\zeta\omega_n s + \omega_n^2} \tag{3-40}$$

By the quadratic formula, the roots of the second denominator can be found to be

$$s_{1,2} = -\zeta\omega_n \pm j\omega_n\sqrt{1-\zeta^2}$$

so Eq. (3-40) can be rewritten as

$$\theta(s) = \frac{\omega_n^2/k}{s\left[(s+\zeta\omega_n)^2 + \left(\omega_n\sqrt{1-\zeta^2}\right)^2\right]}$$

$$= \frac{A}{s} + \frac{Bs + C}{(s+\zeta\omega_n)^2 + \left(\omega_n\sqrt{1-\zeta^2}\right)^2} \tag{3-41}$$

A can now be found by the method of residues used in Sec. 3-6-4. Multiply both sides of the equation by s and then let s equal zero.

$$A = \frac{\omega_n^2/k}{(\zeta\omega_n)^2 + \left(\omega_n\sqrt{1-\zeta^2}\right)^2} = \frac{1}{k}$$

B and C will be found by the algebraic method. The right-hand side of Eq. (3-40) is combined over one denominator, yielding

$$\frac{\omega_n^2/k}{s(s^2 + 2\zeta\omega_n s + \omega_n^2)} = \frac{A(s^2 + 2\zeta\omega_n s + \omega_n^2) + s(Bs + C)}{s(s^2 + 2\zeta\omega_n s + \omega_n^2)}$$

Now if the two quantities are equal and their denominators are equal, their numerators must also be equal. Therefore, set the numerators equal and substitute

$$A = \frac{1}{k}$$

and then multiply both sides of the equation by k.

$$\omega_n^2 = s^2 + 2\zeta\omega_n s + \omega_n^2 + Bks^2 + cks$$

$$(kB + 1)s^2 + (2\zeta\omega_n + kC)s = 0$$

In order for this equation to be valid for all values of s, the coefficient of the s^2 term must be equal to zero and the coefficient of the s term must also be zero. Therefore,

$$B = -\frac{1}{k}$$

$$C = -\frac{2\zeta\omega_n}{k}$$

Substituting these values into Eq. (3-40) yields

$$\theta(s) = \frac{1}{k}\left(\frac{1}{s} - \frac{s + 2\zeta\omega_n}{s^2 + 2\zeta\omega_n + \omega_n^2}\right)$$

This can be rewritten as

$$\theta(s) = \frac{1}{k}\left[\frac{1}{s} - \frac{(s + \zeta\omega_n) + \zeta\omega_n}{(s + \zeta\omega_n)^2 + \left(\omega_n\sqrt{1 - \zeta^2}\right)^2}\right]$$

or

$$\theta(s) = \frac{1}{k}\left[\frac{1}{s} - \frac{s + \zeta\omega_n}{(s + \zeta\omega_n)^2 + \left(\omega_n\sqrt{1 - \zeta^2}\right)^2} - \frac{\zeta\omega_n}{(s + \zeta\omega_n)^2 + \left(\omega_n\sqrt{1 - \zeta^2}\right)^2}\right] \tag{3-42}$$

From a table of Laplace transforms we find that the inverse Laplace transform of Eq. (3-42) is

$$\theta(t) = \frac{1}{k}\left[1 - e^{-\zeta\omega_n t}\cos\left(\omega_n\sqrt{1 - \zeta^2}\, t\right) - \frac{\zeta}{\sqrt{1 - \zeta^2}} e^{-\zeta\omega_n t}\sin\left(\omega_n\sqrt{1 - \zeta^2}\, t\right)\right] \tag{3-43}$$

This equation would be more convenient if it were written with only one sinusoidal term. The transformation is not trivial, so we will demonstrate it with a simple example. Let us find the sum

$$f(t) = A\cos\omega t + B\sin\omega t$$

The easiest way to perform this addition is to define two new quantities by the equations

$$A = C \sin \phi$$
$$B = C \cos \phi \tag{3-44}$$

From Eq. (3.44) and Fig 3-23, the new quantities C and ϕ are determined by

$$C = \sqrt{A^2 + B^2}$$
$$\phi = \mathrm{Tan}^{-1} \frac{A}{B} \tag{3-45}$$

In terms of these quantities, the sum of the two sinusoids is

$$\begin{aligned} f(t) &= C \sin \phi \cos \omega t + C \cos \phi \sin \omega t \\ &= C \sin (\omega t + \phi) \end{aligned} \tag{3-46}$$

The frequency of the resultant wave of Eq. (3-46) is the same as the frequency of the two waves added together to produce it, but the amplitude and the phase angle are different. Using this procedure on Eq. (3-43) yields

$$A = 1$$
$$B = \frac{\zeta}{\sqrt{1 - \zeta^2}}$$

so

$$\phi = \mathrm{Tan}^{-1} \frac{\sqrt{1 - \zeta^2}}{\zeta}$$
$$C = \frac{1}{\sqrt{1 - \zeta^2}}$$

as shown in Fig 3-23.

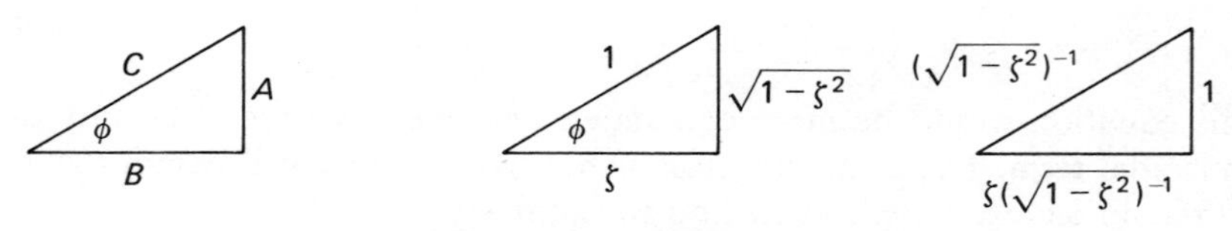

Figure 3-23 Relationship between A, B, C, ϕ, and ζ.

Therefore

$$\theta(t) = \frac{1}{k}\left[1 - \frac{e^{-\zeta\omega_n t}}{\sqrt{1-\zeta^2}} \sin\left(\omega_n\sqrt{1-\zeta^2}\, t + \phi\right)\right] \tag{3-47}$$

For simplicity, let us substitute the damped natural frequency, ω_d, as defined in Sec. 3-6-5.

$$\omega_d = \omega_n\sqrt{1-\zeta^2}$$

Then

$$\theta(t) = \frac{1}{k}\left[1 - \frac{e^{-\zeta\omega_n t}}{\sqrt{1-\zeta^2}} \sin(\omega_d t + \phi)\right] \tag{3-48}$$

The form of this equation is shown in Fig. 3-17 for the systems with ζ less than 1. There are some very interesting properties of the system that can be investigated by manipulating this equation. For example, we can find that the time to peak, T_p, will be the same for any size step input. The time to peak is defined as the time it takes the system to reach its most extreme value, as shown in Fig 3-20. This value is calculated by taking the derivative with respect to time, and finding the values of time that will make this derivative equal zero.

$$\begin{aligned}\frac{d\theta}{dt} &= \frac{-e^{-\zeta\omega_n t}}{k\sqrt{1-\zeta^2}}\left[\omega_d \cos(\omega_d t + \phi) - \zeta\omega_n \sin(\omega_d t + \phi)\right] \\ &= \frac{\omega_n e^{-\zeta\omega_n t}}{k}\left[\cos(\omega_d t + \phi) - \frac{\zeta}{\sqrt{1-\zeta^2}} \sin(\omega_d t + \phi)\right]\end{aligned} \tag{3-49}$$

When the techniques of Eqs. (3-44) to (3-46) are applied, the angles ϕ cancel out and the derivative becomes

$$\frac{d\theta}{dt} = \frac{\omega_n e^{-\zeta\omega_n t}}{k\sqrt{1-\zeta^2}}(\sin\omega_d t) \tag{3-50}$$

This will be equal to zero when

$$t = \frac{n\pi}{\omega_d} = \frac{n\pi}{\omega_n\sqrt{1-\zeta^2}}$$

where $n = 0, 1, 2, 3 \ldots$. The time to peak, T_p, is the smallest value that satisfies the preceding equation:

$$T_p = \frac{\pi}{\omega_n\sqrt{1-\zeta^2}}$$

Substituting this value in either Eq. (3-48) or (3-43) shows that the maximum value for θ is

$$\theta_{\text{max}} = \frac{1}{k}\left(1 + e^{-\zeta\pi/\sqrt{1-\zeta^2}}\right)$$

Some of these results were given without proof in Sec. 3-6-5.

3-7-3.2 IMPLICATIONS OF THE SIMULATION OF THE LINEAR SECOND-ORDER MODEL

Now that we have mathematically analyzed this underdamped linear second-order model, let us see what implications these results have. For the values that Westheimer used for his model, the time to peak becomes

$$T_p = \frac{\pi}{120\sqrt{1-(0.7)^2}} = 37 \text{ msec}$$

It can be immediately noted that the time to peak is independent of the size of the input step; this is not in concert with the experimental data. Figure 3-22 shows the peak velocity as a function of magnitude and the duration as a function of magnitude for normal human saccadic eye movements. The duration increases with saccade size; therefore, the model fails to fit these data. The theoretical values of ζ and ω chosen by Westheimer yielded duration values that were only appropriate for a 10° saccade.

A comparison can also be made between the peak velocities of the model and the human data. To do this, we can differentiate Eq. (3-50), set it equal to zero, and solve for t. The time of maximum velocity becomes

$$t = \frac{n\pi + \phi}{\omega_d}$$

with the first peak occurring at $n = 0$. This value can be substituted into Eq. (3-50) to yield

$$\left.\frac{d\phi}{dt}\right|_{\text{max}} = \frac{\omega_n}{k\sqrt{1-\zeta^2}}(\sin\phi)e^{-\phi/\tan\phi}$$

which is constant depending upon parameter values and the size of the input step. For a step of $\Delta\theta$ with $K = 1°/n, \dot{\theta}_{max} = 55\Delta\theta$.

The second derivative test shows that this is indeed a maximum. This value of the peak velocity is directly proportional to the size of the input step. The human saccadic peak velocity data could be fit with a linear approximation for saccades 15° and smaller, but there is a soft saturation for larger magnitudes. Therefore, the model does not match the physiological data. Westheimer noted this deficiency in his original model.

In summary, the response of a linear second-order system has the same duration for all input magnitudes, and the peak velocity is directly proportional to the size of the input step. Because the human-eye-movement system has neither of these properties, we can conclude that this is a valid model for saccadic eye movements of only one size. In spite of its deficiencies, this model continued to be used (see, e.g., Young and Stark 1963) until Robinson's model supplanted it in 1964.

3-7-3.3 THE PULSE-STEP INPUT

Westheimer assumed that there was a step input to the extraocular muscles. In 1964, Robinson performed an experiment to investigate this assumption. He applied a suction contact lens to the eye and then held the lens so that the eye could not move. He then measured the force required to hold this eye stationary while the other eye executed a saccade. Because the same innervation is sent to both eyes, he could infer the muscle force responsible for the eye movement. This muscle force was a *pulse-step* similar to that shown for the agonist in Fig. 3-24. Robinson also increased the complexity of the model to that of a fourth-order system. His model then consisted of a pulse-step input signal and the following transfer function:

$$\frac{\phi_{eye}}{T} = \frac{0.667(0.02s + 1)}{(0.3s + 1)(0.06s + 1)(1.03 \times 10^{-5}s^2 + 0.004s + 1)}$$

The numerical values were estimated from experimental data obtained from cats and were then adjusted by trial and error to match the human-eye-movement data.

This model could simulate saccades over a range of magnitudes between 5 and 40° by changing the size of the pulse and step. The input to this model was a force that was hypothetically proportional to the difference of the forces generated by the agonist and antagonist muscles. The position-versus-time traces from Robinson's model matched his data fairly well. However, the velocity-versus-time records had abrupt inflection points (see Fig. 3-25) that did not match the physiological data.

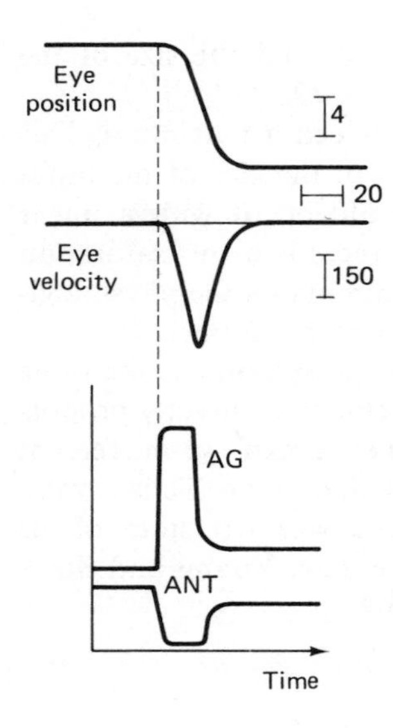

Figure 3-24 (a) Position as a function of time (top) and velocity as a function of time (bottom) for a typical saccade. (b) Presumed motoneural activity of agonist (top) and antagonist (bottom) as functions of time. Calibrations represent 4°, 20 msec, and 150°/sec. No time delay is shown between the start of the controller signal and the beginning of the saccade because this delay depends upon where the controller signal is measured. [From A. T. Bahill, M. R. Clark, and L. Stark, "The Main Sequence, A Tool for Studying Human Eye Movements," *Mathematical Biosciences*, 24 (1975), 191–204. Copyright Elsevier North-Holland, Inc.]

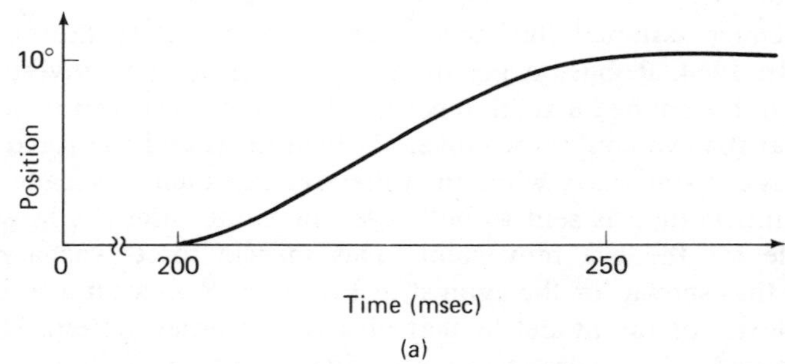

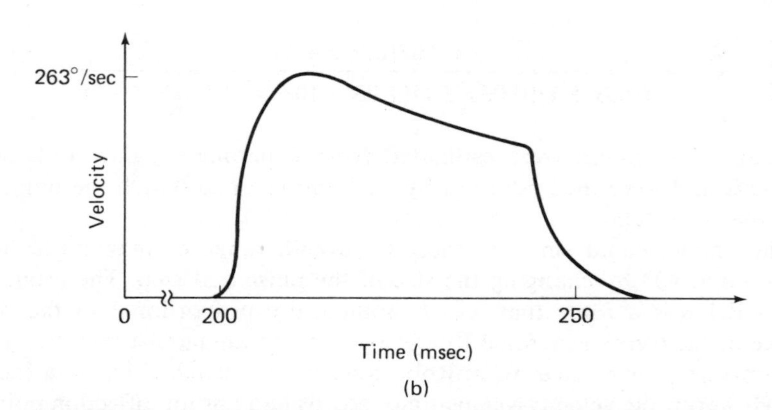

Figure 3-25 (a) Position and (b) velocity curves of Robinson's model. [From G. Cook and L. Stark, "Dynamic Behavior of the Human Eye-positioning Mechanism, *Communications in Behavioral Biology*, Part A, 1 (1968), 197–204.]

3-7-3.4 THE RECIPROCAL INNERVATION MODEL

The model developed by Cook and Stark (1968) and improved by Clark and Stark (1974) differed from Robinson's in two major ways: (1) the duration of their pulse was not as long as the duration of the saccade; and (2) they modeled the antagonist muscle as a separate independent element, implementing Descartes' principle of reciprocal innervation (see Ciuffreda and Stark 1975). The big improvement in their model was its capability of producing realistic records of velocity as a function of time. The following development of the reciprocal innervation model is similar to Cook and Stark's, but it explicitly includes the effects of the length–tension diagram, and it incorporates the results of more recent physiological experiments.

In order to develop a good model of the neuromuscular control of movement, a good model of a single muscle is needed. The Cold Springs Harbor Symposium of 1973 research reports discuss muscular physiology and provide data for such a model. The following muscle-model derivation is based upon Wilkie (1968).

The Passive Elasticity Resting muscle is elastic. It can only be stretched by applying a force: the greater the force, the greater the extension (Fig. 3-26). Human physiological data are available for estimating this parameter for human eye movements (PM curve of Fig. 3-32a) (Robinson, O'Meara, Scott, and Collins 1969; Collins, O'Meara, and Scott 1975). Passive muscle is nonlinear because it becomes more and more resistive as it is stretched: the curve becomes steeper and steeper. For simplicity, this *passive elasticity* will be modeled with an ideal linear spring. The coefficient K_{PE} will be chosen so that this linear approximation yields a good fit to the data in the region of interest for our model.

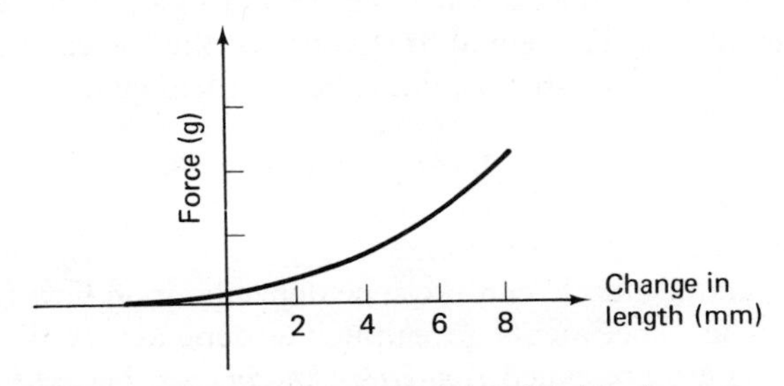

Figure 3-26 Length–tension relationship for passive muscle.

The Active-State Tension Generator A muscle produces a force when it is stimulated. A typical striated muscle responds to a single adequate stimulus with a twitch, a brief period of contraction followed by relaxation. As indicated in Fig. 3-27a, the time course of the twitch depends upon the temperature and the particular type of muscle.

The size of the twitch response depends on the strength of the stimulus, as shown in Fig. 3-27c. With very weak stimuli, nothing happens; but when the strength passes the threshold, a small response results which increases progressively until the stimulus reaches its maximum value. This occurs because the weak shock stimulates only a few muscle fibers close to the electrodes where the current density is highest, whereas the supramaximal shock stimulates all of them. Most of the experiments described here were performed using supramaximal shocks.

If a second shock is given to the muscle before the response to the first has completely died away, summation occurs (Fig. 3-27c). If the stimuli are repeated regularly at a sufficiently high frequency, a smooth tetanus results with tension maintained at a high level for as long as the stimulus train continues or until the onset of fatigue. The maximum tension developed on tetanic stimulation varies from 1.5 to 4.0 kg/cm^2 (frog, mammal) up to about 10 kg/cm^2 (edible mussel).

Because the most important property of a muscle is the fact that it produces a force, an ideal force generator will be the primary component of our muscle model. Each of the other components in the muscle model will represent physiological properties that modify the force which is available from this ideal force generator. A. V. Hill wanted to emphasize the difference between the force exerted by a muscle on a load and the fundamental force producing mechanisms inside the actual muscle. He called this ideal force-producing element the *active-state tension generator* (Hill 1938, 1950 b). For any particular degree of activation, this active-state tension will be a constant represented by F_0.

The Length–Tension Diagram The muscle model, shown in Fig. 3-28d, has two components in parallel: a force generator and a parallel elasticity. The tension in the muscle, T, is equal to the sum of the forces produced by the active-state tension generator, F_0, and the parallel elasticity, K_{PE}.

$$T = F_0 + LK_{PE}$$

where L is the length of the muscle.

The force that a muscle can produce depends upon how far the muscle is stretched. The experiments revealing the dependence of muscle force upon muscle length are called *isometric experiments*, because the muscle is kept at constant length. Typically, the muscle is fixed at a particular

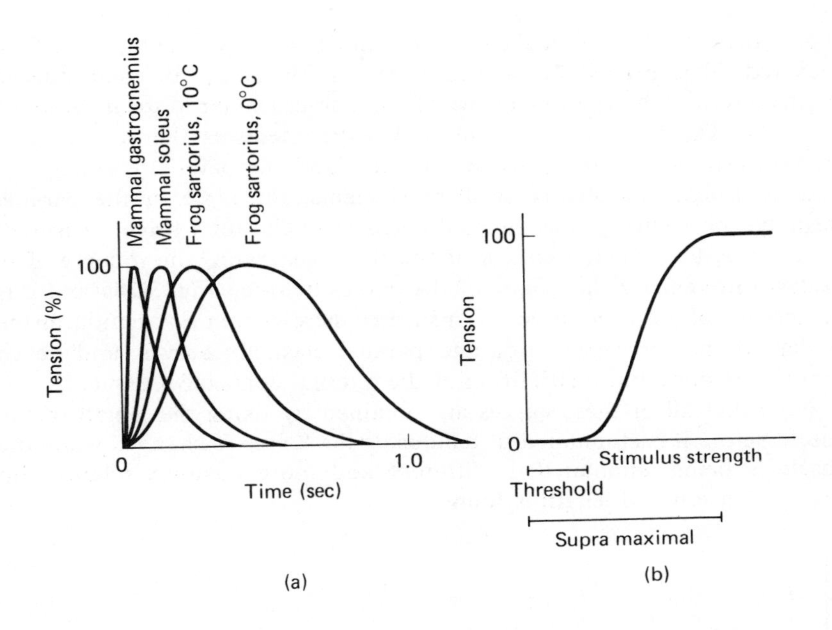

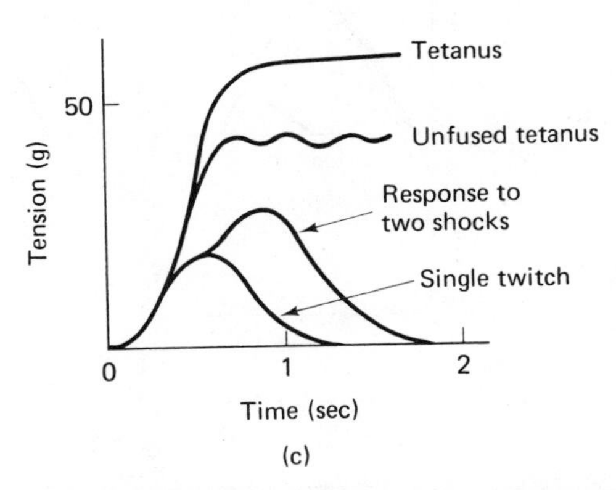

Figure 3-27 (a) Graph showing muscle twitches from different species and temperatures, scaled to the same peak height. (b) Relation between strength of stimulus and size of response. (c) Summation of response following repeated stimulation for frog muscle at °C. [From D. R. Wilke, *Muscle*, Edward Arnold, (Publishers) Ltd., London, 1968.]

length; then it is stimulated tetanically and the tension in the muscle is measured. This procedure is then repeated for many different muscle lengths. From such experiments, we obtain curves similar to the *a* curves of Fig. 3-28. The force recorded in such experiments is the sum of two different effects: the active force generator and the parallel elasticity. In order to isolate the effects of these elements, the force of the parallel elasticity, the *r* curve, must be subtracted from the total force measured, the *a* curve, to yield the effects of the force generator, the *d* curve. The greatest difference in the shape of the curves between Fig. 3-28a and c is the absence of a dip in curve a. This arises merely from a variation in the overlap of the force generator and parallel elasticity curves, and hence from the amount and distribution of the internal connective tissue.

Note that all of these curves are obtained by fixing the length of the muscle before it is stimulated to tetanus. If the length is changed while the muscle is being stimulated, a different and more complex relationship between tension and length is found.

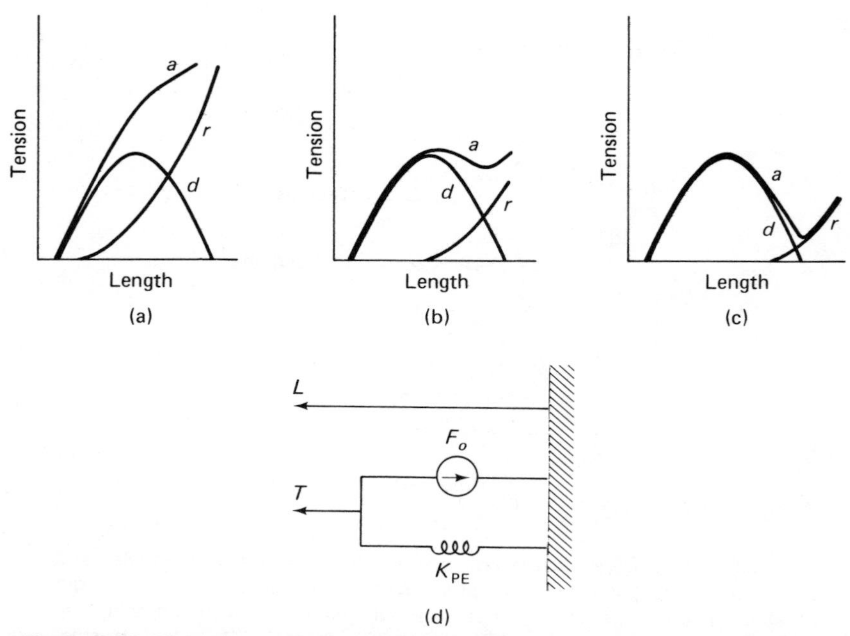

Figure 3-28 Length–tension diagrams for different types of frog muscle: (a) gastrocnemius; (b) sortorius; (c) semitendinosus. In each case r shows the length–tension curve of the resting muscle, a shows that of the active muscle and d $(= a - r)$ shows the extra tension developed on stimulation. (d) Simple model of muscle. F_0 is the force-generating contractile component and K_{PE} is the parallel elastic component. [From D. R. Wilkie, *Muscle*, Edward Arnold, (Publishers) Ltd., London, 1968.]

There is a rational explanation for the shape of these length–tension diagrams. To explain their shape, let us closely examine the microstructure of a muscle shown in Fig. 3-29.

Muscle is composed of anatomical units that can be subdivided into increasingly smaller functional units. A whole muscle (A) can be subdivided into fiber bundles called fasciculi (B), which can be further broken down into single muscle cells (C). The cells contain myofibril (D), which are composed of a series arrangement of repeating structures, the sarcomeres, that extend from Z disc to Z disc. Every sarcomere is composed of complex protein filaments, the myofilaments (E), which are in turn composed of large molecules of the protein myosin (thick filaments) or the smaller protein actin (thin filaments), along with other protein components. f, g, h, and i are cross sections at the levels indicated.

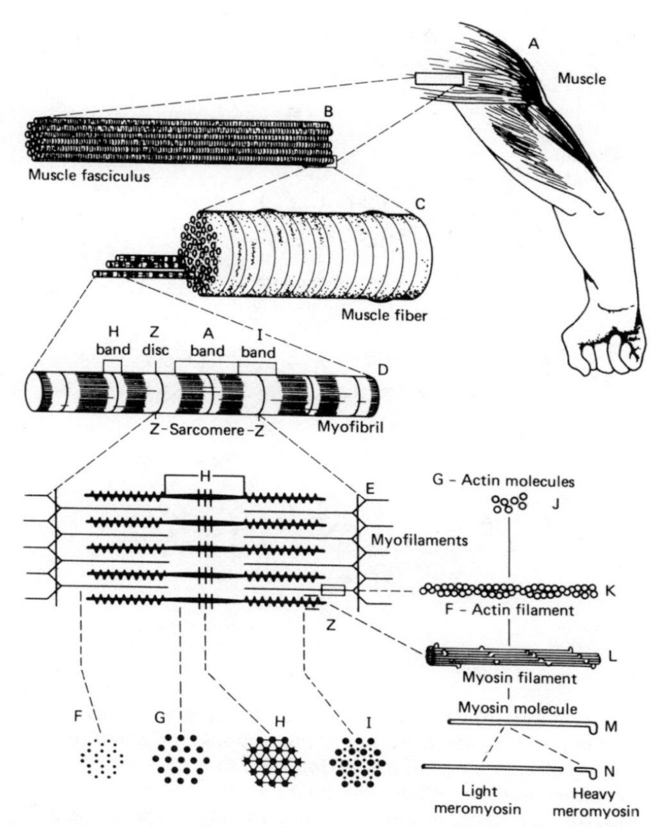

Figure 3-29 Skeletal muscle organization. (Drawing by S. C. Keene; from W. Bloom and D. W. Fawcett, *A Textbook of Histology*, 9th ed., W. B. Saunders Company, Philadelphia, 1968.)

Macroscopic movements of the muscle can be explained by the movements of the microscopic cross bridges in the sliding filament model of a muscle. Figure 3-30 suggests the molecular transformations that are responsible for this movement.

Electrical impulses are transmitted from the muscle membrane into the interior of the cell by a specialized structure called the *sarcoplasmic reticulum*. These impulses stimulate a rapid release of calcium which starts the biochemical processes.

Figure 3-30a shows the biochemical reactions involved in the contraction process. This model was derived from studies in which the long tails of the thick myosin molecules had been chemically removed; only the enzymatically active head portions, the cross bridges of intact muscle, were used. The thin helical actin filaments were separated from the *Z* lines, but

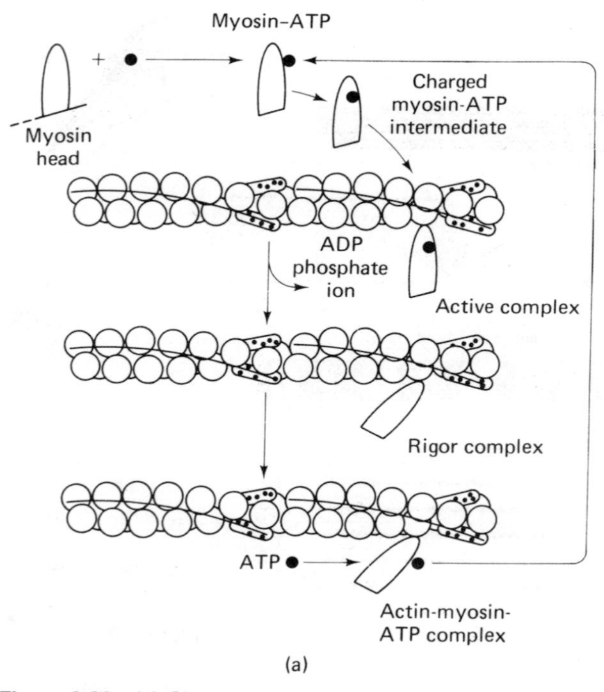

Figure 3-30 (a) Chemical events of the contraction cycle. Calcium ions (small dots) cover the troponin molecule (the oblong structure covered with dots) and thereby supress its inhibitory activity. When the charged myosin–ATP intermediate is formed, it combines with an actin molecule and the ATP is split. When sufficient ATP is present the complex separates and the myosin goes through the process again.

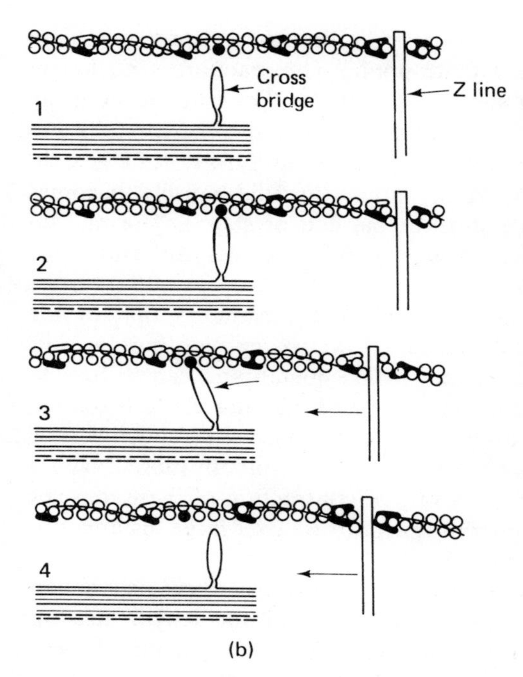

Figure 3-30 (b) The sequence of events in the contraction cycle of an intact muscle. The myosin head is permanently attached to the thick filament. When ATP is split, the myosin head converts the chemical energy into mechanical energy by rolling along the actin fiber; this action moves the thin filament parallel to the thick filament, and a small shortening of the muscle results. When the myosin head releases, it returns to its original position and is free to repeat the whole cycle. Many such cycles, when added together, can produce a large movement of the muscle. (From J. M. Murray and A. Weber, "The Cooperative Action of Muscle Proteins." Copyright 1974 by Scientific American, Inc. All rights reserved.)

they were intact and contained their normal complement of troponin. An excess amount of calcium was present in order to fully activate the process. This preparation was used because it mimics the actual process in intact muscle and is experimentally convenient. Figure 3-30b shows the mechanical counterpart in intact muscle of the biochemical reactions.

The cycle begins when a myosin head combines with a molecule of ATP, forming a *charged* myosin–ATP intermediate (step 1). This intermediate binds to an actin molecule (step 2); the actin myosin combination forms an ATPase system that splits ATP into ADP and an inorganic phosphate ion (step 3). This is the energy-liberating and motion-producing step in the contraction of intact muscle. After the ATP molecule is split, the myosin head is bound to the actin molecule. When the myosin head combines with a new ATP molecule, it becomes detached from the actin (step 4). It then becomes a new charged myosin–ATP intermediate (as in step 1) and repeats the sequence for as long as the calcium concentration and the ATP supply allow.

Calcium ions control the contraction process. Unless calcium ions are present the troponin molecule will not allow the myosin ATP-charged intermediate to combine with the actin molecule, and the sequence halts at the end of step 1. The muscle is ready to contract, but it is inhibited from further action; this is the normal state of resting muscle. When a voltage spike causes calcium ions to be released from the sarcoplasmic reticulum in intact muscle, they combine with troponin. The reaction is no longer inhibited, and it cycles through steps 2, 3, 4, and 1. The calcium pump in the sarcoplasmic reticulum removes calcium ions and thereby reestablishes the troponin inhibition of the contraction process; no further ATP is split, and the muscle relaxes. If the ATP supply for the muscle becomes exhausted, the reaction sequence stops at the end of step 3. The myosin remains bound to the actin and the whole muscle becomes stiff and rigid; this is the cause of rigor mortis.

Contraction of cardiac muscle is basically the same as that of the model just developed for skeletal muscle. However, cardiac muscle cells are smaller, so calcium released at the surface membrane can diffuse to more myofilaments. Therefore, the twitch of a cardiac muscle lasts longer. This same model can also be used for smooth muscle, although less is known of the physiological process. Many types of smooth muscle lack a sarcoplasmic reticulum: the very slow rate of contraction of these muscles suggests that the activator calcium ions must diffuse over large distances to reach the bulk of the myofilaments.

With this understanding of the movement of the cross bridges, we can now return to our attempt to explain the shape of the length–tension diagrams. The sliding-filament model predicted that when a muscle was lengthened, the area of overlap of thick and thin filaments diminished, so that the tension also diminished. This does happen, but it was difficult to make an accurate quantitative comparison between tension development and the area of overlap. Such a comparison demands (1) accurate measurements of the lengths of the thick and thin filaments and of their overlap at various sarcomere lengths, and (2) length–tension curve measurements which are accurately related to the sarcomere length (from Z line to Z line in Fig. 3-29) rather than to the length of the muscle as a whole.

The accurate length measurements were made by taking electronmicrographs of muscles that had been fixed and sectioned with special care to avoid artifacts from shrinkage and other causes (see Wilkie 1968) and also by using laser diffraction techniques (Iwazumi and Pollack 1979). The results are summarized in Fig. 3-31a.

Accurate length–tension curves were obtained from a series of experiments by Gordon, Huxley, and Julian (1966). For whole muscle, the shape of the length–tension curve depends on the connective tissue present. So they worked with a single fiber freed of connective tissue. The sarcomeres at the ends of the fiber were shorter than those in the middle, so they made their measurements only on the middle part of a single fiber by using an electromechanical feedback system to eliminate the unwanted contributions of the ends of the fiber. This yielded the length–tension curve of Fig. 3-31b, which consists of straight segments with fairly sharp corners be-

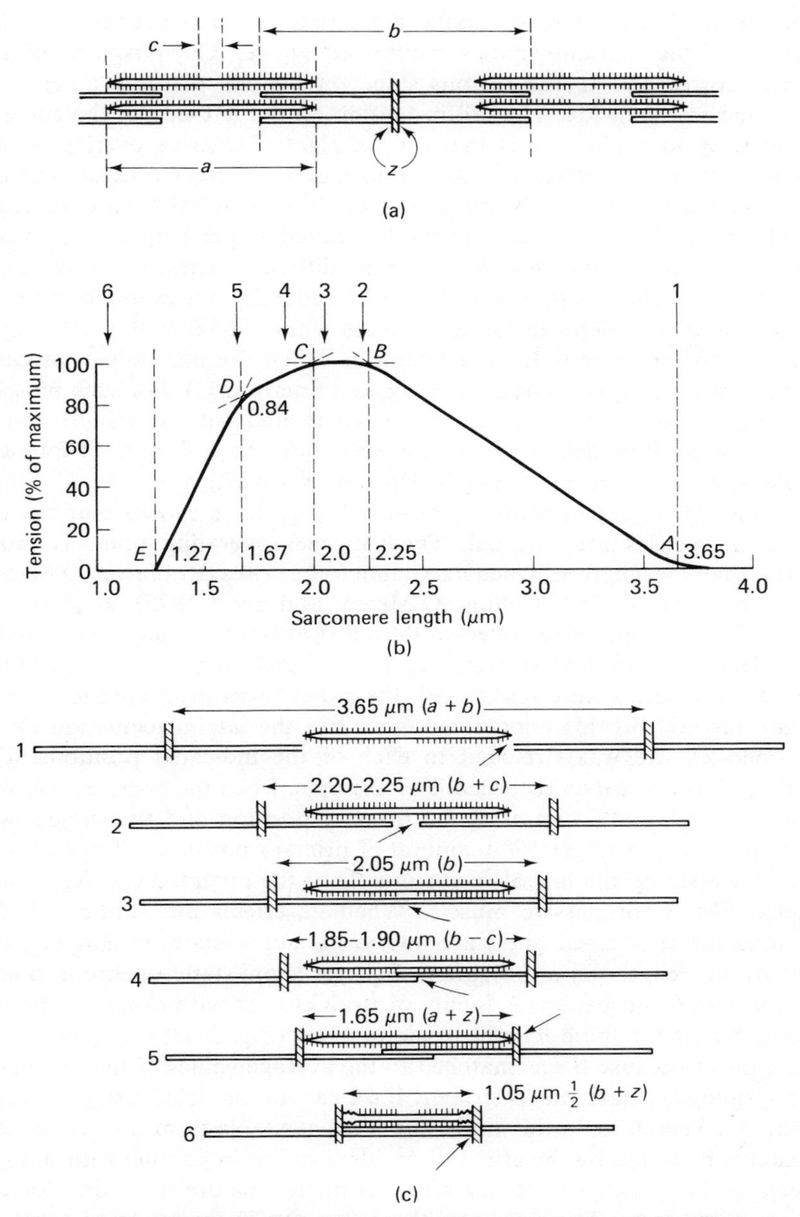

Figure 3-31 (a) Explanation of length–tension curves using sliding-filament model. (b) Length–tension curve from part of a single muscle fiber. The arrows along the top show the various critical stages of overlap . (c) Critical stages in the increase of overlap between thick and thin filaments as a sarcomere shortens. [From A. M. Gordon, A. F. Huxley, and F. J. Julian, *Journal of Physiology*, 284 (1966), 170–92.]

tween them. For the whole muscle, these sharp corners become rounded because of the nonuniformities mentioned above. The positions of the corners correspond to the various stages of overlap of the filaments, as illustrated in Fig. 3-31c. The fall in tension on the left side of the curve is not as easy to explain as is that on the right. Extensive overlap of the filaments probably interferes with the formation of cross bridges, and the thick filaments probably absorb part of the force that has been developed.

The exact form of a muscle model should depend upon the neuromuscular system being modeled because different types of muscle have different properties, as shown in Figs. 3-27 and 3-28. For example, many of the locomotory muscles in the leg operate within $\pm 5\%$ of the rest length, which is, for many muscles, the length at which the muscle can produce maximum force, L_0 (Goslow, Reinking, and Stuart 1973). For such muscles the length–tension diagram can be approximated with a straight line of zero slope. Such models would produce the same force for any length and would appear to ignore the length–tension relationships.

Physiological experiments on human beings have shown that the extraocular muscles are unusual. Their normal operating point is much shorter than the length at which maximum force arises (Robinson, O'Meara, Scott and Collins 1969; Collins, O'Meara, and Scott 1975), as shown in Fig. 3-32a. The data were collected during strabismus surgery: the muscles were detached and then reattached to the eyeball in a different position. Before the muscles were reattached, the patient was instructed to look at certain targets with his unoperated eye while the lateral rectus muscle of the operated eye was stretched to each of the indicated positions. The resulting force was then recorded. For example, when the operated eye was held in a position 10° nasal of the primary position and the subject was asked to look at a target 15° temporal of primary position, a force of 50 g was developed by the lateral rectus muscle of the operated eye. The curve labeled PM is for passive muscle. When its effects are subtracted, the resulting forces (dashed lines) are due to the active-state tension as modified by the length-tension diagram. Squares show static operating points for normal human beings. A family of straight lines with slope K' yields a reasonable fit for these length–tension curves (Fig. 3-32b). Figures 3-32a and b differ because b was matched to the average values of the two pools of physiological data. Such straight lines can be modeled using an ideal spring K'. Therefore, in the model the force available from the active-state tension is modified by the effect of an ideal spring in parallel with it. (The effects of the parallel elasticity were subtracted before the value for the spring was chosen.) There are now three elements in the model of a muscle (Fig. 3-33) and the tension in the muscle is

$$T = F_0 + LK_{PE} + LK'$$

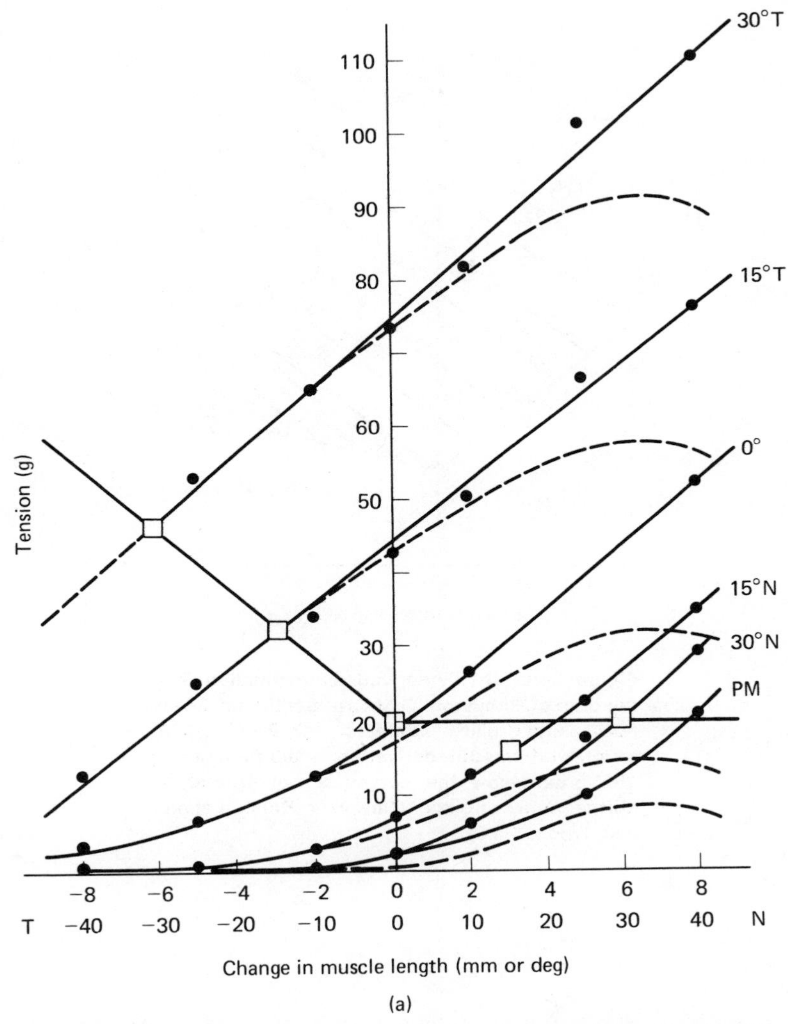

Figure 3-32 (a) Length–tension curves for human extraocular muscle. [Based on D. A. Robinson, D. M. O'Meara, A. B. Scott, and C. C. Collins, "Mechanical Components of Human Eye Movements," *Journal of Applied Physiology*, 26 (1969), 548–53.]

The Series Elasticity The next element we wish to add to our model is the series elasticity. The presence of this element was suggested by quick-release experiments of Levin and Wyman (1927): (1) a weight is hung from a muscle, (2) the muscle is stimulated tetanically, (3) the weight is quickly released, and (4) the muscle length and force are recorded as functions of time. Typical curves are shown in Fig. 3-34.

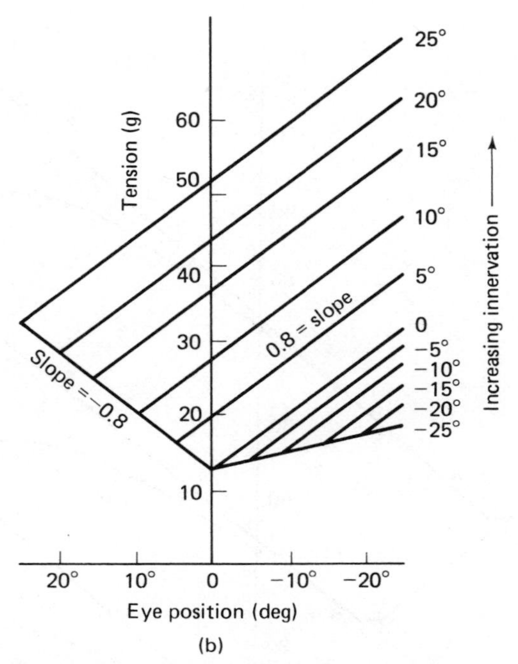

Figure 3-32 (b) Linearized approximation fit to the data of Robinson, O'Meara, Scott, and Collins (1969) and Collins, O'Meara, and Scott (1975), which were used to derive the model parameters. The area below the curves is not entered by normal human beings. (Based on Bahill, Latimer, and Troost 1980).

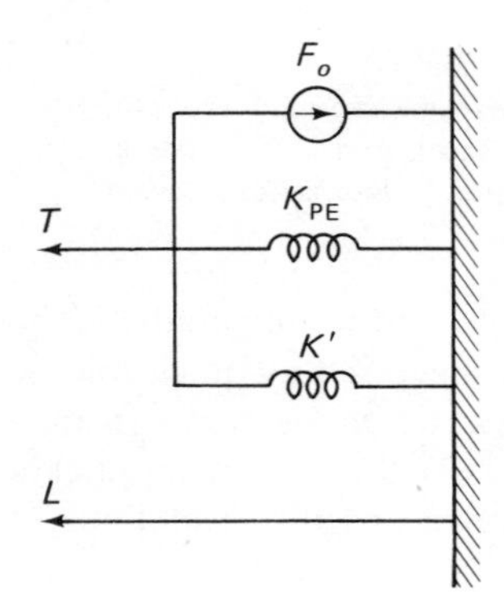

Figure 3-33 Muscle model with the additional affect of the length–tension diagram, K'. The active-state tension generator is F_0 and the total force exerted by the muscle is T.

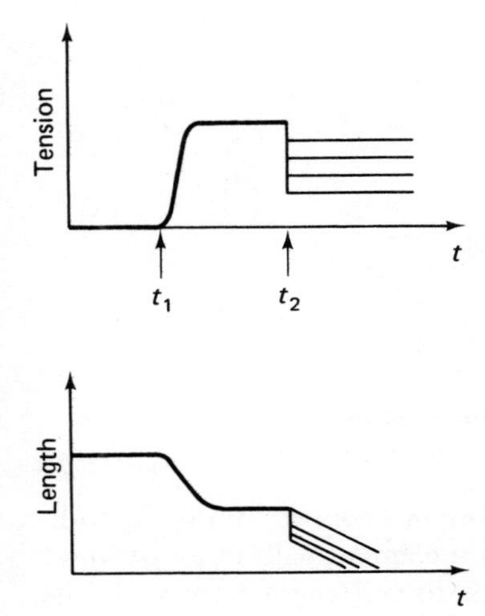

Figure 3-34 Typical data from quick-release experiments used to derive parameters for the series elasticity.

Two things happen after the weight is released at time t_2. The force abruptly decreases (because there is less of a load to support) and then the muscle shortens. There is an ideal mechanical element which will change its length instantaneously in response to an instantaneous change of force: a spring. For the model, this spring must be in series with the force generator; therefore, it is called the *series elasticity*. In the body, most of this elasticity is located in the tendon and in the actin and myosin cross bridges of the muscle (Rack and Westbury 1974). The spring should be nonlinear (Hill 1950a; Bahler 1968), but for small movements it can be approximated as a linear spring.

At time t_2 there is an abrupt change in force due to removal of the weight and a concomitant change in length due to the series elasticity. After t_2 there is a steady reduction in length until the new force, as determined by the length–tension diagram, is equal to the force required for the new weight. Adding the series elasticity produces the muscle model of Fig. 3-35. Numerical data from human extraocular muscles show that K_{SE} is approximately 125 N/m (2.5 g tension/deg) (Collins 1975).

The parallel elasticity has been connected beyond the series elasticity. It could have been connected on the other side of the series elasticity (Bahler 1968). The form shown will allow for simplifications later.

It now becomes important to define precisely where the force is being measured. The force exerted by the tendon on the mass will be called the *muscle force* or *muscle tension* and will be given the symbol T; the force from the ideal force generator will be called the *active-state tension* and will be given the symbol F. The physiological literature is not so unambiguous.

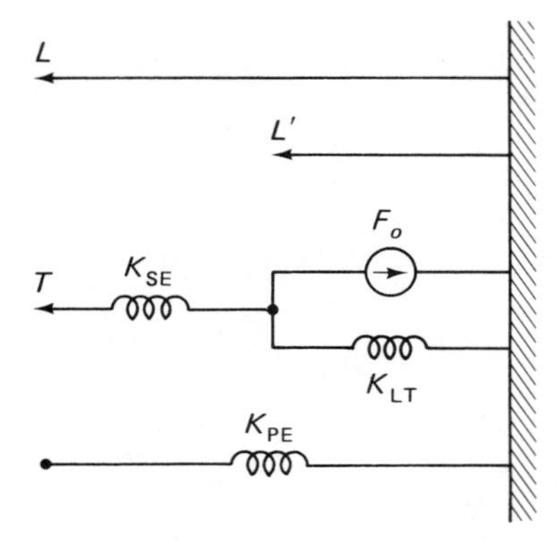

Figure 3-35 Muscle model with addition of series elasticity, K_{SE}.

Derivation of Numerical Value for Length–Tension Element With the addition of the series elasticity, the length–tension element will no longer have a numerical value equal to the slope of the force–length curves of Fig. 3-32. To calculate the value of K_{LT}, we must define the distance L', which is a hypothetical reference length (see Fig. 3-35), and L, which is the length of the muscle. Both lengths are positive from the base of the model. By temporarily neglecting the passive elasticity and defining T as the muscle force exerted on the eye, the static equations for equilibrium become

$$T = F_0 + K_{LT} L' \tag{3-51}$$

and

$$T = K_{SE}(L - L') \tag{3-52}$$

From Eq. (3-52),

$$L' = L - \frac{T}{K_{SE}}$$

Substituting back into Eq. (3-51), we obtain

$$T = F_0 + K_{LT}\left(L - \frac{T}{K_{SE}}\right)$$

and rearranging yields

$$T = \frac{K_{LT} K_{SE} L}{K_{LT} + K_{SE}} + \frac{K_{SE} F_0}{K_{LT} + K_{SE}} \tag{3-53}$$

Equation (3-53) will produce the length–tension curves of Fig. 3.32b, which approximate those of Fig. 3.32a, except for the differences in the ordinate intercept. The slope of these curves, K', is given in the length–tension equation (3-53) as

$$K' = \frac{K_{LT} K_{SE}}{K_{LT} + K_{SE}}$$

K_{SE} is approximately 125 N/m (2.5 g tension/deg) (Collins 1975) and K' is read from the graph of Fig. 3.32a as roughly 40 N/m (0.8 g tension/deg), so K_{LT} can be calculated to be

$$\begin{aligned} K_{LT} &= \frac{K_{SE} K'}{K_{SE} - K'} \\ &= 60 \text{ N/m} = 1.2 \text{ g tension/deg} \end{aligned}$$

The Force–Velocity Relationship The nonlinear force–velocity relationship is revealed by *isotonic* experiments. In these experiments, first performed by Fenn and Marsh in 1935, the muscle length is adjusted to be near rest length, then it is stimulated (usually to 100% activation). The load is then applied and the muscle is allowed to shorten. In this way, records similar to Fig. 3-36a are obtained. It can be seen that the greater the load, the smaller the total shortening. In fact, force and final length follow the left-hand part of the length–tension curve for active muscle, which was shown in Figs. 3-28, 3-31, and 3-32. Because the experiments are done isotonically, the length of the series elasticity will remain constant and the changes in the force will be due to changes in the active-state tension. For this discussion, the most important effect is that as the load gets larger, the maximum speed of shortening gets smaller; that is, the maximum slope of the curve gets smaller. If force or load is plotted against slope, a graph similar to Fig. 3-36b is obtained. Each curve in Fig. 3-36a yields one data point for Fig. 3-36b.

The effects of this force–velocity relationship are quite general. Curves similar to Fig. 3-36b have been obtained from cardiac, smooth, and striated muscle and even from contracting actomyosin threads. The only exceptions seem to be those insect muscles that work by small-scale vibrations rather than by gross changes in length. Therefore, many models have been developed to explain the behavior. One early theory was that the force developed by the muscle was actually constant at all speeds, but when the muscle shortened, part of the force was absorbed by internal viscosity, so that less force was exerted externally. If the internal viscosity were a nonlinear one, this theory would account for the purely mechanical properties of muscle; but it was abandoned because it did not seem to fit

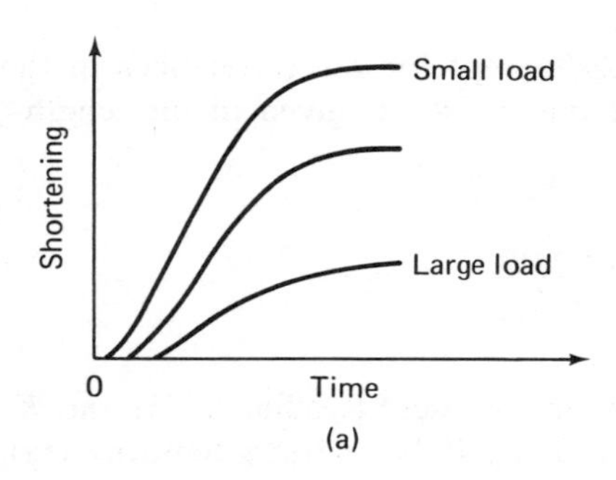

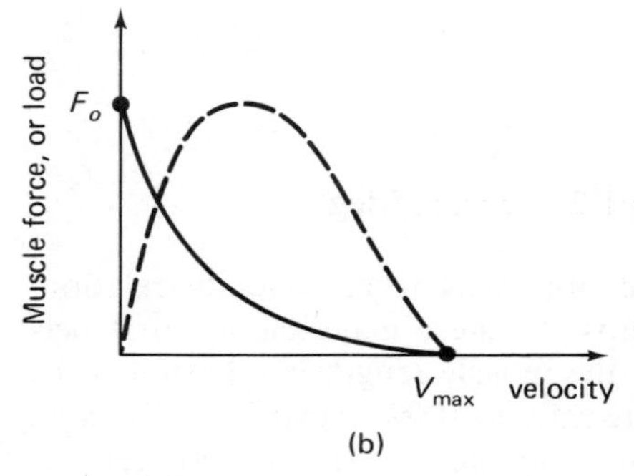

Figure 3-36 (a) Records of shortening against time from a muscle lifting various loads. Tetanic stimulation started at time zero. (b) Force–velocity relationship. The interrupted line shows how mechanical power produced varies according to the force on the muscle. (After Hill 1951, and Wilkie 1968.)

the observed energetic properties of contraction. The force–velocity relationship is probably due to the fact that the rate of the chemical reactions in the muscle is linked to the force on it. That is, larger forces require more cross bridges to be active, and this takes more time.

Many different equations can be fit to the force–velocity data. The most interesting one is Hill's equation (Hill 1938), which fits part of a hyperbola to the curve:

$$V = \frac{(F_0 - T)b}{(T + a)} \tag{3-54}$$

where V is the velocity, T the muscle force, F_0 the isometric force, and a and b are parameters that are related to measurable quantities. Hill's experimental data suggested values of $a = F_0/4$ and $b = V_{max}/4$. The symbol V_{max} is used to represent the maximum muscle velocity. The symbol F_0 is used to represent the muscle force, available at zero velocity. For these isotonic experiments F_0 is the active-state tension. This parameter depends upon the degree of activation of the muscle; that is F_0 for 100 activation is larger than F_0 for 50% activation. Although in the experiments described, muscle force, or load, is the independent variable, the curves are usually plotted with muscle force on the ordinate. Equation (3-54) can be rewritten as

$$T = F_0 - \left[\frac{F_0 + a}{b + V} \right] V \tag{3-55}$$

and therefore

$$T = F_0 - BV \tag{3-56}$$

where

$$B = \frac{F_0 + a}{b + V} \tag{3-57}$$

This suggests that the force–velocity relationship may be modeled with a dashpot, as shown in Fig. 3-37. The muscle force available at the tendon is decreased by a velocity-dependent term.

In the preceding section we derived an equation, Eq. (3-53), representing the force available at the tendon after the active-state tension was modified by the effects of the length–tension diagram. To make this equation appropriate for rotations, a new variable, θ_1, will be defined (see Fig. 3-39). Its value will be zero when the eye is in primary position (looking straight ahead). For variations about this operating point, $\theta_1 = -L$. Thus Eq. (3-53) becomes

$$T = \frac{K_{SE} F_0}{K_{LT} + K_{SE}} - \frac{K_{LT} K_{SE} \theta_1}{K_{LT} + K_{SE}}$$

We must now decrement this force to account for the effects of the force–velocity relationship. The muscle force available at the tendon

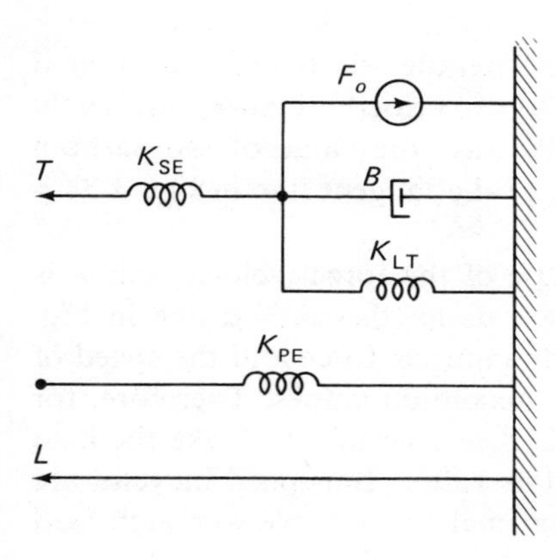

Figure 3-37 Muscle model with addition of nonlinear force–velocity relationship, *B*.

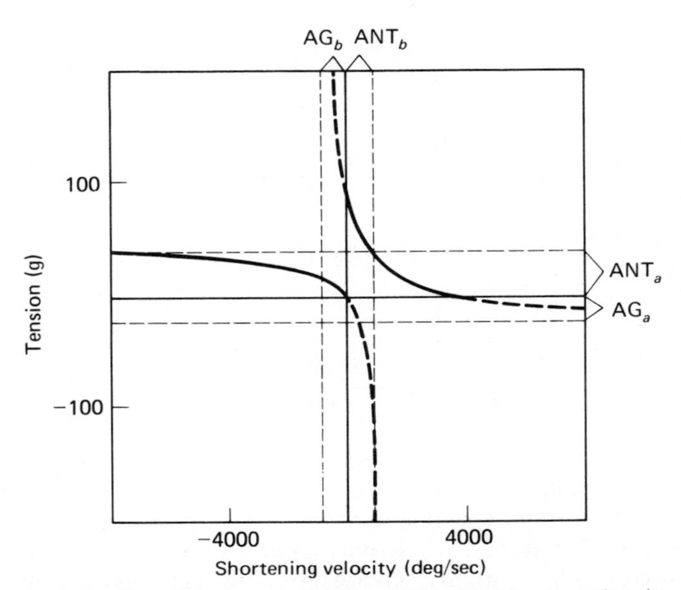

Figure 3-38 The force–velocity relationship for muscles is hyperbolic. The curve in the first quadrant is for the agonist muscle, and the curve in the second quadrant is for the antagonist muscle. The constants *a* and *b* of the Hill equation shift the asymptotes away from the origin. There are, of course, a family of hyperbolic curves for different activation states. These curves show the force–velocity relationships for 10° saccades.

becomes

$$T = \frac{K_{SE} F_0}{K_{LT} + K_{SE}} - \frac{K_{LT} K_{SE} \theta_1}{K_{LT} + K_{SE}} - B\dot{\theta}_2 \tag{3-58}$$

This equation also defers treatment of the parallel elasticity. Note that if the dashpot were linear, the slope of the force–velocity relationship would be the value of the dashpot. However, because the value of the dashpot actually depends upon velocity, the slope of the tangent line in Fig. 3-38 is not the dashpot value.

One practical consequence of the shape of the force–velocity curve is that the mechanical power output of the muscle (the dashed line in Fig. 3-36b) passes through a maximum when the muscle force and the speed of shortening have about one-third of their maximum values. Therefore, for efficient performance of mechanical work, it is necessary to make the load presented to the muscles approximately this value. Ten-speed bicycles are excellent examples of practical devices that make it possible to match load and speed to the properties of the muscle. Wilkie (1968) notes that the maximum power output is about $0.1F_0 \times V_{max}$.

The Antagonist Muscle So far our discussions have been limited to muscles that shorten when stimulated. But for every muscle that is shortened, there is, roughly speaking, an antagonist muscle that is being lengthened. This antagonist muscle has a great effect on the dynamics of the resulting movement. In the model, the inhibition of the antagonist has just as great an effect on the velocity and duration of a saccade as the high-frequency burst of activity for the agonist. Scott (1975) described saccadic eye movements with almost normal velocities which resulted from the action of only the antagonist (the agonist was paralyzed and could contribute no force). The force–velocity relationship of a muscle which is being lengthened is quite different from that of a muscle which is allowed to shorten (Katz 1939; Huxley 1957; Joyce, Rack, and Westbury 1969; Joyce and Rack 1969; Julian 1971; Julian and Sollins 1973; Partridge 1978).

The antagonist muscle is usually activated only slightly, yet it offers a large resistance to being stretched. It is possible to use the same function, a hyperbola, to model the antagonist force–velocity relationship as shown in Fig. 3-38. Using these curves, and replacing F_0 with the active-state tensions, F_{AG} and F_{ANT}, the functions for the agonist and antagonist dashpots become

$$B_{AG} = \frac{F_{AG} + AG_a}{\dot{\theta}_2 + AG_b} \tag{3-59}$$

and

$$B_{ANT} = \frac{F_{ANT} - ANT_a}{-\dot{\theta}_3 - ANT_b} \tag{3-60}$$

These equations were derived by Hsu, Bahill, and Stark (1976) for their reciprocal innervation eye-movement model.

The Passive Tissues and the Eyeball The rotational inertia of the eyeball is represented by J. The eyeball was modeled as a solid sphere of ice. A much more complicated model using several concentric spheres connected by viscoelastic elements could have been used, but a sensitivity analysis of the model showed that this was unnecessary (Sec. 3-7-4.3).

The optic nerve, the other extraocular muscles, orbital fat, check and suspensory ligaments, and other tissues contribute to the viscoelasticity that limits the movement of the eyeball. Human physiological data are available for evaluating these parameters (Robinson, O'Meara, Scott, and

Collins 1969; Collins, O'Meara, and Scott 1975). They are modeled by the spring and dashpot labeled K'_p and B_p. Although some modelers have used two groups of springs and dashpots in series for such elements, the sensitivity analysis proved this added complication unnecessary.

The new model is shown in Fig. 3-39. It is now obvious why we chose the form of model where K_{PE} is connected outside K_{SE}: it is now possible to combine the three parallel springs $K_{PE_{AG}}$, $K_{PE_{ANT}}$, and K'_p into one larger spring, K_p. This greatly simplifies the overall model. This model is said to be a homeomorphic model because there is a one-to-one relationship between the elements of the model and the elements of the physiological system. Such a similarity of form makes evaluation of a model easier.

Activation and Deactivation Time Constants As shown in Fig. 3-27, the muscle force does not rise to its maximum value instantaneously. However, even the active-state tension does not rise to its maximum value instantaneously.

A pulse-step controller signal is used to produce saccadic eye movements. In the reciprocal innervation model, the idealized pulse-steps were modified by first-order lag circuits (low-pass filtering) in order to produce the active-state tensions of the muscles. These tensions were then modified by the nonlinear force–velocity relationships, by the effects of the length–

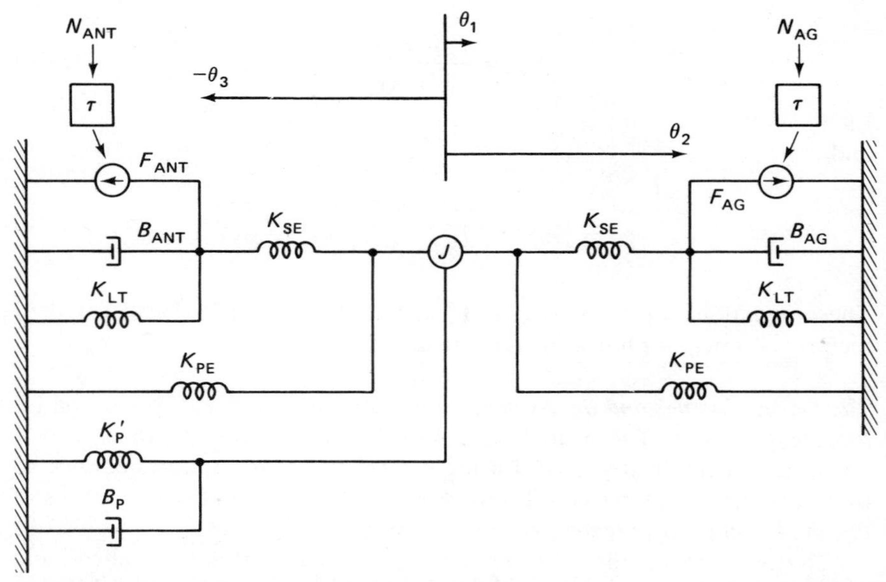

Figure 3-39 Reciprocal innervation model of the extraocular plant. θ_1 is the distance of the eye from primary position (0°). θ_2 is the distance of the agonist node from primary position; it is separated from θ_1 by the agonist series elasticity. θ_3 is the distance of the antagonist node from primary position.

tension diagrams, and by the series elasticities to produce muscle forces (see Figs. 3-40 and 3-41b).

In the physiological system, there is both a delay and a low-pass filtering in the transition between the pulse-step and the active-state tensions. The reciprocal innervation model does not include the delay between the onset of the pulse and the start of the saccade, because this delay is not fixed: it depends upon where the pulse-step is measured. The model does include the filtering. Physiologically, this filtering is due to both spreading in time and rate-limiting factors. The spreading is due to variations between cells in synchronization, synaptic transmission delays, motoneuronal firing frequency acceleration, neuronal conduction velocity, depolarization, spread of activity in the sarcoplasmic reticular formation, and acceleration of actinomysin cross bridges. The rate-limiting processes include the synaptic transmissions, the release and reuptake of the Ca^{2+} and its modification of the actinomysin fibers. For simplicity, we have accounted for all of these with simple first-order time constants (τ_{AC} and τ_{DE}).

Most of this low-pass filtering is probably due to the Ca^{2+} activation process. A clue for a physiological value of this time constant can be taken from Rudel's work on the frog with a Ca^{2+}-sensitive bioluminescent protein which emitted light in the presence of Ca^{2+} (Taylor, Rudel, and Blinks 1975; Rudel, personal communication). The muscle was electrically stimulated and the light flux versus time was plotted. From these data, a 10 to 90% rise time of 7 to 20 ms was calculated (for frog muscle at 15° C). Adjusting this value for 25° C (division by 2.2) yields 3 to 9 msec. Changing this into time constants (division by 2.2) yields 2 to 4 msec. These values were fine-tuned by using them in the model and by comparing model data to human physiological data. The time constants that were finally used ranged between 0.2 and 13 msec.

At least one of these time constants, τ_{AG-AC}, is a function of motoneuronal firing frequency. This hypothesis is supported by two separate arguments. The first is that the rate of muscular tension development increases with higher motoneuronal firing frequencies. This has been offered as an explanation for oculomotoneurons firing at frequencies that were higher than the muscle tetanus frequency. Although maximum tension will not continue to rise with increasing frequency, the rate of tension rise will increase. This proportionality between rate of tension development and motoneuronal firing frequency has been demonstrated for lateral rectus muscles (Barmack, Bell, and Rence 1971), for fast units of the inferior oblique muscles (Lennerstrand 1972), and for slow units of the soleus muscle (Henneman 1974). Larger movements are produced by, among other things, higher motoneuronal firing frequencies. Therefore, larger movements should have a faster rate of tension rise. This phenomenon is probably related to the facilitation of the second of a pair of closely

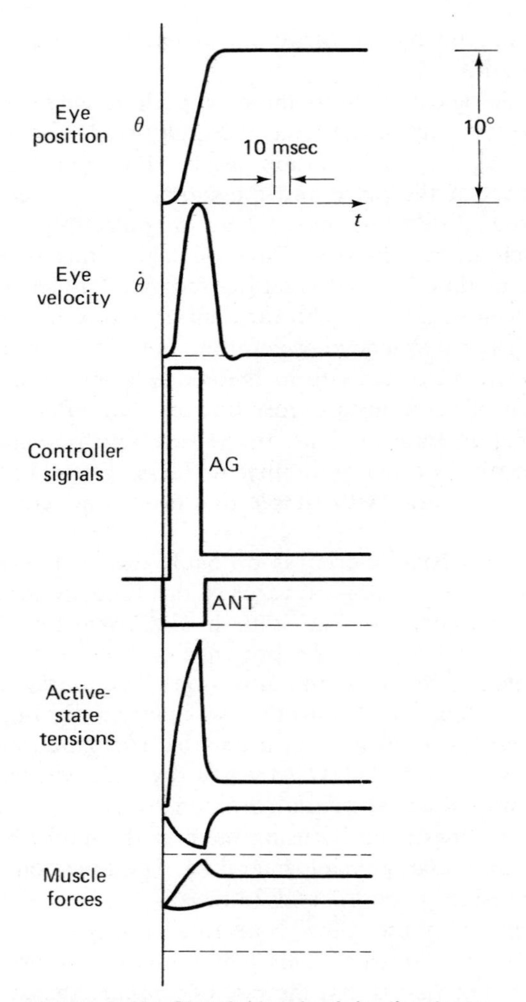

Figure 3-40 Signals involved in the transformation of input commands into eye movements. The pulse-step controller signal has abrupt transitions which are filtered out by the activation and deactivation time constants to produce the active-state tensions. These are, in turn, filtered by the series elasticity and the nonlinear force–velocity relationship to produce the forces that are applied to the globe. These forces produce the eye movements.

spaced neuronal spikes and is similar to the catch property of certain invertebrate muscles (Burke, Rudomin, and Zajac 1970; Barmack, Bell, and Rence 1971). The way this relates to the time constant is that in large saccades, the motoneurons fire at higher frequencies and yield faster rates of tension rise.

To show the second reason for time-constant variability, we must digress to discuss muscle physiology. There are two types of muscle fibers: fast and slow. Well, at least two types; some researchers claim there are three, some claim there are five, and some claim there is a continuum of fiber types between the fast and slow types.

The slow fibers contract slowly but resist fatigue, they are the tonic fibers. Each fiber is small and contributes a small amount to the total muscle force. The fast fibers contract rapidly and fatigue quickly; they are the phasic fibers. Each fast fiber is larger and contributes much more force than a slow fiber. Some muscles are composed predominately of fast fibers, some predominately of slow fibers, and some of an intermediate mixture.[1]

In muscles of mixed fiber types, there is an orderly recruitment of fibers. This is called the *Henneman size principle*(Henneman, Somjen, and Carpenter 1965; Goslow, Cameron, and Stuart 1977; Binder, Cameron, and Stuart 1978). For small muscle forces, only the small, slow fibers will be recruited. For larger forces, the large, fast fibers will also be recruited. As the required muscle force increases, fibers will be recruited in a fixed order from the smallest to the largest. As the muscle force is relaxed, the fibers will drop out in inverse order. This size principle applies over a wide range of muscle types and animal species.

Finally, let us go back and explain the second reason for time-constant variability in the reciprocal innervation model. Large saccadic eye movements will recruit the large, fast, pale, fast-fatiguing, global fibers (those nearest the globe), whereas small saccades will only recruit the slower orbital fibers (those nearest the orbit) (Lennerstrand 1974; Collins 1975). Therefore, because the larger saccades recruit the faster motoneurons, the rate of rise of muscle force should be greater for larger saccades.

Lacking physiological evidence to the contrary, the agonist deactivation time constant and both of the antagonist time constants were fixed for all sizes of saccades. Only the agonist activation time constant was made a function of saccadic size.

[1] We take advantage of the differences in muscle types every day. When walking home from school you probably carry your books primarily on your ring and little fingers. The muscles controlling these fingers are composed predominately of slow fibers. To see the difference, try carrying your books on your middle and index fingers, which are composed predominately of fast fibers. Or merely try carrying your lunchbag between your thumb and index finger and see how long you last.

The following equation demonstrates the functional relationship of the agonist activation time constant:

$$\tau_{\text{AG}-\text{AC}} = (13 - 0.1\Delta\theta) \quad \text{msec} \tag{3-61}$$

where $\Delta\theta$ is the size of the saccade.

Reciprocal Innervation Model Figure 3-41a shows the nonlinear reciprocal innervation model of Hsu, Bahill, and Stark (1976). The top figure shows Descartes' basic concept of reciprocal action of extraocular muscles: Descartes thought that muscles were like balloons; when inflated they would be short and fat, when drained of fluid they would be long and skinny. The pipes were to pump fluid reciprocally in and out of the muscles. Shortening of the agonist, together with lengthening of antagonist, produces eye movements. The bottom figure shows the ideal mechanical elements used for modeling the plant. The globe and surrounding tissues were modeled by the inertia (J), a viscous element (B_p), and a passive elasticity (K_p'). Each muscle was modeled by an active-state tension generator (F_{AG} and F_{ANT}), a nonlinear dashpot (B_{AG} and B_{ANT}) representing the nonlinear force-velocity relationship, a series elasticity (K_A), and a parallel elasticity that was combined with passive elasticity of the globe to form (K_p). The active-state generator converts motoneuronal firing into force through a first-order activation-deactivation process. The controller signals (CS_{AG} and CS_{ANT}) represent the aggregate activity of all the motoneurons in the agonist and antagonist motoneural pools. This model is similar to the model shown in Fig. 3-39 except for its treatment of the length–tension diagram: it was assumed that the muscles worked at the peak force, and this region was approximated with a horizontal line. Thus K_{LT} equaled zero and was removed from the model.

The system equations for the model are

$$F_{\text{AG}} = K_{\text{SE}}(\theta_2 - \theta_1) + B_{\text{AG}}\dot{\theta}_2$$

$$F_{\text{ANT}} = K_{\text{SE}}(\theta_1 - \theta_3) - B_{\text{ANT}}\dot{\theta}_3$$

$$K_{\text{SE}}(\theta_2 - \theta_1) - K_{\text{SE}}(\theta_1 - \theta_3) = K_p\theta_1 + B_p\dot{\theta}_1 + J_{\text{G}}\ddot{\theta}_1 \tag{3-62}$$

F_{AG} is the muscle force exerted by the agonist. F_{ANT} is the muscle force exerted by the antagonist. θ_1 is eye position; θ_2 and θ_3 are the nodal positions separated from θ_1 by the series elasticities of the agonist and antagonist muscle, respectively, as shown in Fig. 3-39. $\dot{\theta}_1$ and $\ddot{\theta}_1$ indicate eye velocity and acceleration, respectively.

The Newtonian equation is

$$F_{\text{AG}} = B_{\text{AG}}\dot{\theta}_2 + K_p\theta_1 + B_p\dot{\theta}_1 + J_{\text{G}}\ddot{\theta}_1 + F_{\text{ANT}} + B_{\text{ANT}}\dot{\theta}_3 \tag{3-63}$$

This model is a sixth-order nonlinear system. For the six state variables, let us choose the three θ's shown in Fig. 3-39: the eye velocity and the two

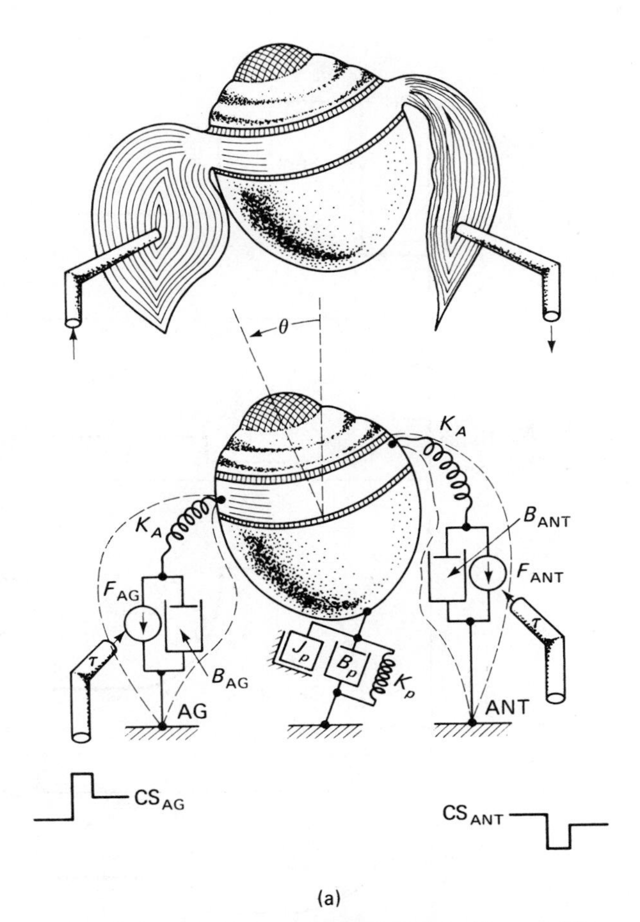

Figure 3-41 (a) Two reciprocal innervation models for the human eye movement system—Descartes' 1626 model and the 1976 model. [From A. T. Bahill, F. K. Hsu, and L. Stark, "Glissadic Overshoots Are Due to Pulse Width Errors," *Archives of Neurology*, 35 (1978), 138–42. Copyright 1978, American Medical Association.]

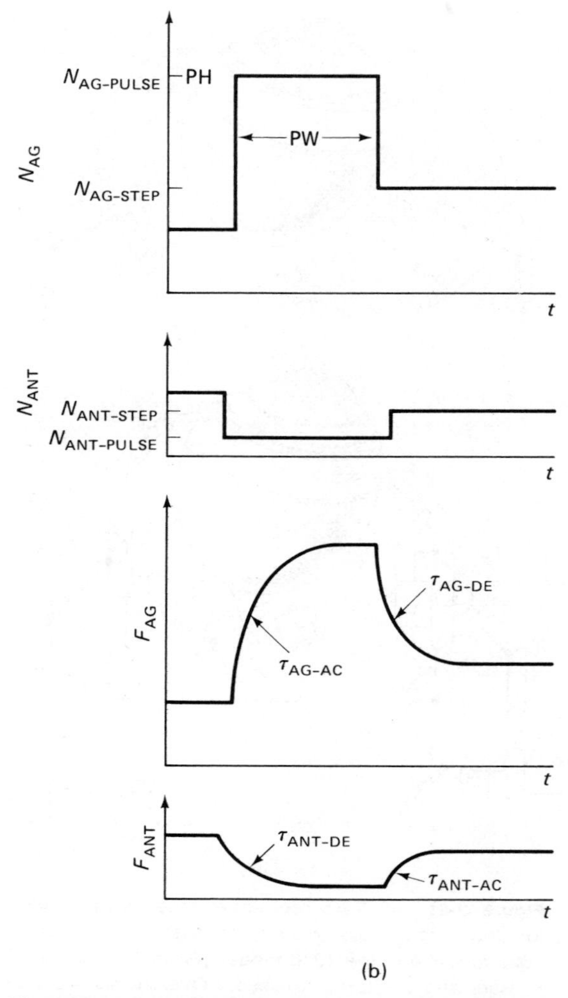

(b)

Figure 3-41 (b) Sketches showing how the controller signals are transformed into active-state tensions by first-order activation and deactivation processes.

active-state tensions. The six state equations and initial conditions are as follows:

State Equations	*Initial Conditions*
$\dot{x}_1 = \dot{\theta}_1 = x_4$	$x_1 = 0$
$\dot{x}_2 = \dot{\theta}_2 = \dfrac{AG_b[K_{SE}(x_1 - x_2) + x_5]}{AG_a + K_{SE}(x_2 - x_1)}$	$x_2 = \dfrac{16}{K_{SE}}$
$\dot{x}_3 = \dot{\theta}_3 = \dfrac{ANT_b[K_{SE}(x_1 - x_3) - x_6]}{ANT_a - K_{SE}(x_1 - x_3)}$	$x_3 = \dfrac{-16}{K_{SE}}$
$\dot{x}_4 = \ddot{\theta}_1 = \dfrac{K_{SE}x_2 + K_{SE}x_3 - x_1(2K_{SE} + K_p) - B_p x_4}{J}$	$x_4 = 0$
$\dot{x}_5 = \dot{F}_{AG} = \dfrac{N_{AG} - x_5}{\tau_{AG}}$	$x_5 = 16$
$\dot{x}_6 = \dot{F}_{ANT} = \dfrac{N_{ANT} - x_6}{\tau_{ANT}}$	$x_6 = 16$

The following parameters values were used in this model for saccadic and vergence eye movements from 0.1 to 50°. The size of the eye movement is $\Delta\theta$.

$$K_{SE} = K_A = 1.8 \text{ g tension/deg} = 91.9 \text{ N/m}$$
$$K_P = 0.86 \text{ g tension/deg} = 43.9\ N\text{/m}$$
$$B_P = 0.018 \text{ g tension-sec/deg} = 0.919 \text{ N-sec/m}$$
$$J = 4.3 \times 10^{-5} \text{ g tension-sec}^2\text{/deg}$$
$$= 2.192 \times 10^{-3} \text{N} - \text{sec}^2\text{/m}$$
$$N_{\text{AG-pulse}} = PH \quad \text{(see Table 3-4)}$$
$$N_{\text{ANT-pulse}} = (0.5 + 16e^{-\Delta\theta/2.5}) \text{ g tension}$$
$$N_{\text{AG-step}} = (16 + 0.8\Delta\theta) \text{ g tension}$$
$$N_{\text{ANT-step}} = (16 - 0.06\Delta\theta) \text{ g tension}$$
$$\tau_{\text{AG-AC}} = (13 - 0.1\Delta\theta) \text{ msec}$$
$$\tau_{\text{AG-DE}} = 2 \text{ msec}$$
$$\tau_{\text{ANT-AC}} = 3 \text{ msec}$$
$$\tau_{\text{ANT-DE}} = 11 \text{ msec}$$

PW = see Table 3-4 (antagonist pulse always circumscribes agonist pulse by 3 msec on each side)

TABLE 3-4 Values of PW and PH used for simulation.

Magnitude (deg)	*Pulse Width, PW (msec)*	*Pulse Height, PH (g)*
0.1	10	17.6
0.5	10	20
1	11	22
5	15	53
10	20	87
20	31	124
30	40	155
40	54	160
50	70	160

From Eq. (3-58), the nonlinear force–velocity relationships yield

$$B_{\mathrm{AG}} = \frac{F_{\mathrm{AG}} + \mathrm{AG}_a}{\dot{\theta}_2 + \mathrm{AG}_b} = \frac{1.25 F_{\mathrm{AG}}}{\dot{\theta}_2 + 900}$$

for the agonist muscle, and from Eq. (3-59),

$$B_{\mathrm{ANT}} = \frac{F_{\mathrm{ANT}} - \mathrm{ANT}_a}{\dot{\theta}_3 - \mathrm{ANT}_b} = \frac{F_{\mathrm{ANT}} - 40}{\dot{\theta}_3 - 900}$$

For a 10° vergence movement,

$$\tau_{\mathrm{AG}} = \tau_{\mathrm{ANT}} = 50 \text{ msec}$$

(Note: 1 g tension $= 9.806 \times 10^{-3}$ N; $1° = 1.92 \times 10^{-4}$ m. $\Delta\theta$ represents the absolute value of the eye-movement size.)

The antagonist activity circumscribes the agonist activity (see Figs. 3-24, 3-40, and 3-41). Electromygraphic (EMG) studies have shown that the antagonist resumes its activity after the agonist ceases its burst of activity (Collins 1975). It has been reported that the pause of the antagonist motoneuronal pool starts before the agonist motoneuronal pool begins its high-frequency burst of firing in human lateral and medial rectus muscles (Sindermann, Geiselmann, and Fischler 1978), the abducens motoneurons (Fuchs and Luschei 1970; Maeda, Shimazu, and Shinoda 1971), the trochlear motoneurons (Fuchs and Luschei 1971; Baker and Berthoz 1974), the pausing and bursting units within the reticular formation that are associated with saccades (Keller 1974), the arm muscles (Kirsner 1973; Evarts 1974), and the jaw muscles (Clark and Luschei 1974).

Figure 3-41 also compares the reciprocal innervation model to a much older model drawn by Descartes in 1626. This comparison of the two models prompts the questions: Which model is better? How can you tell if a model is a good model?

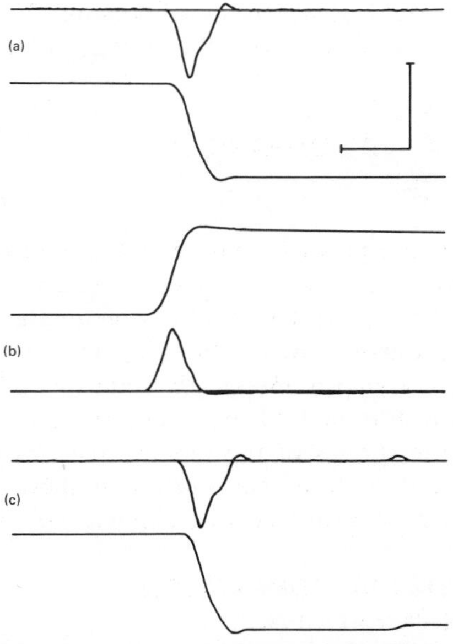

Figure 3-42 Three types of overshoot in saccadic eye movements: dynamic overshoot (a), glissadic overshoot (b), and both dynamic and static overshoot (c). The calibration bars represent 18.2 degrees, 750 deg/sec, and 80 msec. [From A. T. Bahill, M. R. Clark, and L. Stark, "Glissades — Eye Movements Generated by Mismatched Components of the Saccadic Motoneuronal Control Signal," *Mathematical Biosciences*, 26 (1975), 303–18. Copyright Elsevier North-Holland, Inc.]

3-7-4 Techniques for Validating Models

Although everyone makes models, few people validate their models. One question that can be used to assess the validity of a model is: Is the development of the model logical and scientific? Each element of the newly developed reciprocal innervation model was based on a physiological experiment. Every experiment was explained and the sources of the data were presented. Most bioengineering models will pass this test for logical development.

The reciprocal innervation model is one of the best justified models in the bioengineering literature, and therefore it was chosen as an example of the further validation of a model. First, the individual subsystems of the model must be valid. For this model that means that the model of an agonist muscle must match data from physiological experiments. A logical model development should ensure valid subsystems. The second validation technique used was to see how well the outputs of the model matched physiological data.

3-7-4.1 QUALITATIVE COMPARISON OF SHAPES

Figure 3-24 shows the position and velocity, as functions of time, of a typical human saccadic eye movement. This figure also shows the pulse-step motoneuronal controller signal that is used to generate saccades. The pulse is the high-frequency burst of neuronal firing that drives the eye rapidly between two points: the step is the steady-state firing level that holds the eye in its new position. Figure 3-43 shows eye movements from the model. It can be seen that the shapes of the model saccades match those of the human saccades. But is there a quantitative method with which we can measure how well the model fits? Yes, we can use *main-sequence* diagrams.

3-7-4.2 QUANTITATIVE COMPARISON WITH MAIN SEQUENCE

One of the first scientists to study eye movements quantitatively was Raymond Dodge (see Dodge and Cline 1901). His data fit neatly on the main-sequence diagrams of Bahill, Clark, and Stark (1975c). These main-sequence diagrams (Fig. 3-44), which plot peak velocity and duration of human eye movements as functions of their magnitudes, summarize the dynamic characteristics of saccadic eye movements. These data are representative of human saccades, although there are inter- and intrasubject variabilities, and fatigue and instrumental low-pass filtering will diminish the velocities. The dots are for individual saccades of normal unfatigued human beings. The V's represent normal vergence eye movements. G's represent glissadic eye movements. O's represent dynamic overshoots. L's represent overlapping saccades: the single L's are the two individual dynamic saccades, and the double L represents the abnormal condition created by trying to treat the movement as one large dynamic saccade. D's represent the data of Dodge and Cline (1901). Single crosses represent the saccadic simulations on the reciprocal innervation model. The stars represent vergence simulations. Solid and dashed lines represent analytic functions that fit the experimental data.

The greatest upper bound of the peak-velocity main sequence shows the maximum peak velocity (PV) for various-sized eye movements of normal unfatigued humans, while the least lower bound of the duration main sequence shows the shortest duration (DUR) for various magnitudes of human saccadic movements. These bounds are the most important,

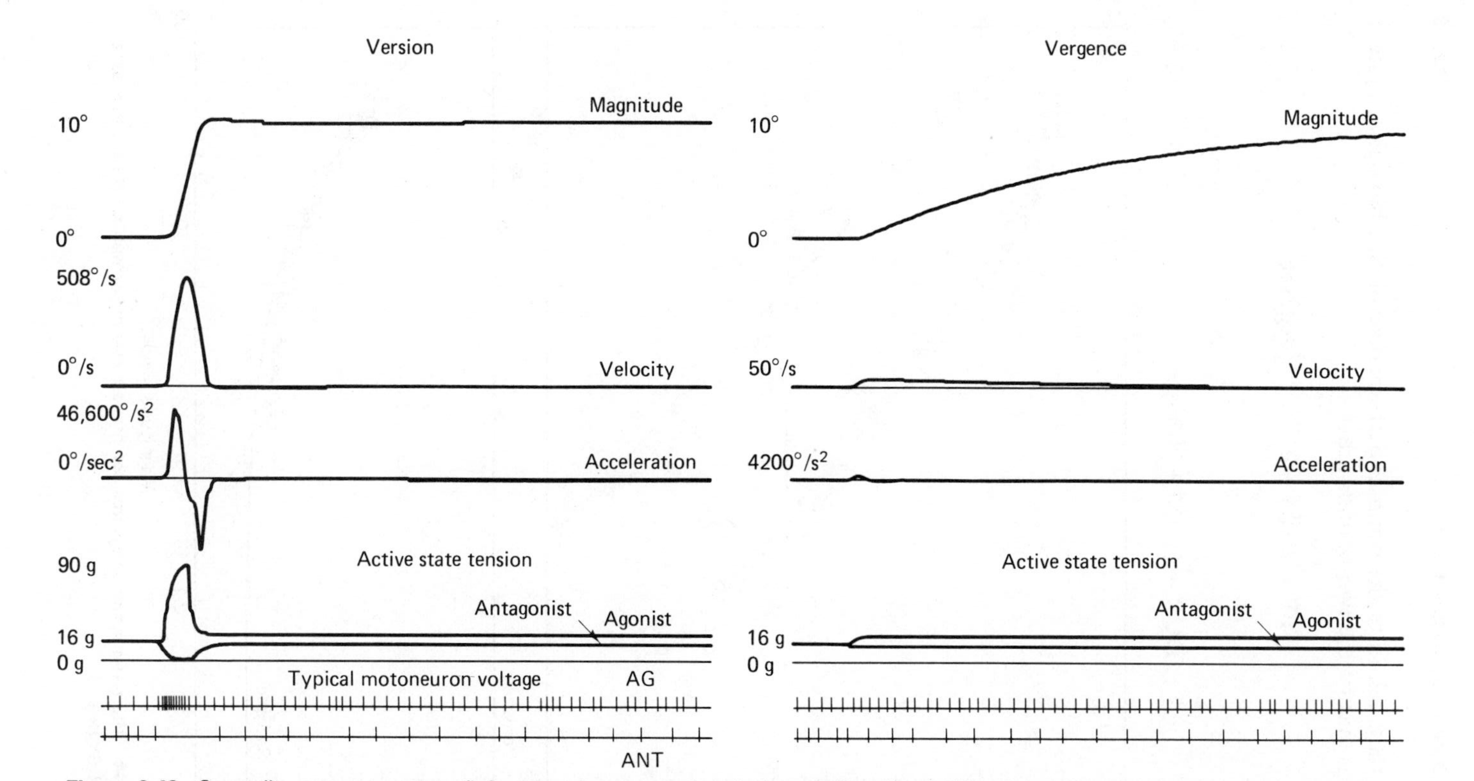

Figure 3-43 Saccadic eye movements (left column) use a pulse-step controller signal. They are therefore faster than vergence eye movements (right column), which use only a step-controller signal. [From F. K. Hau, A. T. Bahill, and L. Stark, "Parametric Sensitivity of a Homeomorphic Model for Saccadic and Vergence Eye Movements," *Computer Programs in Biomedicine*, 6 (1976), 108–16.]

because they delineate the optimal performance of this biological system. Equations that fit the two bounds are

$$PV = 850\left[1 - e^{-\Delta\theta/10.6}\right]\text{deg/sec} \tag{3-64}$$

and

$$DUR = (1.7\Delta\theta + 20)\text{msec} \tag{3-65}$$

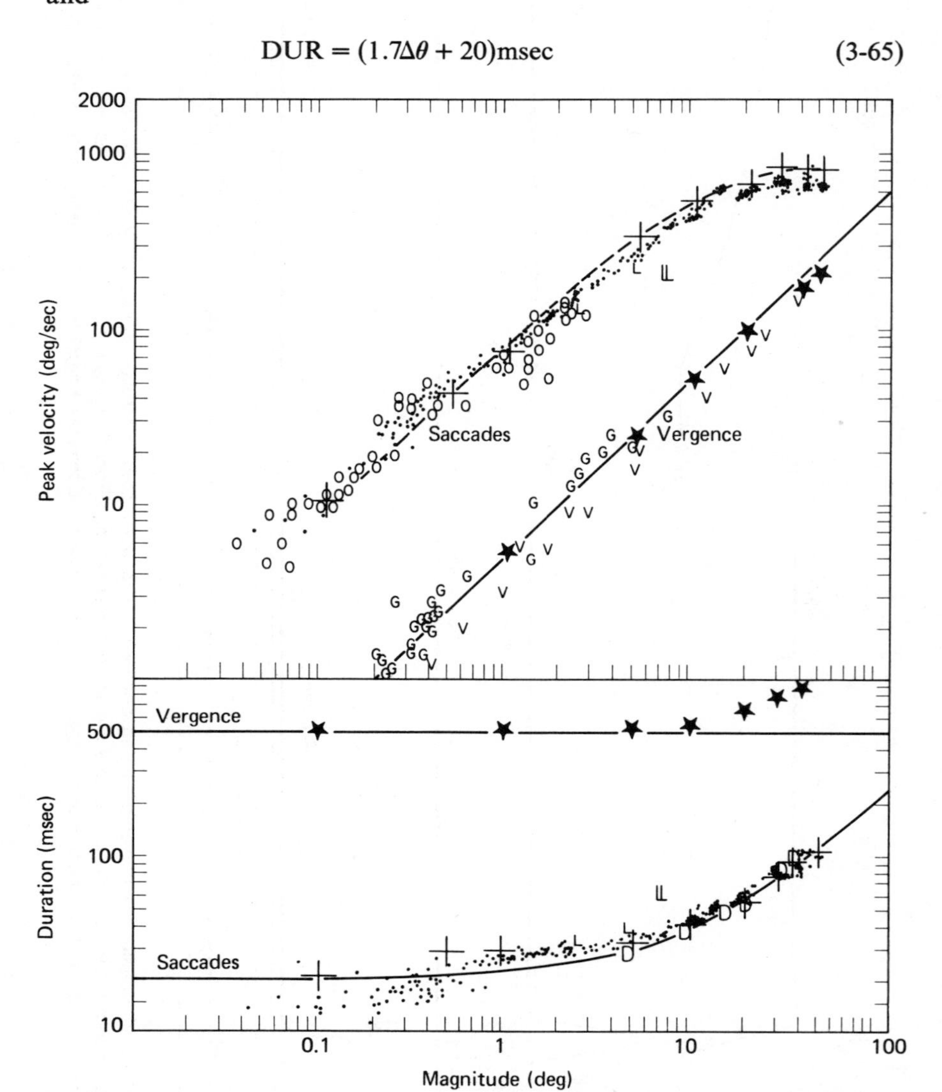

Figure 3-44 Main-sequence diagrams for human eye movements. (From the data of Bahill and Stark.)

To evaluate the validity of our model, we simulated saccades over the whole range of magnitudes. The large crosses in Fig. 3-44 demonstrate the good fit between our model and the actual human data.

So far we had validated our model qualitatively by comparing the shapes of model and human saccades, and quantitatively by comparing the main sequence diagrams of model and human saccades. But many models can emulate physiological data; therefore, we were looking for an analytical method of evaluating our model. This led us to do a sensitivity analysis.

3-7-4.3 SENSITIVITY ANALYSIS

A sensitivity analysis shows how a model's outputs change with variations in its parameters. The sensitivity analysis may be analytic or experimental. The results of a sensitivity analysis can be used (1) to validate a model; (2) to warn of strange or unrealistic model behavior; (3) to suggest new experiments or guide future data collection efforts; (4) to point out important assumptions of the model; (5) to guide the formulation of the structure of the model, pointing out which unimportant elements can be treated simply; and (6) to select numerical values for the parameters. The sensitivity analysis tells which parameters are the most important and most likely to affect predictions of the model. Values of critical parameters can then be refined while parameters that have little effect can be simplified or ignored.

A traditional root-locus plot graphically displays the results of a sensitivity analysis: it shows the movement of the system's closed-loop poles as a function of the system gain.

A partial derivative is a sensitivity function. In a system described by analytic functions, for example, calculating partial derivatives constitutes a sensitivity analysis. There are two common definitions for sensitivity functions: absolute sensitivity and relative, or logarithmic, sensitivity.

The *absolute sensitivity* of the function F (which is a function of α and time) to variations in the parameter α evaluated at the nominal parameter value α_0 is given by

$$S_\alpha^F = \left.\frac{\partial F}{\partial \alpha}\right|_{\alpha_0} \tag{3-66}$$

Absolute-sensitivity functions are useful for calculating output errors due to parameter variations and for assessing the times at which a parameter has its greatest or least effect.

The *relative sensitivity* of the function F to the parameter α evaluated at the nominal value of that parameter, α_0, is given by

$$\bar{S}_{\alpha}^{F} = \left.\frac{\partial \ln F}{\partial \ln \alpha}\right|_{\alpha_0}$$

$$\bar{S}_{\alpha}^{F} = \left.\frac{\partial F/F}{\partial \alpha/\alpha}\right|_{\alpha_0} = \left.\frac{\partial F}{\partial \alpha}\right|_{\alpha_0} \frac{\alpha_0}{F_0} \tag{3-67}$$

Relative-sensitivity functions are ideal for comparing parameters, because they are dimensionless, normalized functions.

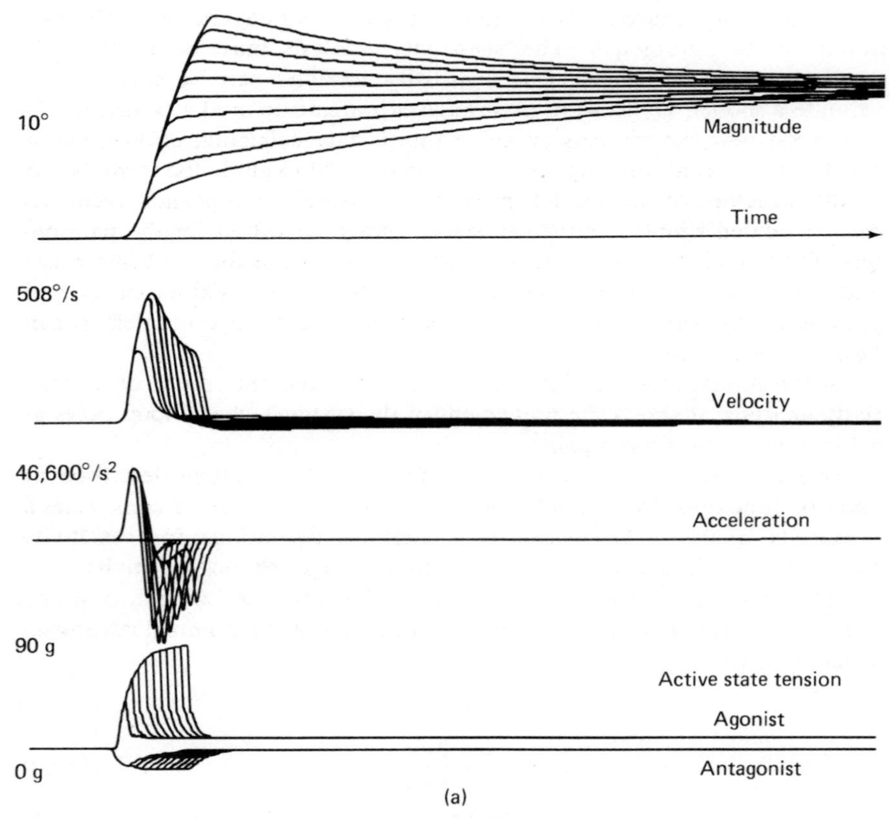

Figure 3-45 Trajectories of position, velocity, and acceleration as (a) pulse width (PW) and (b) pulse height (PH) are varied from 20 to 200% of the value for producing a good 10° saccade. [From F. K. Hsu, A. T. Bahill, and L. Stark, "Parametric Sensitivity of a Homeomorphic Model for Saccadic and Vergence Eye Movements," *Computer Programs in Biomedicine*, 6 (1976), 108–16.]

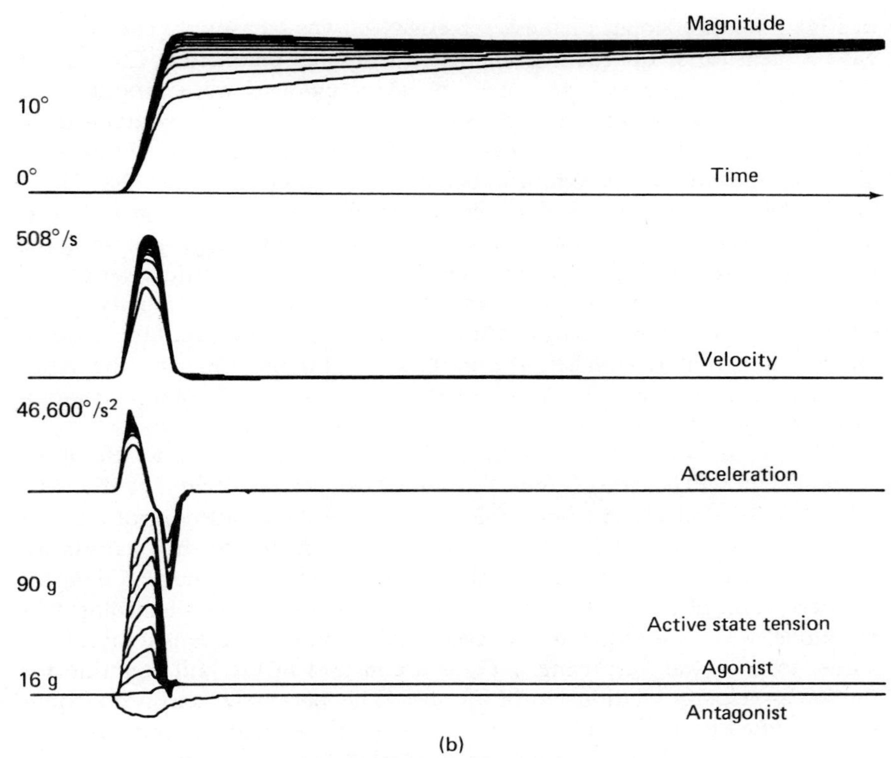

Figure 3-45 Continued.

Both of these sensitivity functions are evaluated at the operating point where the parameters take on their nominal values; and both are functions of time.

Because the reciprocal innervation model was a sixth-order nonlinear model, the analytic partial derivative technique for sensitivity analysis was not feasible. An empirical sensitivity analysis was performed for a 10° saccade, because this is a normal physiological magnitude and data were both abundant and relatively noise-free. Each parameter was varied from 20% to 200% of the value used for producing a good 10° saccade while the other 19 parameters were held constant. Figure 3-45 shows examples of trajectory variations produced when pulse width (PW) and pulse height (PH) were varied. These simulations were the method used to numerically estimate the partial derivatives of the eye position and the eye velocity with respect to each of the parameters. The sensitivity of eye position with respect to each parameter was evaluated for one instant of time near the end of the saccade; the sensitivity of eye velocity was evaluated near the middle of the saccade. These were plotted as functions of each parameter in Fig. 3-46. The slopes of these sensitivity curves (sensitivity coefficients) gave a measure of the relative importance of each parameter. The steeper the slope, the more eye-movement behavior changes were produced by variation of that parameter. For instance, when the pulse width was 20 msec, the saccadic magnitude was 10°. When pulse width was doubled to 40 msec, the saccadic magnitude became 17.3° increasing by 73%. However, when another parameter, the series elasticity of the agonist, was doubled (from 1.8 g tension/deg to 3.6 g tension/deg) the saccadic magnitude increased by only 7%. The slope of the magnitude sensitivity curve for pulse width was 0.54 compared to 0.1 for series elasticity of the agonist. This sensitivity analysis showed which parameters had the greatest effects on the model. When the parameters did not describe the input controller signals, we had to search for several sources of good physiological data to justify our values.

The model was most sensitive to four parameters pulse width (PW), pulse height (PH), AG_b, and K_p, all of whose sensitivity coefficients exceeded 0.4. Pulse width and pulse height are the input control signals; therefore, it was reasonable that the model was sensitive to their variations. K_p is a large lumped parameter and there was ample physiological data for the derivation of its value; so it was neither surprising nor disturbing that the model was so sensitive to its variations. However, the sensitivity of the model to AG_b was surprising. AG_b is a constant in the Hill equation for the force–velocity relationship of muscles. This parameter had been extensively studied by Hill, and he estimated that AG_b should be approximately one-fourth V_{max}. But V_{max} must be evaluated for each particular muscle studied. Because this model was so sensitive to AG_b, experiments were performed to investigate it in greater detail. The treatment of this agonist dashpot was revised in subsequent models.

In contrast, the model behavior has very little dependence on many other parameters (see Fig. 3-46). For instance, the slope of the sensitivity curves for the inertia, J, were almost zero. This means that the inertia, J, had very little effect on the output. This, then, was the justification for modeling the inertia of the eyeball as a globe of ice rather than as a series of concentric shells connected with viscoelastic elements. The sensitivity analysis provided us with direct information concerning the importance of each parameter, helped us choose numerical values for the parameters, and suggested new experiments for further understanding of the system and for updating the model.

Sensitivity functions are usually functions of time. In the preceding analysis the sensitivity functions were evaluated at only two points in time: near the middle and near the end of the saccade. However, the analysis covered a broad range of parameter perturbation sizes: from 20 to 200% of the nominal values. If one parameter perturbation size is sufficient, as is the case with a linear system, it may be advantageous to look at sensitivity

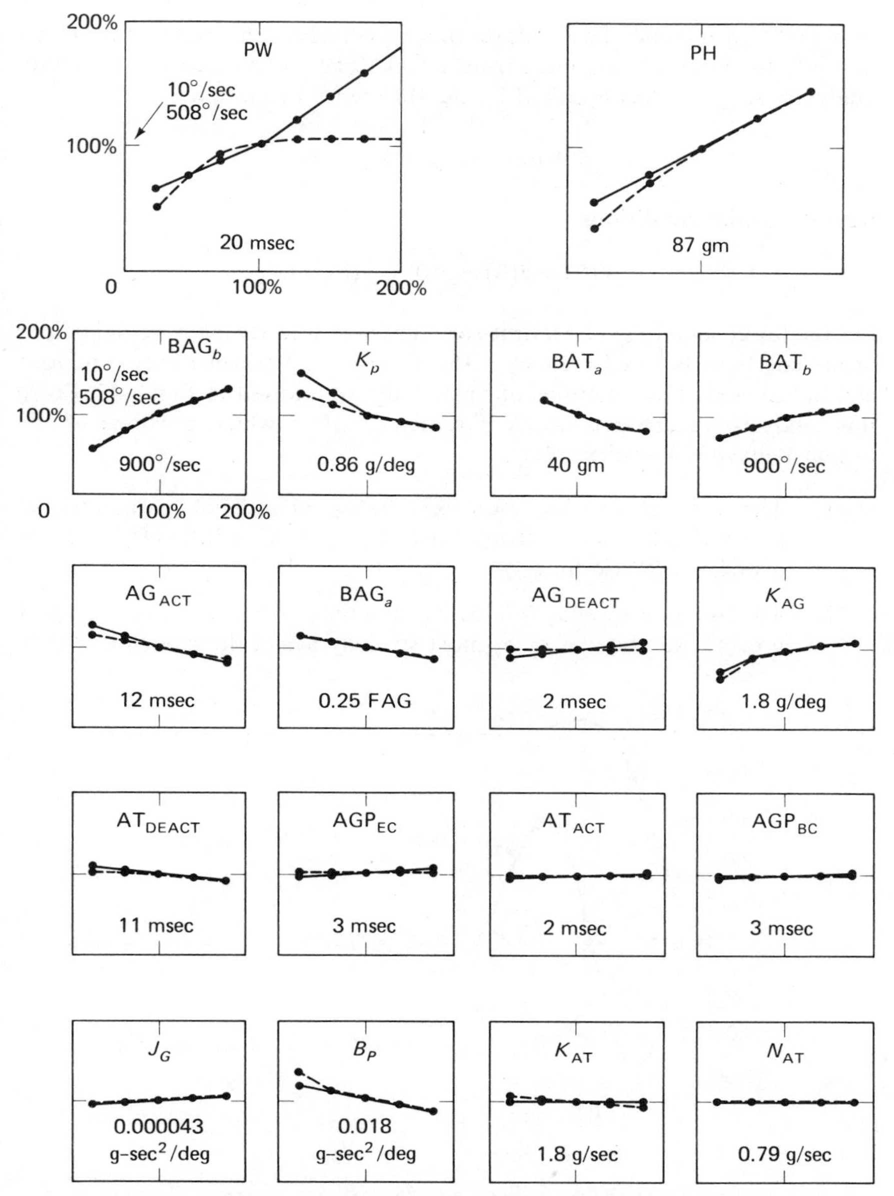

Figure 3-46 Rank-ordered relative sensitivity curves for 16 of the internal parameters. The actual value of each parameter used for stimulating a 10° saccade is given along the *X* axis and normal output values along the *Y* axis. The solid line is the magnitude relative sensitivity; the dashed line is the peak-velocity relative sensitivity. [From F. K. Hsu, A. T. Bahill, and L. Stark, "Parametric Sensitivity of a Homeomorphic Model for Saccadic and Vergence Eye Movements," *Computer Programs in Biomedicine*, 6 (1976), 108–16.]

as a function of time. To illustrate this aspect of a sensitivity analysis, let us study the following example from Frank's (1978) textbook on sensitivity analyses. A system is described by the differential equation

$$\dddot{y} + a_2\ddot{y} + a_1\dot{y} + a_0 y = u$$

with the initial conditions

$$\ddot{y}(0) = \ddot{y}(0) = \dot{y}(0) = y(0) = 0$$

Let the input signal $u(t)$ be a unit-step function, and let the nominal values of the coefficients be $a_0 = 20$, $a_1 = 15$, and $a_2 = 5$. The sensitivity functions can be calculated as functions of time. They are plotted in Fig. 3-47. From this analysis we can see which parameters affect which portions of the output behavior. For example:

1. The plot of $\partial y/\partial a_0$ indicates that a change of parameter a_0 primarily affects the steady state of $y(t)$, having little effect on the overshoot or rise time.

2. The function $\partial y/\partial a_1$ is largest at the times where the overshoots of y occur, so changes of a_1 most strongly affect the overshoots of y.

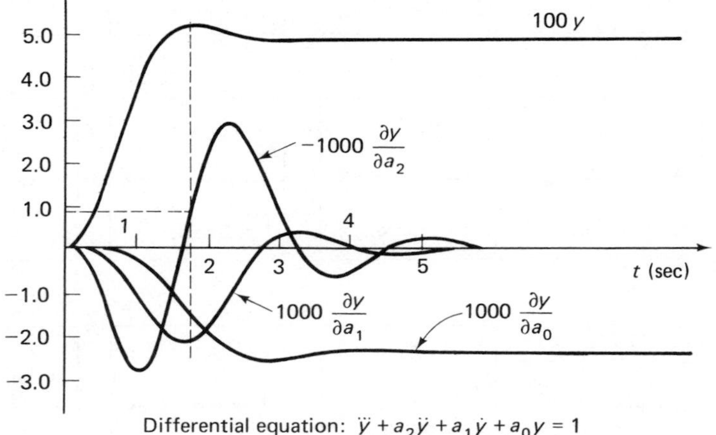

Figure 3-47 Step response of a third-order system and the absolute sensitivity functions for each of the three parameters. Scale factors were used so that the curves would not overlap. (From P. M. Frank, *Introduction to System Sensitivity Theory*, Academic Press, Inc., New York; 1978.)

3. The plot of $\partial y/\partial a_2$ indicates that a change of a_2 most strongly affects the slopes of y, having no effect on the overshoots or the steady state.

The absolute sensitivity function is useful for seeing when the parameters have their maximum effects, but not for comparing parameters. For these comparisons we want a relative sensitivity function such as,

$$\bar{S} = \frac{\partial \ln y}{\partial \ln \beta} = \frac{\partial y}{\partial \beta}\frac{\beta_0}{y_0}$$

β_0 and y_0 are the values of β and y at the nominal operating point. However, for a ten degree saccadic eye movement the nominal output value, y_0, varies from 0 to 10 degrees, and division by zero is frowned upon. Furthermore, the relative sensitivity function gives undue weight to the beginning of the saccade when y_0 is small. Therefore, the semirelative sensitivity function was used by Bahill, Latimer, and Troost (1980b). The semirelative sensitivity function is

$$\tilde{S} = \frac{\partial y}{\partial \ln \beta} = \frac{\partial y}{\partial \beta}\beta_0$$

For small parameter changes in their linear system this became

$$\tilde{S} = \frac{\Delta y}{\Delta \beta}\beta_0$$

To perform the sensitivity analysis, a ten degree saccade was simulated (solid line labeled θ_n in Fig. 3-48). Then one parameter was changed by a set amount $\Delta\beta$, +5% for this figure, and the model was run again, producing the perturbed saccade (dashed line labeled θ_p in Fig. 3-48). The difference between the nominal and perturbed saccades (Δy) was calculated for each millisecond and this difference was divided by the change in the parameter value ($\Delta\beta$). This ratio was then multiplied by the nominal parameter value (β_0). This process was repeated for each of the 18 parameters in the model. Some of the results are shown in Fig. 3-48. The pulse width (PW) and pulse height (PH) primarily effect the dynamic saccade and the behavior immediately following; the steady state neural firing levels, $N_{AG\text{-step}}$ and $N_{ANT\text{-step}}$ primarily effect the static behavior of the eye. The effect of $N_{ANT\text{-pulse}}$ is too small to be seen on this scale. Its sensitivity function looks like noise on the absissa. The perturbed saccade is that produced by increasing $N_{AG\text{-step}}$ by 5%. The model was most sensitive to the input control signals for the agonist muscle, $N_{AG\text{-step}}$, PH and PW. The control signals for the antagonist, $N_{ANT\text{-step}}$ and $N_{ANT\text{-pulse}}$ were not as important.

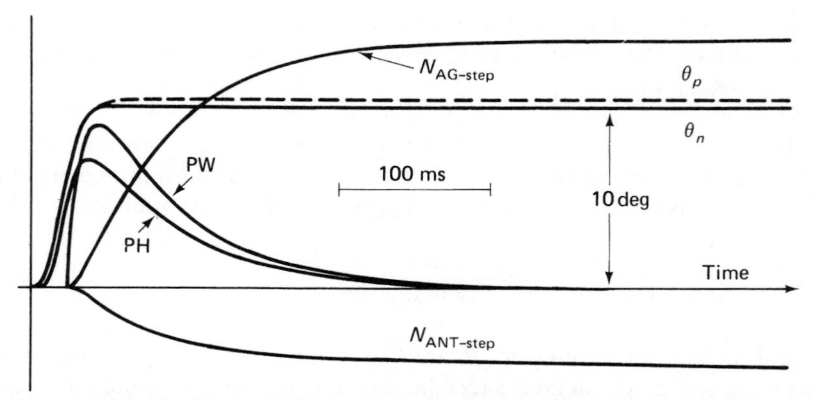

Figure 3-48 Nominal (solid line labeled θ_n) and perturbed (dotted line labeled θ_p) ten degree saccadic eye movements and the *semirelative sensitivity functions* for the parameters describing the input controller signals. The sensitivity functions have units of degrees; the calibration bar is applicable to the eye position records and the sensitivity functions. Record length is 490 ms. Based on Bahill, Latimer, and Troost (1980a).

The sensitivity analysis demonstrated that:

1. The parameters $N_{AG\text{-step}}$, $N_{ANT\text{-step}}$, $K_{AG\text{-LT}}$, and K_p affect the steady state of $\theta(t)$ and have little affect on the overshoot or rise time.

2. The parameter $\tau_{ANT\text{-DE}}$ attained its peak value at the beginning of the saccade, as shown in Fig. 3-49. No other parameter was as large at the beginning of the saccade. This parameter has its greatest effect on the initial rapid rise.

3. The sensitivity functions for the other three time constants and for the three dashpots had shapes that were similar to each other and that peaked near the end of the saccade. This means that trade-offs could be made between these six parameters without affecting the precision of the model.

4. Pulse height (PH) and pulse width (PW) primarily affect the output near the end of the saccade and immediately following the saccade. Variations in both of these parameters produce slow drifts after the saccades. These drifts are called *glissades*. However, the shapes of these two sensitivity functions differ. The pulse-height function rises gradually, starting at the beginning of the saccade; whereas the pulse width sensitivity function is zero until near the end of the saccade, where it abruptly rises to its peak, as shown in

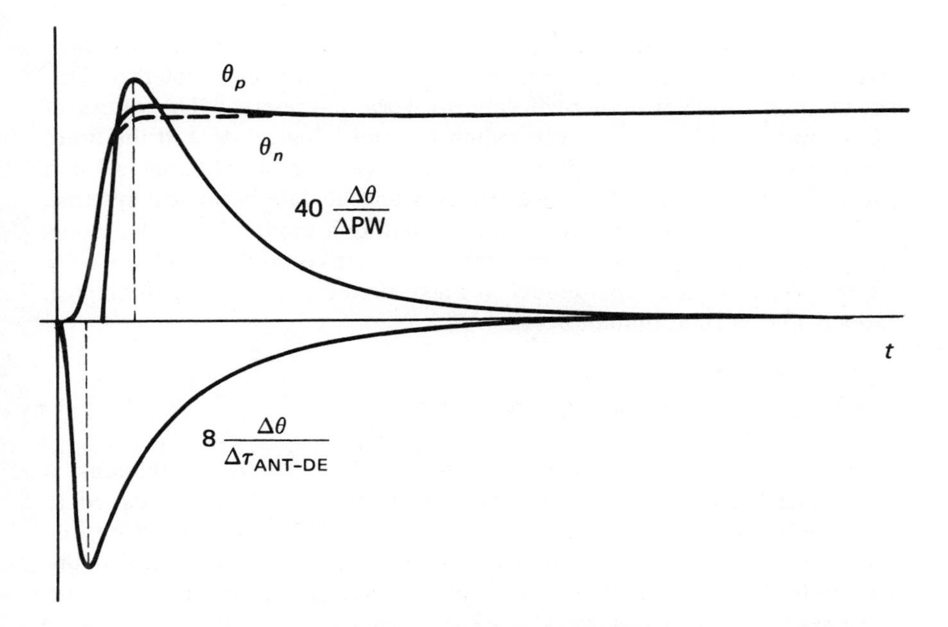

Figure 3-49 A 10° saccade and the *absolute sensitivity functions* for pulse width (PW) and the antagonist deactivation time constant ($\tau_{ANT\text{-}DE}$). The time constant primarily affects the initial rise. The pulse width has no effect on the first half of the saccade. Variations in pulse width produce glissades, which are slow drifts appended to the end of the saccades. Each record is 490 mscc. The perturbed saccade (the solid line labeled θ_p) is that produced by increasing PW by 5%.

Fig. 3-48. This means that increasing either parameter would produce a glissade, a slow drift. However, increasing the pulse height would also increase the peak velocity of the saccade; whereas increasing the pulse width would not affect the peak velocity, because the peak velocity occurs while this sensitivity function is still zero. This result will be seen again in the section on glissades, specifically in Fig. 3-61.

Sensitivity analyses are an important method of validating social models (Ford and Gardner 1979), engineering models (Frank 1978), and physiological models (Hsu, Bahill, and Stark 1976; Lehman and Stark 1980; Bahill, Latimer and Troost 1980b). They show which parameters of the model have the most effect on the model behavior. They allow a simplified treatment of parameters that are not important. If the sensitivity coefficients are calculated as functions of time, it can be seen *when* each parameter has the greatest effect on the output function of interest. This can be used to select numerical values for the parameters. The values of the parameters will be chosen to match the physiological data at the times when they have the most effect upon the output. Sensitivity analyses can also be used to suggest future experiments to elucidate biological systems.

This suggests another method for evaluating a model: using the model to simulate new types of eye movements, to make predictions about the system. We have used this model to make predictions about pathological patients and normal human beings.

3-7-4.4 PREDICTIONS OF THE MODEL

A counterexample of a model being able to make predictions is Westheimer's model, which we have already studied. It was designed to simulate 20° saccades, but it could not simulate 30° saccades. Often, the new phenomenon to be simulated is something that is well known but ignored in the development of the model. Sometimes the new phenomenon is the result of a new discovery.

An eye movement has overshoot when the eye travels beyond its final position and then returns, finally coming to rest on the target. There are three distinct types of overshoot of saccadic eye movements: dynamic, glissadic, and static (see Fig. 3-42). They are named according to their most distinctive feature: the behavior of the eye immediately after the primary saccade. Dynamic overshoot has a very fast saccadic return to the true target position. For example, the return phase of a 1° dynamic overshoot lasts about 20 msec and has a peak velocity of about 60°/sec. Glissadic overshoot has a slow gliding return phase which, for 1° of overshoot, lasts approximately 300 msec with a peak velocity of 5°/sec. In static overshoot the eye remains steady in the incorrect position for 150 to 200 msec until feedback instigates a corrective saccade, eliminating the error (Bahill, Clark, and Stark 1975a, b). Dynamic overshoot was a new discovery that was used to test the model. Glissadic overshoot was a previously known phenomenon that was ignored in the development of the model: it was also used to test the model. Static overshoot, because of its simplicity, could not be used to test the model.

Dynamic Overshoot Dynamic overshoot is the most common type of saccadic overshoot. Two-thirds of human saccades have dynamic overshoot; however, this varies greatly from subject to subject. The occurrence of dynamic overshoot is very capricious. On one day most of a subject's saccadic eye movements will have dynamic overshoot, while on another

Figure 3-50 Dynamic overshoots are variable as shown by these successive saccadic eye movements between two targets 10° apart. This variability is often obscured by filtering or averaging to remove noise from the data. Most subjects exhibit similar variability, although few have dynamic overshoots as large as this. Calibration bar represents 1 sec.

day very few will. Short-term variability is illustrated by the sequence of 10° saccades shown in Fig. 3-50. Some of these saccadic eye movements have dynamic overshoot, and some do not. Although the dynamic overshoots are very obvious in the figure, they are seldom so prominent in published eye-movement records because most published data have been low-pass-filtered in order to remove noise. This removes the evidence of the dynamic overshoot; therefore, the existence of dynamic overshoot was a recent discovery.

Dynamic overshoot averages 0.25° for 10° saccades: the overshoot size increases with saccadic size. The existence of dynamic overshoot is not limited to certain initial conditions or direction of travel, because saccades have been shown with dynamic overshoot for temporal and nasally directed saccades both at primary position and with the eye abducted 35° from the primary position.

Studies of the peak velocity—magnitude—duration relationships (main-sequence diagrams) for normal saccades and for the return phases of dynamic overshoots suggested that the return phases of dynamic overshoots were small saccades. The peak velocity-versus-magnitude main-sequence diagram showed that the return phases of dynamic overshoots were just as fast as normal saccades, and that both of these were much faster than vergence eye movements (Bahill, Clark, and Stark 1975a). See Fig. 3-44.

Because the return phases of dynamic overshoots had saccadic dynamics, they were expected to have saccadic motoneuronal controller signals. To investigate this possibility, saccades with dynamic overshoots were simulated on the reciprocal innervation model. Dynamic overshoots were not used in the formulation of the model, so this served as an important test for the model.

A simulated saccade with dynamic overshoot is illustrated in Fig. 3-51. This figure shows the eye position, eye velocity, active-state tensions of the agonist and antagonist muscles, the motoneuronal signals required for the agonist and antagonist in order to produce dynamic overshoot, and the firing pattern of typical agonist and antagonist motoneurons. The reversal of motoneuronal activity at the end of the main pulse requires a high degree of synchronization between the motonuclei. For instance, the firing rate of the pool of motoneurons supplying the agonist muscle rises from the original level to the high-frequency level, remains there until about the middle of the saccade, drops to a very low value, and then rises again up to its new step level. The pool of motoneurons innervating the antagonist muscle perform an analogous, but opposite task. Because of this motoneuronal innervation, the active-state tensions of the muscles reverse. The active-state tension of the agonist first becomes larger, then smaller, and then equal to all of its opposing forces. These active-state tensions are not the muscle forces measured at the tendon of a muscle. The changes in force at the tendon were slower than the changes in active-state tension shown in this figure due to the series elasticity and the force–velocity relationship of the muscle (see Fig. 3-40). Dynamic overshoots must be caused by motoneuronal control signal reversals. However, these control signal reversals cannot be due to random noise because as many as a dozen consecutive saccades can have nearly identical overshoot.

The modeling results enabled us to make the prediction that there should be pauses for the agonist motoneurons and bursts for the antagonists motoneurons in saccades with dynamic overshoot. Such pausing and bursting behavior could not easily be found in the neurophysiological literature. Once the prediction was made, investigators specifically looked for such behavior and found it in single-cell recordings of the brain stem of monkeys (Zee and Robinson 1979b) and in human electromyographic recordings (Stark 1977).

Thus the model was able to simulate a type of movement that was not known when the model was designed. Furthermore, predictions were made based on the modeling results, and these predictions were subsequently confirmed by neurophysiological experiments.

Glissades Glissades are the slow, gliding eye movements that are often appended to the end of normal human saccadic eye movements. Their frequency of occurrence is increased by fatigue and pathology.

Several papers (Dell'Osso, Robinson, and Daroff 1974; Kommerell 1975; Fricker and Sanders 1975; Metz 1976; Baloh, Yee and Honrubia 1978) have suggested that the existence of glissades could be used to diagnose internuclear ophthalmoplegia: a syndrome that is usually caused by multiple sclerosis. In these patients there is a glissadic undershooting of the adducting eye and a concommitant overshooting of the abducting eye. There may also be abduction nystagmus. The data of Fig. 3-52, recorded by Bahill, Ciuffreda, Kenyon, and Stark (1976) with the photoelectric technique for eye-movement recording, show that normal subjects can exhibit similar saccadic tracking patterns.

The model was not built to simulate these glissadic eye movements; therefore, we tried producing glissades in the model in order to test it and

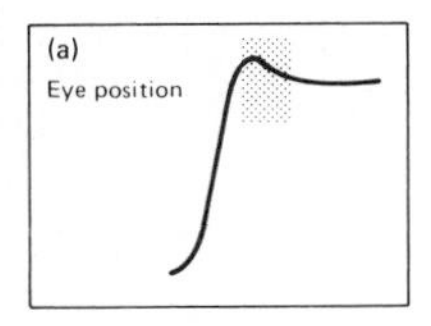

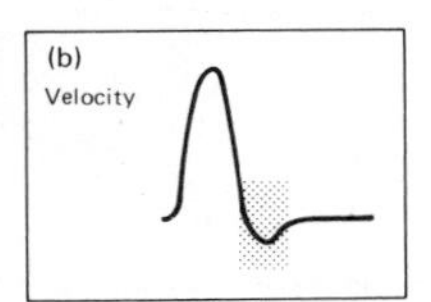

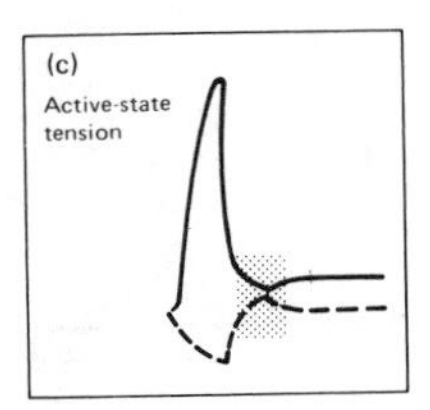

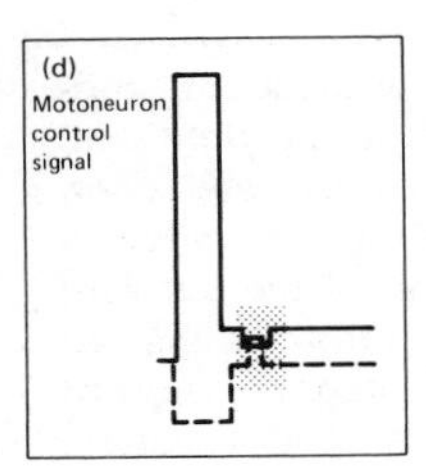

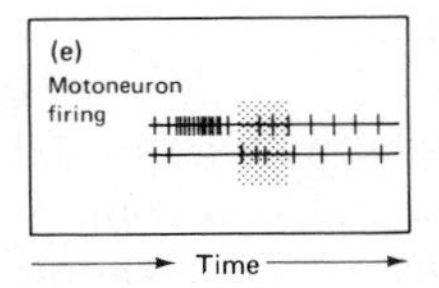

Figure 3-51 (a) Eye position, (b) velocity, (c) active-state tensions, (d) motoneuronal control signals, and (e) motoneuron firing for saccades with dynamic overshoot. (From A. T. Bahill and L. Stark, "The Trajectories of Saccadic Eye Movements." Copyright 1979, Scientific American, Inc. All rights reserved.)

to try to understand the CNS errors that produce glissades. It was found that glissades could be caused by errors in either the pulse or the step components of the motoneuronal controller signal. In this section only those glissades caused by errors in the pulse component will be considered. If the pulse was too small for its accompanying step, the result was glissadic undershoot, similar to the human glissadic undershoot of Fig. 3-53d. If the pulse was too large for the accompanying step, glissadic overshoot would result, as shown in Fig. 3-53c.

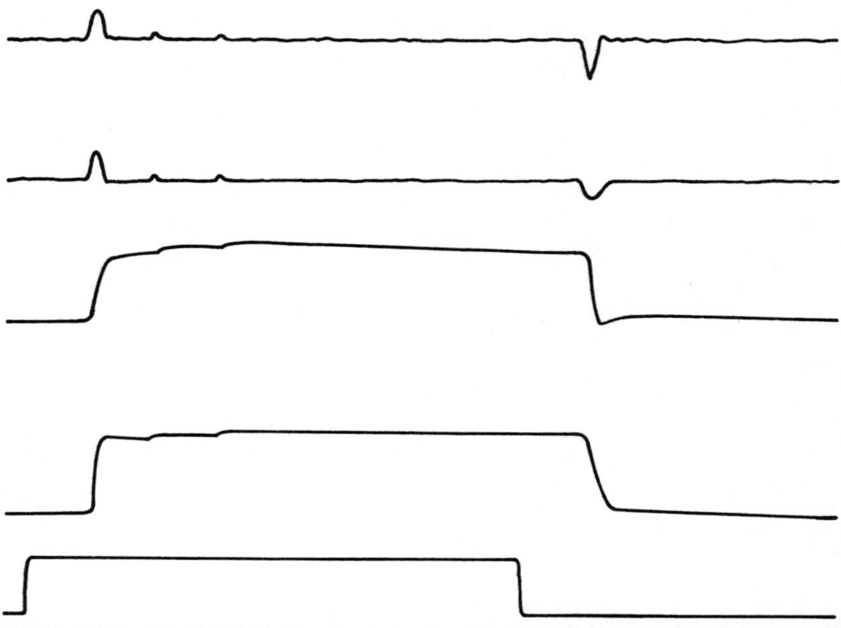

Figure 3-52 Pseudo-INO records showing abductor glissadic undershoot, concomitant adductor glissadic overshoot, and adductor nystagmus in the right eye. Although this is similar to tracking patterns described for patients with abduction INO, these eye movements were executed by a normal subject. Shown from top to bottom are left-eye velocity, right-eye velocity, horizontal position of the left eye, horizontal position of the right eye, and the target position. The target jumps are physiologically large: 20°. Leftward movements are represented by upward deflections. [From A. T. Bahill, K. R. Ciuffreda, R. V. Kenyon, and L. Stark, "Dynamic and Static Violations of Hering's Law of Equal Innervation," *American Journal of Optometry and Physiological Optics*, 53 (1976), 786–96. Copyright 1976, American Academy of Optometry.]

There are two ways of making the pulse portion of the motoneuronal controller signal too large, thereby mismatching the pulse and step components of the controller signal and producing this type glissadic overshoot: the pulse could be either too wide or too high. Both of these possibilities were tried in the model. The resulting mismatched saccades had similar qualitative shapes, but different quantitative main-sequence parameters.

Figure 3-54 shows (with dashed lines) ideal 10° saccades with no mismatch of the pulse and step components and (with solid lines) mismatched 12° saccades with 2° glissades appended. These mismatched saccades were produced by making the pulse portions of the agonist controller signals too large. For simplicity, the antagonist controller signals were not altered. The final eye positions were the same because the tonic firing rates of the motoneurons (the step portions of the controller signals)

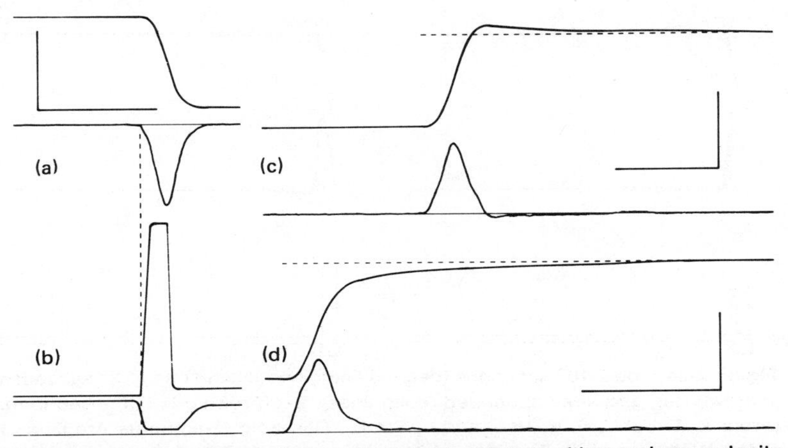

Figure 3-53 Three types of glissadic behavior. Eye position and eye velocity as functions of time for saccadic eye movements with (a) no glissade, (c) glissadic overshoot, and (d) glissadic undershoots. The hypothesized firing frequencies for the agonist (top) and the antagonist (bottom) motoneuronal pools are shown in (b). The calibrations represent (a) 13° and 640°/sec, (b), 14.6° and 600°/sec, and (c), 10° and 500°/sec. The time calibration in each represents 100 msec. [From A. T. Bahill, M. R. Clark, and L. Stark, "Glissades —Eye Movements Generated by Mismatched Components of the Saccadic Motoneuronal Control Signal," *Mathematical Biosciences*, 26 (1978), 303–18. Copyright Elsevier North-Holland, Inc.]

were the same. The mismatched saccade (solid lines) of Fig. 3-54PW was produced by increasing the pulse width of an ideal 10° saccade from 20 to 26 msec. This pulse width was the only parameter that was different in the controller signals for the 10° and 12° saccades of Fig. 3-54PW. The mismatched saccade (solid lines) of Fig. 3-54PH was produced by increasing the pulse height of an ideal 10° saccade from 87 to 120. This pulse height was the only parameter that was different in the two controller signals of Fig. 3-54PH. Therefore, Fig. 3-54 shows that eye movements, which qualitatively look like human glissades, could be created by making either the pulse width or the pulse height too large. However, quantitatively these eye movements differed: the peak velocity of the pulse-width mismatched saccade (Fig. 3-54PW) was smaller than the peak velocity of the pulse-height mismatched saccade (Fig. 3-54PH), although they were both 12° saccades. In order to gain further insight into the neurological control of saccades, these mismatched 12° saccades were next compared to ideal 12° saccades. Figure 3-55 demonstrates this comparison.

Figure 3-55 shows (with dashed lines) ideal 12° saccades, and (with solid lines) the same movements as the solid lines of Fig. 3-54 (i.e., mismatched 12° saccades with 2° glissades appended. The important

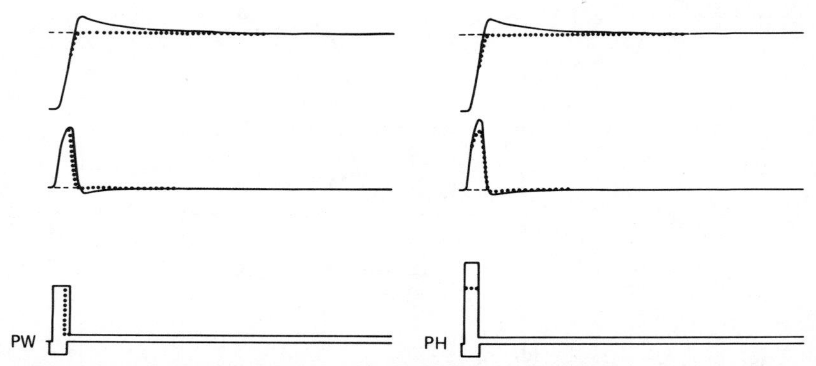

Figure 3-54 Ideal 10° saccades (dashed lines) and mismatched 12° saccades with two deg glissades appended (solid lines). Each record is 500 msec long. [From A. T. Bahill, F. K. Hsu, and L. Stark, "Glissadic Overshoots Are Due to Pulse Width Errors," *Archives of Neurology*, 35 (1978), 138–42. Copyright 1978, American Medical Association.]

traces in Fig. 3-55 are the velocity traces. The peak velocity of the 12° pulse-width mismatched saccade (Fig. 3-55PW) was 518°/sec, which was smaller than the 560°/sec peak velocity of an ideal 12° saccade. (The solid line is below the dashed line.) In contrast, the peak velocity of the 12° pulse-height mismatched saccade (Fig. 3-55PH) was 612°/sec, which was larger than the 560°/sec peak velocity of an ideal 12° saccade. (The solid line is above the dashed line.)

These peak velocities are indicated with the symbols PW, PH, and *I* on the linear peak-velocity main-sequence diagram in the inset of Fig. 3-56.

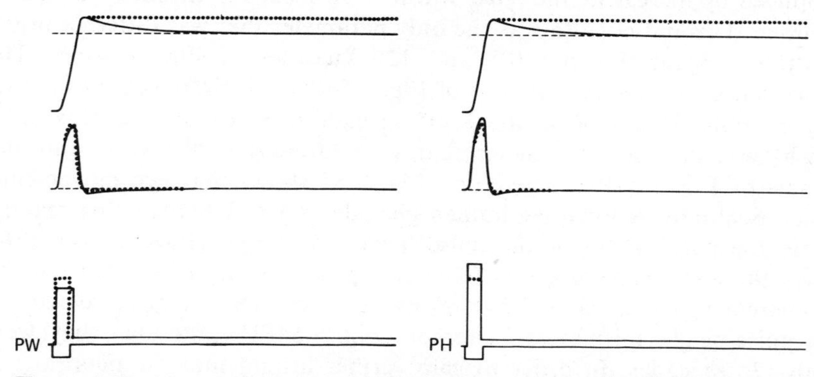

Figure 3-55 Ideal 12° saccades (dashed lines) and (solid lines) the same mismatched 12° saccades with 2° glissades appended as Fig. 3-54. The velocity traces show that pulse width errors (PW) produce smaller saccadic peak velocities, while pulse height errors (PH) produce larger saccadic peak velocities. [From A. T. Bahill, F. K. Hsu, and L. Stark, "Glissadic Overshoots Are Due to Pulse Width Errors, *Archives of Neurology*, 35 (1978), 138–42. Copyright 1978, American Medical Association.]

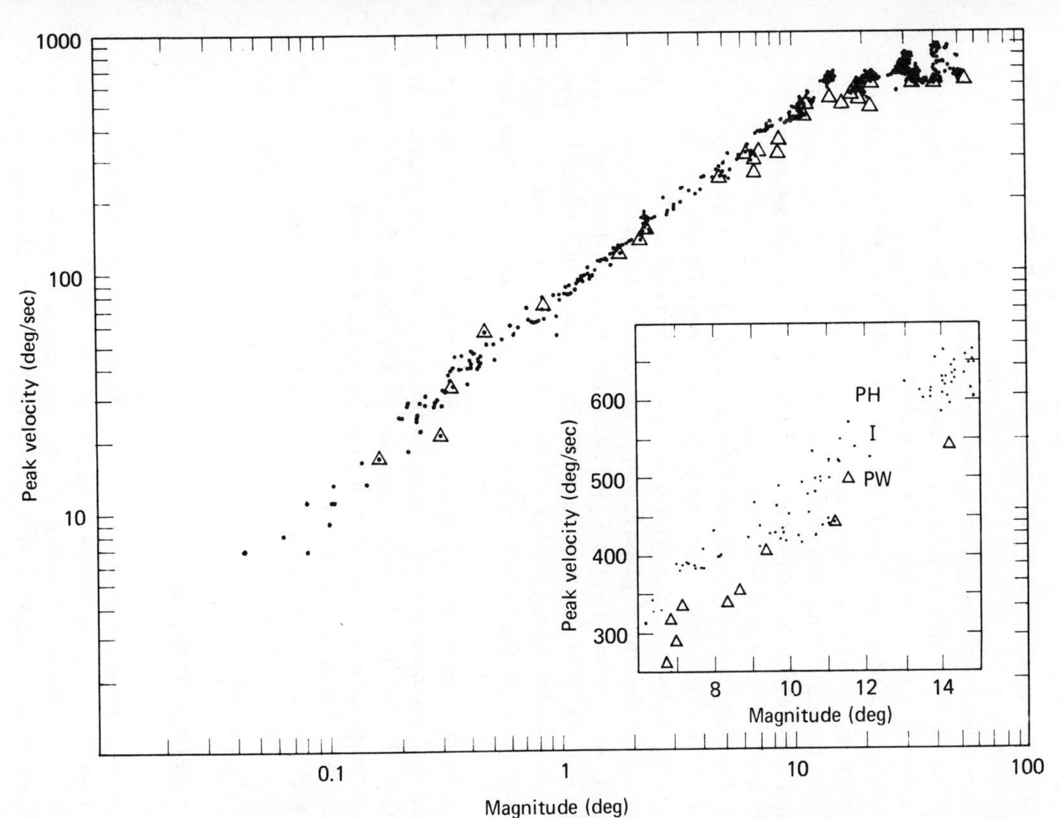

Figure 3-56 Main-sequence diagrams showing that naturally occurring human saccades with glissadic overshoot have reduced saccadic peak velocities just as pulse-width mismatched saccades from the model. Shown are normal human saccades (dots), human mismatched saccades (triangles), model pulse-width mismatched saccades (PW), model pulse-height mismatched saccades (PH), and model ideal 12° saccades (I). [Based on A. T. Bahill, F. K. Hsu, and L. Stark, "Glissadic Overshoots Are Due to Pulse Width Errors," *Archives of Neurology*, 35 (1978), 138–42. Copyright 1978, American Medical Association.]

Normal human saccades are indicated with dots in Fig. 3-56. A comparison of the model and the human data allowed us to predict that actual human mismatched saccades with this type glissadic overshoot should have either unusually low peak velocities, if they were produced by pulse-width errors, or else exceptionally high peak velocities, if they were produced by pulse-height errors.

After the modeling studies were completed, human physiological data were gathered to test this prediction. Human saccades with glissadic overshoot actually had unusually low peak velocities. Figure 3-56 shows normal human saccades indicated with dots, and human mismatched saccades with glissadic overshoot due to oversized pulse components indicated with triangles. The peak velocities of these mismatched saccades were smaller than the peak velocities of normal saccades, just as in the pulse-width error mismatched saccades of the model.

This implies that in normal eye movements in which glissadic overshoot is the result of a too-large pulse, the error is caused primarily by the brain's mistake in computing the pulse width, not the pulse height. This, in turn, implies that normal human glissades are not caused by peripheral disturbances but are caused by the CNS networks that produce the pulse width. It also implies that in controlling a neuronal burst, it is easier for the central nervous system to regulate pulse height than pulse width. Perhaps this is because pulse-height control is hardwired (recall the Henneman size principle).

Figure 3-56 also shows the more usual log-log main-sequence diagram with normal and mismatched human saccadic eye movements plotted. It confirms that the healthy human saccades in the range 5 to 50° with glissadic overshoot have smaller-than-normal saccadic peak velocities. For smaller saccades, this may not be true.

The model prediction that glissades are due to pulse-width errors lays the groundwork for analyzing the central nervous system's adaptation for certain specific pathological states.

3-7-4.5 VALIDATION SUMMARY

We used physiological data to build a homomorphic reciprocal innervation model for the eye-movement system. We validated our model qualitatively by comparing shapes of human and model saccades, quantitatively by comparing main-sequence relationships of human and model saccades, analytically by performing a sensitivity analysis and heuristically by simulating eye movements that the model was not designed to simulate. Use of this model has enabled the following predictions to be made: that dynamic overshoots are caused by reversals in the neurological control signals, and that glissades, which are part of a diagnostic syndrome for certain diseases, are produced primarily by errors in computation of the pulse width of the motoneuronal controller signal.

The validation of the reciprocal innervation model was presented as an example of model validation. There were five major components of this validation scheme: logical development, qualitative comparisons, quantitative comparisons, simulation of data not used in the model formulation, and sensitivity analysis. These techniques were all helpful in validating the reciprocal innervation model; however, use of all of these techniques may not be appropriate for other models. Let us now examine the validation of some other physiological models.

3-7-5 Validation of Other Physiological Models

As the first example of model validation, let us reexamine the Hodgkin–Huxley model of a nerve axon that we studied in Chap. 1. Their model is composed of a set of nonlinear differential equations. The solutions to these equations match the experimental data both qualitatively and quantitatively. All of their experimental data were derived from voltage-clamp experiments. But they used the model to simulate other phenomena, such as real action potentials, thresholds, the refractory period, anode break responses, subthreshold oscillations, and accommodation. This large number of phenomena that were not used to construct the model, but could be simulated by the model, greatly contributed to the rapid acceptance of the Hodgkin-Huxley neuronal model.

Some models are not usually recognized or described as models. The validation techniques presented in this chapter are seldom utilized or applicable to these models. For example, the description of simple cells, complex cells, and hypercomplex cells by Hubel and Weisel (1965, 1974, 1979) is a model for the function of the human visual system from the retina to the cortex. This model was mentioned in the introductory chapter of this book. There is a qualitative agreement between animal and model data, but quantification would be difficult. Predictions of new types of cells not in the original data base, for example a grandmother cell, have not been substantiated. There are also techniques for validating models that were not appropriate for the reciprocal innervation model. For example, the principles of controllability and observability could be applied to the model if it could be represented in state-variable form.

The validation techniques used must depend on the particular model being studied. The more techniques that are used, the more confidence can be placed in the model.

3-7-6 Parameter Estimation

In developing a model of a system, the parameters of the model must be assigned numerical values. Sometimes this assignment can be made based upon knowledge of the physical system. For example, the mass of an object being controlled is often known. However, some physical parameters, such as the viscosities in the reciprocal innervation model, can only be estimated. Choosing parameter values can be done manually or under computer control.

So far in this text all parameter fitting has been done by eye. For example, the experimental passive length–tension diagram of Fig. 3-32 was approximated with a hand-drawn straight line. This approximation was then used in the construction of the model. Other parameters, such as the agonist activation time constant, were adjusted after the model was constructed. The model was run and the output was visually compared to human eye movements. Then the time constant was changed and the model was run again. After many iterations, a satisfactory value of the parameter was obtained.

A digital computer can perform this parameter estimation more efficiently. Figure 3-57 shows the estimation scheme. The criterion function is the mean squared error between the model and plant outputs. It is a function of the parameter values, $\boldsymbol{\beta}$.

$$h(\boldsymbol{\beta}) = \int_0^t (\mathbf{x} - \boldsymbol{\theta})^T \mathbf{G}(\mathbf{x} - \boldsymbol{\theta})\, dt \qquad (3\text{-}68)$$

where the superscript T represents the matrix transpose operation and $\mathbf{G}$ is

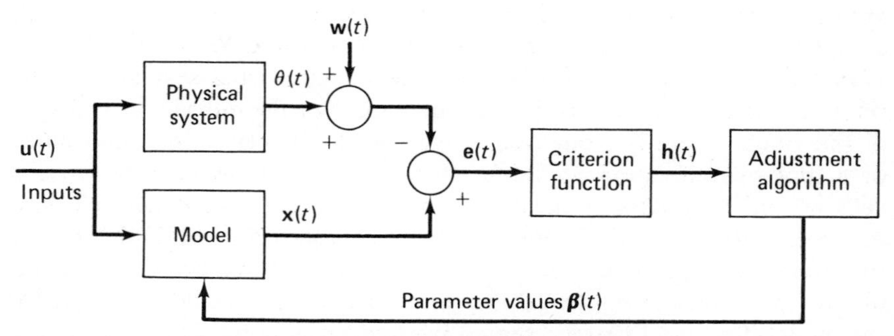

Figure 3-57 To find the optimal model parameters, both the model and the physical system are excited with the same input. The two outputs are compared and the difference between the two, the error e, is used to form the criterion function. The adjustment algorithm iteratively changes the parameter values until the criterion function is minimized. Appropriate steps may be taken to account for measurement noise, w. (Based on Latimer and Bahill, 1979.)

a weighting matrix. For the simple example presented in the next section, $\mathbf{G} = 1$, and only one variable, the model output, is studied; the criterion function becomes

$$h(\boldsymbol{\beta}) = \int_0^t (x_1 - \theta_1)^2 \, dt$$

The purpose of the algorithm is to adjust the parameters so as to minimize this function. Many function-minimization programs can be used for this purpose (Eykoff 1974; Bekey and Yamashiro 1976). The Davidson–Fletcher–Powell method is perhaps the most powerful, but it is also difficult to use. Latimer and Bahill (1979) discussed several classical techniques and developed a modified steepest-descent technique for function minimization. This technique allowed several parameters to be varied simultaneously and also allowed several outputs to be compared.

This technique was used to determine the values of the agonist and antagonist dashpots and of the time constants that would yield the least-squared error between the model output and eye position during a human saccade. The method was very powerful: almost all changes in the saccadic trajectories that were not produced by input signal variations could be matched by small parameter changes. The model could match either the right eye or the left eye better than the two eyes matched each other (Fig. 3-60). Parameter estimation by minimizing the difference between the model behavior and the behavior of the physical system is an important step in building a model. As an example of using the parameter-estimation routine, we will find values for agonist and antagonist dashpots of a linear homeomorphic eye movement model.

3-7-7 Linearizing the Model

Modeling is a dynamic process and the models should be continually changed. Models for human neuromuscular control should be reevaluated every time there are new physiological data or new engineering analysis techniques. Changing parameters to match new human data is relatively simple. The much more difficult changes are those that require a change in the form of the model. Two directions for changes in the form of the model have been suggested: (1) make the whole system more complicated by making most elements nonlinear time-varying elements (Hatze 1977, 1978); and (2) make the model simpler by linearizing the elements. Most of our recent efforts have gone into linearizing the model.

Nonlinearities are primarily due to the nonlinear force–velocity relationships characterized by the Hill equation (Hill 1938), Figs. 3-36 and 3-58. Latimer, Troost, and Bahill (1978), in an effort to linearize the model, plotted numerical values for the dashpots B_{AG} and B_{ANT} as functions of time. Straight-line approximations of these functions closely fit the actual values of the dashpots for a 10° saccade. The nonlinear problem was thereby transformed into a time-variant problem. However, time-varying problems are also difficult, so several other linearization techniques were tried.

Clark and Kamat (discussed in Clark, Jury, Krishnan, and Stark 1975) tried to linearize the reciprocal innervation model by using the first partial derivatives of Taylor series expansions. Their linearized model did not match physiological data: the poles fell into an unacceptable region of the *s*-plane. This failure was probably the result of the oversimplification of omitting the activation and deactivation time constants.

An obvious method of linearizing the model is to approximate the force–velocity curves with straight lines. Stark (1968), in formulating the BIOSIM simulation language, approximated these curves with straight lines through the point V_{max}. This linearized the force–velocity relationship but did not linearize the model because the dashpot parameters became functions of the model-state variables (Latimer, Troost, and Bahill 1978). Moreover, newer physiological data have shown that V_{max} varies with the percentage of activation (Joyce, Rack, and Westbury 1969; Joyce and Rack 1969; Julian 1971; Julian and Sollins 1973; Partridge 1978). Thus a

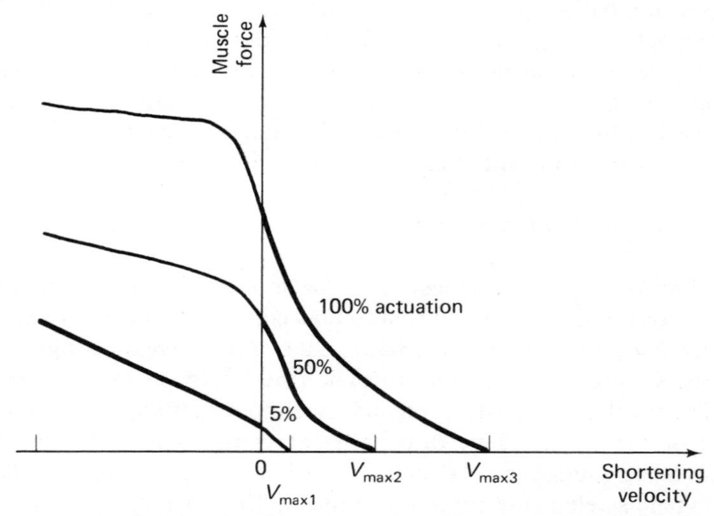

Figure 3-58 Force–velocity relationships for muscle stimulated with neural signals appropriate for 5%, 50%, and 100% activation. Agonist activation is typically between 50 and 100% activation. Antagonist activity is typically less than 5% activation. For large rapid movements, the force–velocity relationship for the agonist muscle would be represented by the upper curves in the first quadrant, while the antagonist muscle would be represented by the lower curve in the second quadrant. (Based on Latimer and Bahill, 1979.)

better linearization is a family of piecewise linear curves with equal, constant slopes (Fig. 3-59).

When a muscle is stimulated and quickly stretched, it offers a high resistance to the external force. This antagonist force–velocity relationship can be modeled by a two-piece linear approximation as shown in left half of Fig. 3-59. The intersection of the two lines is a linear, force-dependent function. During normal saccadic movements the antagonist muscle force is reduced, corresponding to the 5% innervation curve of Figs. 3-58 and 3-59. The parameter estimation algorithm was used in an attempt to find the intersection and slopes of the piecewise linear approximations of the force–velocity curves using a 10° saccade. The results showed that the best fit to the data was obtained by using only one line for the force–velocity approximation of the 5% innervation curve.

Eye-movement records were sampled at 1000 Hz and stored in the computer. These records, together with the model simulations, were used to form the error criterion. The minimization algorithm was then used to make the following parameter estimations:

$$B_{AG} = 2.36 \text{ N-sec/m}$$

$$B_{ANT} = 1.12 \text{ N-sec/m}$$

These values resulted in excellent model simulations (Fig. 3-60). The minimization was repeated for various different initial guesses of B_{AG} and B_{ANT} to ensure that the estimated parameters corresponded to a global minimum instead of a local minimum.

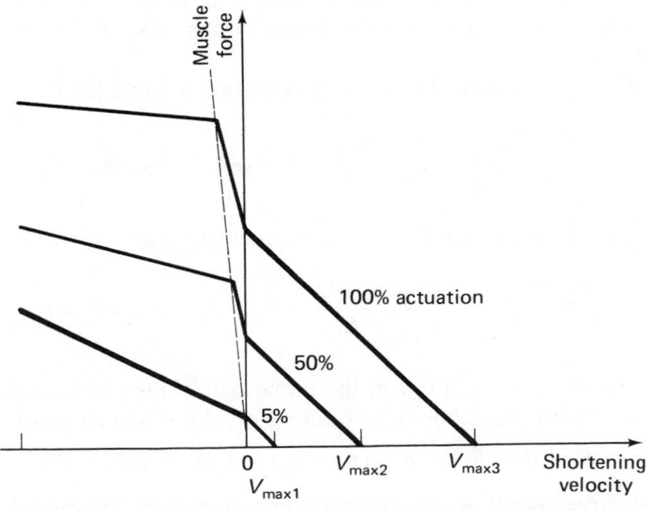

Figure 3-59 Linearized force–velocity relationships for muscle with 5%, 50%, and 100% activation. The antagonist curve for 5% activation is best fit with only one straight line. (Based on Latimer and Bahill, 1979.)

The linear homeomorphic model derived by Latimer and Bahill has the same form as that shown in Fig. 3-39. It differs from the old reciprocal innervation model of Fig. 3-41a, in that it contains the effects of the length–tension diagram, it is linear, the parameters were updated with recent physiological data, and the fine tuning of the parameters was done with the parameter-estimation routine.

This linear model can be derived with reference to Fig. 3-39. The three parallel elasticities, K'_p, K_{PE}, and K_{PE}, are combined into one spring, K_p. Two muscles are pulling in opposite directions on the globe, J. Their forces are

$$T_{AG} = K_{SE}(\theta_2 - \theta_1) \tag{3-69}$$

$$T_{ANT} = K_{SE}(\theta_1 - \theta_3) \tag{3-70}$$

From Eq. (3-58), noting that $L = -\theta_1$,

$$T_{AG} = \frac{K_{SE}F_{AG}}{K_{LT} + K_{SE}} - \frac{K_{LT}K_{SE}\theta_1}{K_{LT} + K_{SE}} - B_{AG}\dot{\theta}_2 \tag{3-71}$$

and

$$T_{ANT} = \frac{K_{SE}F_{ANT}}{K_{LT} + K_{SE}} + \frac{K_{LT}K_{SE}\theta_1}{K_{LT} + K_{SE}} + B_{ANT}\dot{\theta}_3 \tag{3-72}$$

The minus signs of Eq. (3-71) are plus signs in Eq. (3-72) because the antagonist dashpot force adds to the antagonist active-state tension, to resist eye movement. Also, as the muscle gets longer, the length–tension diagram prescribes more muscle force, which also helps to resist eye movement.

Now Eqs. (3-71) and (3-69) can be combined to yield

$$\frac{K_{SE}F_{AG}}{K_{LT} + K_{SE}} - \frac{K_{LT}K_{SE}\theta_1}{K_{LT} + K_{SE}} - B_{AG}\dot{\theta}_2 = K_{SE}(\theta_2 - \theta_1) \tag{3-73}$$

and Eqs. (3-72) and (3-70) can be combined to yield

$$\frac{K_{SE}F_{ANT}}{K_{LT} + K_{SE}} + \frac{K_{LT}K_{SE}\theta_1}{K_{LT} + K_{SE}} + B_{ANT}\dot{\theta}_3 = K_{SE}(\theta_1 = \theta_3) \tag{3-74}$$

The two muscle forces acting on the globe equations (3-69) and (3-70) can be combined with the other forces acting on the globe to yield

$$K_{SE}(\theta_2 - \theta_1) - K_{SE}(\theta_1 - \theta_3) = K_p\theta_1 + B_p\dot{\theta}_1 + J\ddot{\theta}_1 \tag{3-75}$$

These last three equations describe the movements of the model. However, the model is a sixth-order system, so it will take six differential equations

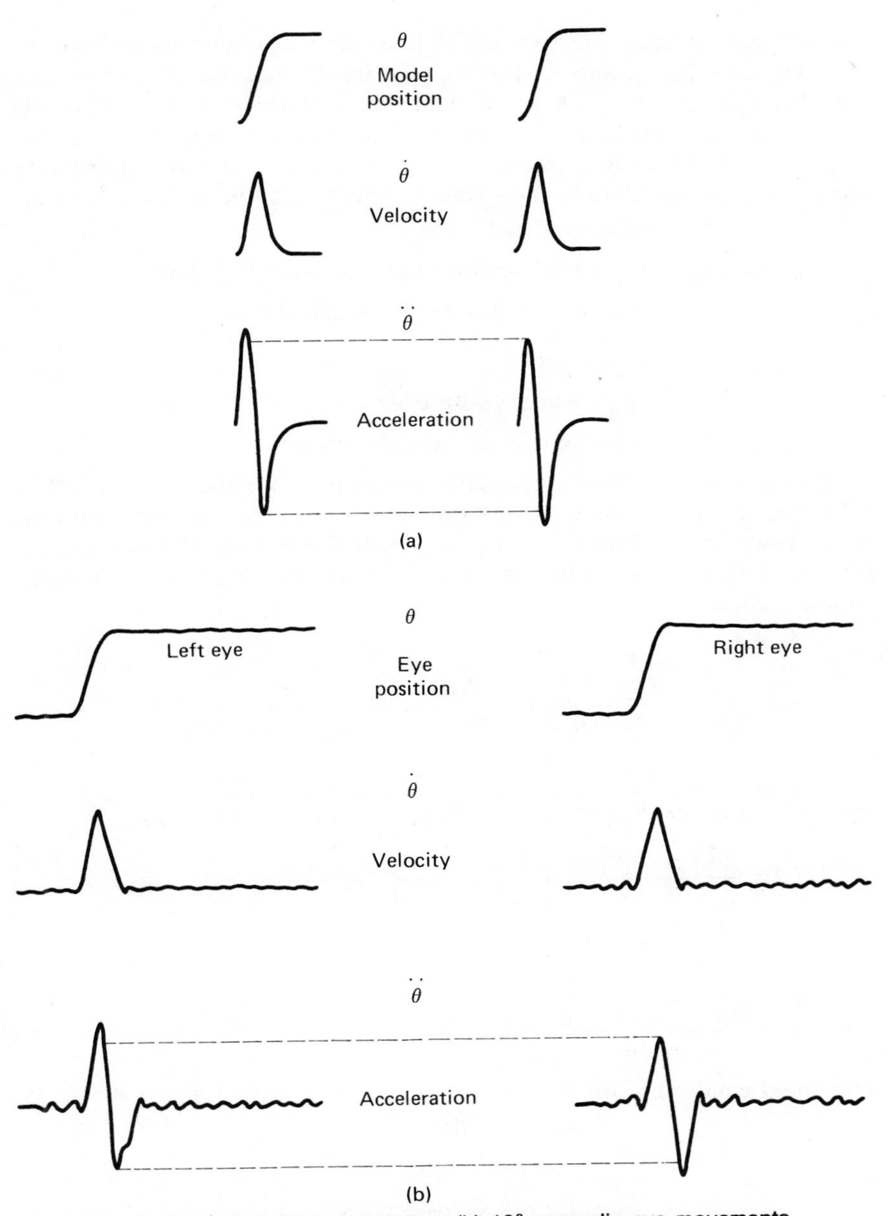

Figure 3-60 Model (a) and human (b) 10° saccadic eye movements with small differences between the simultaneous saccades of the left (left column) and right (right column) eyes. Each human record is 340 msec in length. Differences show up best in the acceleration traces.

to completely describe the system. For these state equations we will use the three positions θ_1, θ_2, and θ_3; the eye velocity $\dot{\theta}_1$; and the two active-state tensions, F_{AG} and F_{ANT}. A conversion factor of 0.004 N·sec/pulse was used to convert the firing frequency of a typical motoneuron, N_{AG} or N_{ANT}, into active-state tension. To conform with standard engineering procedures, we will identify these state variables with the symbols x_1 to x_6.

$$x_1 = \theta_1 = \text{position of eye}$$
$$x_2 = \theta_2 = \text{position of agonist node, shown in Fig. 3-39}$$
$$x_3 = \theta_3 = \text{position of antagonist node, shown in Fig. 3-39}$$
$$x_4 = \dot{\theta}_1 = \text{eye velocity}$$
$$x_5 = F_{AG} = \text{agonist active-state tension}$$
$$x_6 = F_{ANT} = \text{antagonist active-state tension}$$

This is but one of many possible assignments for the state variables. This assignment happens to be intuitive and convenient. The three simultaneous equations (3-73) to (3-75) can be solved for each of the variables, and the three auxiliary equations can be formed to yield the following six state equations:

$$\dot{x}_1 = x_4$$
$$\dot{x}_2 = \frac{K_{SE}^2}{(K_{LT} + K_{SE})B_{AG}} x_1 - \frac{K_{SE}}{B_{AG}} x_2 + \frac{K_{SE}}{(K_{LT} + K_{SE})B_{AG}} x_5$$
$$\dot{x}_3 = \frac{K_{SE}^2}{(K_{LT} + K_{SE})B_{ANT}} x_1 - \frac{K_{SE}}{B_{ANT}} x_3 - \frac{K_{SE}}{(K_{LT} + K_{SE})B_{ANT}} x_6$$
$$\dot{x}_4 = \frac{-2K_{SE} - K_p}{J} x_1 + \frac{K_{SE}}{J} x_2 + \frac{K_{SE}}{J} x_3 - \frac{B_p}{J} x_4$$
$$\dot{x}_5 = \frac{0.004 N_{AG} - x_5}{\tau_{AG}}$$
$$\dot{x}_6 = \frac{0.004 N_{ANT} - x_6}{\tau_{ANT}}$$

The initial conditions are

$$x_1(0) = x_4(0) = 0$$
$$x_2(0) = -x_3(0) = 5.6°$$
$$x_5(0) = x_6(0) = 0.2\,\text{N}$$

These state equations completely describe the behavior of the model. It is sometimes more convenient to write these equations using matrix notation.

$$\dot{\mathbf{x}} = \mathbf{Ax} + \mathbf{Bu}$$

In this equation $\dot{\mathbf{x}}$, $\mathbf{x}$, and $\mathbf{u}$ are vectors, $\mathbf{B}$ may be a vector or a matrix, and $\mathbf{A}$ is a square matrix. Using this notation, our six state equations become

$$
\begin{bmatrix} \dot{x}_1 \\ \dot{x}_2 \\ \dot{x}_3 \\ \dot{x}_4 \\ \dot{x}_5 \\ \dot{x}_6 \end{bmatrix}
=
\begin{bmatrix}
0 & 0 & 0 & 1 & 0 & 0 \\
\dfrac{K_{\mathrm{SE}}^2}{(K_{\mathrm{LT}}+K_{\mathrm{SE}})B_{\mathrm{AG}}} & \dfrac{-K_{\mathrm{SE}}}{B_{\mathrm{AG}}} & 0 & 0 & \dfrac{K_{\mathrm{SE}}}{(K_{\mathrm{LT}}+K_{\mathrm{SE}})B_{\mathrm{AG}}} & 0 \\
\dfrac{K_{\mathrm{SE}}^2}{(K_{\mathrm{LT}}+K_{\mathrm{SE}})B_{\mathrm{ANT}}} & 0 & \dfrac{-K_{\mathrm{SE}}}{B_{\mathrm{ANT}}} & 0 & 0 & \dfrac{-K_{\mathrm{SE}}}{(K_{\mathrm{LT}}+K_{\mathrm{SE}})B_{\mathrm{ANT}}} \\
\dfrac{-2K_{\mathrm{SE}}-K_{\mathrm{P}}}{J} & \dfrac{K_{\mathrm{SE}}}{J} & \dfrac{K_{\mathrm{SE}}}{J} & \dfrac{-B_P}{J} & 0 & 0 \\
0 & 0 & 0 & 0 & \dfrac{-1}{\tau_{\mathrm{AG}}} & 0 \\
0 & 0 & 0 & 0 & 0 & \dfrac{-1}{\tau_{\mathrm{ANT}}}
\end{bmatrix}
\begin{bmatrix} x_1 \\ x_2 \\ x_3 \\ x_4 \\ x_5 \\ x_6 \end{bmatrix}
+
\begin{bmatrix} 0 \\ 0 \\ 0 \\ 0 \\ 0.004\dfrac{N_{\mathrm{AG}}}{\tau_{\mathrm{AG}}} \\ 0.004\dfrac{N_{\mathrm{ANT}}}{\tau_{\mathrm{ANT}}} \end{bmatrix}
$$

The following parameter values are given to complete the description of this model. Most of these parameters have been derived in previous sections. They are treated in greater detail in Bahill, Latimer, and Troost (1981a), and Bahill (1980).

Length–tension and quick-release experiments on the isolated globe by Robinson et al. (1969) and Collins (1975) were used to determine the viscosity B_p and the elasticity K_p of the globe orbit. The inertial mass J was calculated based on a spherical radius of 11 mm and a density of 1 g/cm³. Initial values for pulse width, pulse height, and the four time constants were based upon the values used for 10° saccades in the old model. The parameter-estimation program was then run, and these six parameters were adjusted to yield the least-mean squared error between the model and the human responses. These values were then fixed for 10° saccades. The parameter-estimation routine was then run on a different-size saccade adjusting pulse width, pulse height, and $\tau_{\text{AG-AC}}$ to minimize the error between model and human saccades. This procedure was repeated for saccades between 1 and 40°. Straight-line approximations were then fit to these data points to yield the following equations:

$$K_{\text{SE}} = 125\,\text{N/m}$$

$$K_{\text{LT}} = 60\,\text{N/m}$$

$$K_p = 25\,\text{N/m}$$

$$B_p = 3.1\,\text{N}\cdot\text{sec/m}$$

$$B_{\text{AG}} = 2.36\,\text{N}\cdot\text{sec/m}$$

$$B_{\text{ANT}} = 1.12\,\text{N}\cdot\text{sec/m}$$

$$J = 2.2(10^{-3})\text{N}\cdot\text{sec}^2/\text{m}$$

$$N_{\text{AG-PULSE}} = \text{PH} = (135 + 27\Delta\theta)\ \text{pulses/sec for } \Delta\theta \leqslant 11°$$

$$= (392 + 5\Delta\theta)\ \text{pulse/sec} \qquad \text{for } \Delta\theta > 11°$$

$$N_{\text{ANT-PULSE}} = 1.2\ \text{pulses/sec}$$

$$\text{PW}_{\text{AG}} = (10 + \Delta\theta)\text{msec}$$

$$\text{PW}_{\text{ANT}} = \text{PW}_{\text{AG}} + 6\ \text{msec}$$

Antagonist pulse circumscribes agonist pulse by 3 msec on each end.

$$N_{\text{AG-STEP}} = (50.1 + 5.8\Delta\theta)\ \text{pulses/sec}$$

$$N_{\text{ANT-STEP}} = (50.1 - 0.2\Delta\theta)\ \text{pulses/sec}$$

$$\tau_{\text{AG-AC}} = (11.7 - 0.2\Delta\theta)\ \text{msec}$$

$$\tau_{\text{AG-DE}} = 0.2\ \text{msec}$$

$$\tau_{\text{ANT-AC}} = 2.4\ \text{msec}$$

$$\tau_{\text{ANT-DE}} = 1.9\ \text{msec}$$

To validate this linear homeomorphic model and to evaluate the goodness of fit of existing eye movement models, the differences between the outputs of several models and some typical human saccadic eye movements were computed. Ten degree saccades of a normal unfatigued human were recorded and the first 16 of these were used to test the models. Each model was run for 60 msec, and the resulting record was compared point by point to the human saccadic eye movement. The mean squared error between the model and human responses was calculated. This process was repeated 50 times as the model was shifted forward and backward in time. The shift with the minimum mean squared error was chosen as the best possible fit for that model. This process was then repeated for each of the 16 saccades and the mean and standard deviation of the mean squared errors were computed. These are the results: step input driving a plant model with a transfer function of unity, $3500 \pm 200 \times 10^{-6}$ deg^2 (Selhorst, Stark, Ochs, and Hoyt 1976); pulse input and an integrator driving a second-order, linear, overdamped plant model with time constants of 150 and 12 msec, $2760 \pm 820 \times 10^{-6}$ deg^2 (Zee and Robinson 1979a, b); step input driving a second-order, linear, underdamped plant model with $\omega_n =$ 120 rad/sec and $\zeta = 0.7$, $250 \pm 150 \times 10^{-6}$ deg^2 (Westheimer 1954); pulse-step input driving a sixth-order, nonlinear plant model (the model of Fig. 3-41), $140 \pm 60 \times 10^{-6}$ deg^2 (Hsu, Bahill, and Stark 1976); pulse-step input driving a sixth-order, linear plant model that includes a length-tension element (the model developed in this section), $50 \pm 25 \times 10^{-6}$ deg^2 (Bahill, Latimer and Troost 1980a); left eye matching the right eye, $45 \pm 20 \times 10^{-6}$ deg^2. The simulations of the linear homeomorphic model match the human eye movements almost as well as the two eyes match each other. Biological variations produce larger differences in saccadic trajectories than those caused by small parameter adjustments. This implies that the model parameters have been selected optimally; further modifications are not likely to be useful.

During movements the eyes follow different trajectories (Bahill, Ciuffreda, Kenyon, and Stark 1976). The records of Fig. 3-60 are typical, showing small differences between the two eyes. Although it is not known what causes these differences, the parameter-estimation program has shown that they could be caused by small (5%) variations in pulse height and pulse width. Small variations in other parameters could also have been used to match these data.

However, certain large variations in saccadic trajectories can be produced only by specific parameter variations. For example, the glissadic undershoot in the left column of Fig. 3-61 could only be produced in the model by decreasing the pulse width by a large amount. When the parameter-estimation routine was allowed to vary PW and PH, the only two physiological parameters that are likely to change between saccades, the best fit was obtained by changing the pulse width by 20% and the pulse

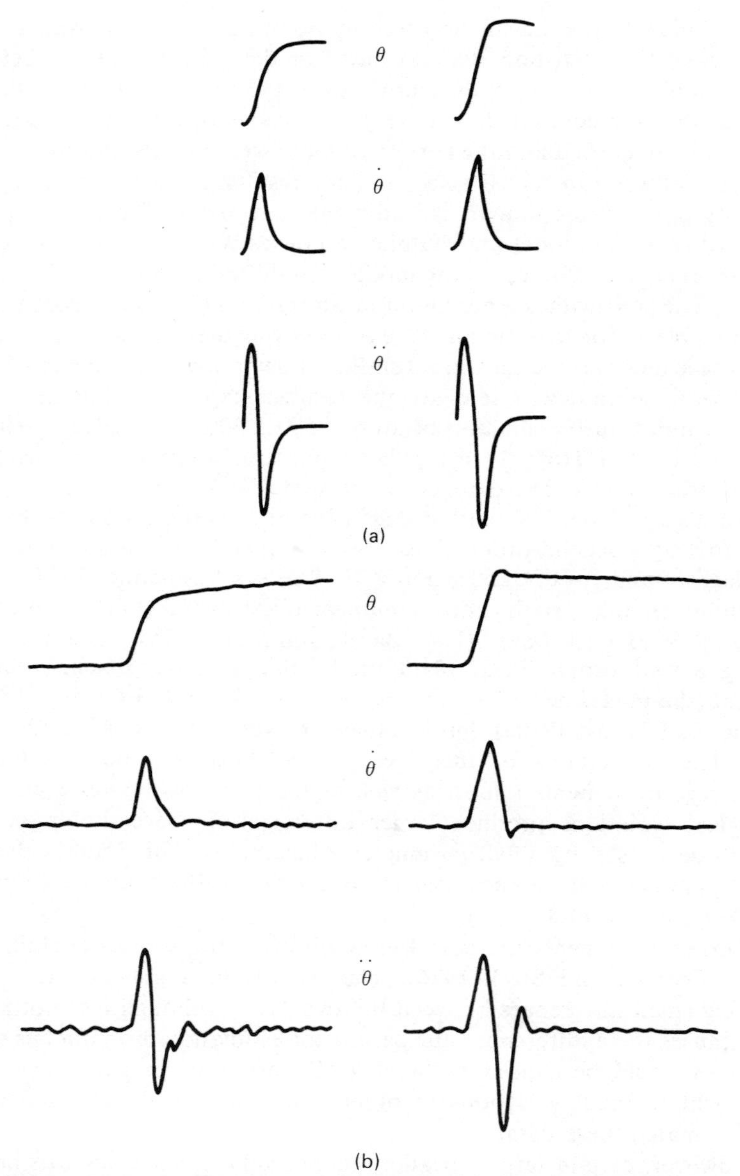

Figure 3-61 (a) Model and (b) human 10° saccadic eye movements with large differences between two sequential saccades of the left eye. Large differences such as these are caused by control-signal variations. In the model, the pulse width had to be decreased by 20% to produce the glissadic undershoot in the left column. The pulse width had to be increased by 25% to produce the overshoot in the right column. In both cases the pulse height had to be held with 2% of the nominal value. Each human record is 340 msec in length. (Unpublished data of Latimer and Bahill.)

height by 2%. Similarly, the overshoot in the right column of Fig. 3-61 was matched with minimal error by a large (25%) increase in pulse width and a small (1%) increase in pulse height.

This once again suggests that human glissadic overshoot and undershoot are caused by pulse-width errors as shown in Sec. 3-7-4.4 on glissades. This implies that the brain can control the height of a motoneuronal burst quite accurately (recall the Henneman size principle), but that it has a much more difficult time accurately controlling the duration of a motoneuronal burst.

It is not easy to linearize a model, but it pays substantial benefits. One of the most important advantages of a linear model is the fact that superposition applies. Superposition is a very powerful tool for studying linear systems. It means that the interaction of model parameters need not be considered. Such interactions were neglected in Hsu, Bahill, and Stark's (1976) sensitivity analysis for the purpose of simplification, not because the system was linear. Linearization of the model allows the use of currently available computer analysis programs from the field of linear control theory, such as the programs by Melsa and Jones (1970). Furthermore, the model had to be simplified before it would be used by other independent investigators.

It is a useful medical technique to construct closed-loop feedback control system models for the oculomotor system and then to remove several regions of the model to discover what parts must be destroyed in order to emulate the eye movements of patients with certain diseases. Previous attempts at doing this (Selhorst, Stark, Ochs, and Hoyt 1976; Zee, Optican, Cook, Robinson, and Engel 1976; Abel, Dell'Osso, and Daroff 1978; Zee and Robinson 1979a, b) have all reverted back to zero-order or second-order systems for the plant model. Better precision could be obtained with a larger model, but these researchers considered the full model to be too hard to use.

3-8 SUMMARY

The development of open-loop-system models depends on an understanding of the physiological processes, and circularly the development of models helps us to understand the physiological systems. Numerical values for the model parameters were initially derived from knowledge of the physiological system. These values were adjusted, either visually or with computer programs, to minimize the error between the model output and the output of the physiological system. Several methods for validating models were shown: qualitative, quantitative, sensitivity analyses, and making predictions based upon novel behavior of the model.

The linear homeomorphic model gradually evolved over a 25-year span, as shown in Table 3-4. It started as a linear second-order system, grew to a fourth-order system, then to a nonlinear sixth-order system, and finally back to a linear sixth-order system. As the model evolved it became more homeomorphic; that is, a one-to-one relationship developed between the elements of the model and the elements of the physical system.

REFERENCES

ABEL, L. A., L F. DELL'OSSO, and R. B. DAROFF, "Analog Model for Gaze-evoked Nystagmus," *IEEE Transactions on Biomedical Engineering*, BME-25 (1978) 71–75.

BAHILL, A. T., "Development, Validation and Sensitivity Analyses of Human Eye Movement Models," *CRC Reviews in Bioengineering*, vol. 4, Issue 4, (1980).

BAHILL, A. T., K. J. CIUFFREDA, R. V. KENYON, and L. STARK, "Dynamic and Static Violations of Hering's Law of Equal Innervation," *American Journal of Optometry and Physiological Optics*, 53 (1976), 786–96.

BAHILL, A. T., M. R. CLARK, and L. STARK, "Dynamic Overshoot in Saccadic Eye Movements is Caused by Neurological Control Signal Reversals," *Experimental Neurology*, 48 (1975a), 95–122.

BAHILL, A. T., M. R. CLARK, and L. STARK, "Glissades—Eye Movements Generated by Mismatched Components of the Saccadic Motoneuronal Control Signal," *Mathematical Biosciences*, 26 (1975b), 303–18.

BAHILL, A. T., M. R. CLARK, and L. STARK, "The Main Sequence, a Tool for Studying Human Eye Movements," *Mathematical Biosciences*, 24 (1975c), 194–204.

BAHILL, A. T., F. K. HSU, and L. STARK, "Glissadic Overshoots Are Due to Pulse Width Errors," *Archives of Neurology*, 35 (1978), 138–42.

BAHILL, A. T., J. R. LATIMER, and B. T. TROOST, "Linear Homeomorphic Model for Human Movement," *IEEE Transactions on Biomedical Engineering*, BME-27, No.11, in press, (1980a).

BAHILL, A. T., J. R. LATIMER, and B. T. TROOST, "Sensitivity Analysis and Validation of Linear Homeomorphic Model for Human Movement," *IEEE Transactions on Systems Man and Cybernetics*, SMC-10, No. 12, in press, (1981b).

BAHILL, A. T., and L. STARK, "The Trajectories of Saccadic Eye Movements," *Scientific American*, 240 (January 1978), 108–17.

BAHLER, A. S., "Modeling of Mammalian Skeletal Muscle," *IEEE Transactions on Biomedical Engineering*, BME-14 (1968), 249–57.

BAKER, R., and A. BERTHOZ, "Organization of Vestibular Nystagmus in Oblique Oculomotor Systems," *Journal of Neurophysiology*, 37 (1974), 195–217.

BALOH, R. W., R. D. YEE and V. HONRUBIA, "Internuclear Ophthalmoplegia," *Archives of Neurology*, 35 (1978), 484–93.

BARMACK, N. H., C. C. BELL, and B. G. RENCE, "Tension and Rate of Tension Development during Isometric Responses of Extraocular Muscle," *Journal of Neurophysiology*, 34 (1971), 1072–79.

BEKEY, G. A., and S. M. YAMASHIRO, "Parameter Estimation in Mathematical Models of Biological Systems," in *Advances in Biomedical Engineering*, Vol. 6, ed. J. H. V. Brown and J. F. Dickson III. New York: Academic Press, Inc., 1976, pp. 1–43.

BINDER, M. D., M. S. CAMERON, and D. S. STUART, "Speed–Force Relations in the Motor Units of the Cat Tibalis Posterior Muscle," *American Journal of Physical Medicine*, 57 (1978), 57–65.

BURKE, R. E., P. RUDOMIN, and F. E. ZAJAC III, "Catch Property in Single Mammalian Motor Units," *Science*, 168 (1970), 122–24.

CAMPBELL, F. W., and R. H. WURTZ, "Saccadic Omission! Why We Do Not See a Gray-out during Saccadic Eye Movement," *Vision Research*, 18 (1978), 1297–1303.

CARPENTER, R. H. S., *Movements of the Eyes*. London: Pion Ltd., 1977.

CIUFFREDA, K. J., and L. STARK, "Descartes' Law of Reciprocal Innervation," *American Journal of Optometry and Physiological Optics*, 52 (1975), 663–73.

CLARK, M. R., E. I. JURY, V. V. KRISHNAN, and L. STARK, "Computer Simulation of Biological Models Using the Inners Approach," *Computer Programs in Biomedicine*, 5 (1975), 263–82.

CLARK, M. R., and L. STARK., "Control of Human Eye Movements," *Mathematical Biosciences*, 20 (1974), 191–265.

CLARK, R. W., and E. S. LUSCHEI, "Short Latency Jaw Movement Produced by Low Intensity Intracortical Microstimulation of the Precentral Face Area in Monkey," *Brain Research*, 70 (1974), 144–47.

Cold Spring Harbor Symposia on Quantitative Biology, *The Mechanism of Muscle Contraction*, Vol. 37, Cold Spring Harbor Laboratory, New York, 1973.

Collins, C. C., "The Human Oculomotor Control System," in *Basic Mechanisms of Ocular Motility and Their Implications*, ed. G. Lennerstrand and P. Bach-y-Rita. Elmsford, N.Y.: Pergamon Press, Inc., 1975, 145–80.

Collins, C. C., D. O'Meara, and A. B. Scott, "Muscle Tension during Unrestrained Human Eye Movements," *Journal of Physiology*, 245 (1975), 351–69.

Cook, G., and L. Stark, "The Human Eye Movement Mechanism: Experiments, Modeling and Model Testing," *Archives of Opthalmology*, 79 (1968), 428–36.

Davson, H., *The Physiology of the Eye*. New York: Academic Press, Inc., 1972.

Dell'Osso, L. F., D. A. Robinson, and B. Daroff, "Optokinetic Asymmetry in Internuclear Ophthalmoplegia," *Archives of Neurology*, 31 (1974), 138–39.

Descartes, René, *Treatise of Man*. Originally published by Charles Angot, Paris, 1664. Published with translation and commentary by T. S. Hall. Cambridge, Mass.: Harvard University Press, 1972.

Dodge, R., and T. L. S. Cline, "The Angle Velocity of Eye Movements," *Psychology Reviews*, 8 (1901), 145–57.

Evarts, E. V., "Precentral and Postcentral Cortical Activity in Association with Visual Triggered Movement," *Journal of Neurophysiology*, 37 (1974), 373–84.

Eykhoff, P., *System Identification—Parameter and State Estimation*. New York: John Wiley & Sons, Inc., 1974, Chaps. 2, 3, and 5.

Fenn, W. O., and B. S. Marsh, "Muscular Force at Different Speeds of Shortening," *Journal of Physiology*, 85 (1935), 277–297.

Ford, A., and P. C. Gardner, "A New Measure of Sensitivity for Social System Simulation Models," *IEEE Transactions on Systems, Man and Cybernetics*, SMC-9 (1979), 105–14.

Frank, P. M., *Introduction to System Sensitivity Theory*. New York: Academic Press, Inc., 1978.

Fricker, J., and J. J. Sanders, "Velocity and Acceleration Statistics of Pseudorandomly Timed Saccades in Humans," *Vision Research*, 15 (1975), 225–29.

Fuchs, A. F., and E. S. Luschei, "Firing Patterns of Abducens Neurons of

Alert Monkeys in Relationship to Horizontal Eye Movement," *Journal of Neurophysiology*, 33 (1970), 382–92.

FUCHS, A. F., and E. S. LUSCHEI, "Development of Isometric Tension in Simian Extraocular Muscle," *Journal of Physiology*, 219 (1971), 155–66.

GORDON, A. M., A. F. HUXLEY, and F. J. JULIAN, "The Variation in Isometric Tension with Sarcomere Length in Vertebrate Muscle Fibers," *Journal of Physiology*, 184 (1966), 170–92.

GOSLOW, G. E., W. E. CAMERON, and D. G. STUART, "The Fast Twitch Motor Units of Cat Ankle Flexors, (2) Speed–Force Relations and Recruitment Order," *Brain Research*, 134 (1977), 47–57.

GOSLOW, G. E., R. M. REINKING, and D. G. STUART, "The Cat Step Cycle: Hind Limb Joint Angles and Muscle Lengths during Unrestrained Locomotion," *Journal of Morphology*, 141 (1973), 1–42.

GOSLOW, G. E., and D. G. STUART, "The Fast Twitch," *Brain Research*, 134 (1977), 35–58.

HATZE, H., "A Myocybernetic Control Model of Skeletal Muscle," *Biological Cybernetics*, 25 (1977), 103–19.

HATZE, H., "A General Myocybernetic Control Muscle," *Biological Cybernetics*, 28 (1978), 143–57.

HENNEMAN, E., "Peripheral Mechanisms Involved in the Control of Muscle," in *Medical Physiology*, ed. V. B. Mountcastle. St. Louis: The C. V. Mosby Company, 1974, pp. 617–35.

HENNEMAN, E., G. SOMJEN, and D. O. CARPENTER, "Functional Significance of Cell Size in Spinal Motoneurons," *Journal of Neurophysiology*, 28, (1965), 560–80.

HILL, A. V., "The Heat of Shortening and Dynamic Constraints of Muscle," *Proceedings of the Royal Society, London (B)*, 126 (1938), 136–95.

HILL, A. V., "The Series Elastic Component of Muscle," *Proceedings of the Royal Society, London (B)*, 137 (1950a), 273–80.

HILL, A. V., "The Development of the Active State of Muscle during the Latent Period," *Proceedings of the Royal Society, London (B)*, 137 (1950b), 320–29.

HILL, A. V., "The Effect of Series Compliance on the Tension Developed in a Muscle Twitch," *Proceedings of the Royal Society, London (B)*, 138 (1951a), 325–29.

HILL, A. V., "The Transition from Rest to Full Activity in Muscles: The Velocity of Shortening," *Proceedings of the Royal Society, London (B)*, 138 (1951b), 329–38.

HSU, F. K., A. T. BAHILL, and L. STARK, "Parametric Sensitivity of a Homeomorphic Model for Saccadic and Vergence Eye Movements," *Computer Programs in Biomedicine*, 6 (1976), 108–16.

HUBEL, D. H., and T. N. WIESEL, "Receptive Fields and Functional Architecture in Two Nonstriate Visual Areas (18 and 19) of the Cat," *Journal of Neurophvsiology*, 28 (1965), 229–89.

HUBEL, D. H., and T. N. WIESEL, "Sequence Regularity and Geometry of Orientation Columns in Monkey Striate Cortex," *Journal of Comparative Neurology*, 158 (1974), 257–318.

HUBEL D. H., and T. N. WIESEL, "Brain Mechanism of Vision," *Scientific American*, 241(3) (1979), 150–62.

HUXLEY, A. F., "Muscle Structure and Theories of Contraction," *Programs in Biophysics and Biochemistry*, 7 (1957), 255–318.

IWAZUMI, T., and G. H. POLLACK, "On-Line Measurement of Sarcomere Length from Diffraction Patterns in Muscle," *IEEE Transactions on Biomedical Engineering*, BME-26 (1979), 86–93.

JOYCE, G. C., and P. M. H. RACK, "Isotonic Lengthening and Shortening Movements of Cat Soleus Muscle," *Journal of Physiology*, 204 (1969), 475–91.

JOYCE, G. G., P. M. H. RACK, and D. R. WESTBURY, "The Mechanical Properties of Cat Soleus Muscle during Controlled Lengthening and Shortening Movements," *Journal of Physiology*, 204 (1969), 461–74.

JULIAN, F. J., "The Effect of Calcium on the Force–Velocity Relation of Briefly Glycerinated Frog Muscle Fibers," *Journal of Physiology*, 218 (1971), 117–45.

JULIAN, F. J., and M. R. SOLLINS, "Regulation of Force and Speed of Shortening in Muscle Contraction," *Cold Springs Harbor Symposium on Quantitative Biology*, 37 (1973), 635–46.

KATZ, B., "The Relation between Force and Speed in Muscular Contraction," *Journal of Physiology*, 46 (1939), 46–64.

KELLER, E. L., "Participation of Medial Pontine Reticular Formation in Eye Movement Generation in Monkey," *Journal of Neurophysiology*, 37 (1974), 316–32.

KIRSNER, R. L. G., "Visual Control of Human Neuromuscular Movement," *Journal of Physiology*, 230 (1973), 28P.

KOMMERELL, G., "Internuclear Ophthalmoplegia of Aduction," *Archives of Ophthalmology*, 93 (1975), 531–34.

LATIMER, J. R., and A. T. BAHILL, "Parameter Estimation by Function Minimization Using a Modified Steepest Descent Method," in *Modeling and Simulation, Proc. Tenth Annual Pittsburgh Conference*, eds. W. Vogt, M. Mickle, Pittsburgh, Instrument Society of America, 1979, 683–90.

LATIMER, J. R., B. T. TROOST, and A. T. BAHILL, "Linearization and Sensitivity Analysis of Model for Human Eye Movements," *Modeling and Simulation*, Proceedings of the Ninth Annual Pittsburgh Conference, eds. W. Vogt and M. Mickle, Instrument Society of America, Pittsburgh, 1978, pp. 365–71.

LEHMAN, S., and L. STARK, "Simulation of Linear and Nonlinear Eye Movement Models: Sensitivity Analyses and Enumeration Studies of Time Optimal Control," *Journal of Cybernetics and Information Science*, 4 (1979).

LENNERSTRAND, G., "Fast and Slow Units in Extrinsic Eye Muscles of Cat," *Acta Physiologica Scandinavia*, 86 (1972), 286–88.

LENNERSTRAND, G., "Electrical Activity and Isometric Tension in Motor Units of the Cat's Inferior Oblique Muscle," *Acta Physiologica Scandinavia*, 91 (1974), 458–74.

LEVIN, A., and J. WYMAN, "The Viscous Elastic Properties of a Muscle," *Proceedings of the Royal Society, London (B)*, 101 (1927), 218–43.

MAEDA, M., H. SHIMAZU, and Y. SHINODA, "Inhibitory Postsynaptic Potentials in the Abducens Motoneurons Associated with the Quick Relaxation Phase of Vestibular Nystagmus," *Brain Research*, 26 (1971), 420–24.

MELSA, J. L., and S. K. JONES, *Computer Programs for Computational Assistance in the Study of Linear Control Theory*. New York: McGraw-Hill Book Company, 1970.

MELSA, J. L., and D. G. SCHULTZ, *Linear Control Systems*. New York: McGraw-Hill Book Company, 1969.

METZ, H., "Saccadic Velocity Measurements in Internuclear Ophthalmoplegia," *American Journal of Ophthalmology*, 81(3) (1976), 296–99.

MILSUM, J. H., *Biological Control Systems Analysis*. New York: McGraw-Hill Book Company, 1966.

MURRAY, J. M., and A. WEBER, "The Cooperative Action of Muscle Proteins," *Scientific American*, 230 (February 1974), 58–71.

PARTRIDGE, L. D., "Muscle Properties: A Problem for the Motor Physiologist," in *Posture and Movement: Prospective for Integrating Sensory and Motor Research on the Mammalian Nervous System*, ed. R. E. Talbott, and D. R. Humphrey. New York: Raven Press, (1979), 189–229.

PETERKA, R. J., "Characterization of Properties of Peripheral Neurons Innervating Semicircular Canals of the Bullfrog." Ph.D. dissertation, Department of Electrical Engineering, Carnegie-Mellon University, 1980.

RACK, P. M. H., and D. R. WESTBURY, "The Effects of Length and Stimulus Rate on Tension in the Isometric Cat Soleus Muscle," *Journal of Physiology*, 204 (1969), 443–60.

RACK, P. M. H., and D. R. WESTBURY, "The Short Range Stiffness of Active Mammalian Muscle and Its Effect on Mechanical Properties," *Journal of Physiology*, 240 (1974), 331–50.

ROBINSON, D. A., "The Mechanics of Human Saccadic Eye Movement," *Journal of Physiology*, 174 (1964), 245–64.

ROBINSON, D. A., D. M. O'MEARA, A. B. SCOTT, and C. C. COLLINS, "Mechanical Components of Human Eye Movements," *Journal of Applied Physiology*, 26 (1969), 548–53.

SCOTT, A. B., "Strabismus—Muscle Forces and Innervations," in *Basic Mechanisms of Ocular Motility and Their Clinical Implications*, eds. G. Lennerstrand and P. Bach-y-Rita. Elmsford, N.Y.: Pergamon Press, Inc., 1975, 181–91.

SEIDEL, R. C., Transfer-Function-Parameter Estimation from Frequency Response Data—A FORTRAN Program. NASA Technical Memorandum NASA TM X-3286, September 1975.

SELHORST, J. B., L. STARK, A. L. OCHS, and W. F. HOYT, "Disorders in Cerebellar Oculomotor Control," *Brain*, 99 (1976), 497–522.

SINDERMANN, F., B. GEISELMANN, and M. FISCHLER, "Single Motor Unit Activity in Extraocular Muscles in Man during Fixation and Saccades," *Electroencephalography and Clinical Neurophysiology*, 45 (1978), 64–73.

STARK, L., *Neurological Control Systems, Studies in Bioengineering*. New York: Plenum Press, 1968.

STARK, L., "Dynamics of the Oculomotor System," in *Control of Gaze by Brain Stem Neurons*, ed. R. Baker, and A. Berthoz. Developments in Neuroscience, Vol. 1. New York: Elsevier North-Holland, Inc., 1977, pp. 73–74.

STARK, L., R. KONG, S. SCHWARTZ, D. HENDRY, and B. BRIDGEMAN, "Saccadic Suppression of Image Displacement," *Vision Research*, 16 (1976), 1185–87.

TAKAHASHI, Y., M. J. RABINS, and D. M. AUSLANDER, *Control and Dynamic Systems*. Reading, Mass.: Addison-Wesley Publishing Co., Inc., 1970.

TAYLOR, S. R., R. RUDEL, and J. R. BLINKS, "Calcium Transients in Amphibian Muscle," *Federation Proceedings*, 34 (1975), 1379–81.

WESTHEIMER, G., "Mechanism of Saccadic Eye Movements," *AMA Archives of Ophthalmology*, 52 (1954), 710–24.

WILKIE, D. R., *Muscle*: *Studies in Biology*, Vol. 11. London: Edward Arnold (Publishers) Ltd., 1968.

YOUNG, L. R. and L. STARK, "Variable Feedback Experiments Testing a Sampled Data Model for Eye Tracking Movements," *IEEE Transactions on Human Factors in Electronics*, HFE-4 (1963), 38–51.

ZEE, D. S., L. M. OPTICAN, J. D. COOK, D. A. ROBINSON, and W. K. ENGEL, "Slow Saccades in Spinocerebellar Degeneration," *Archives of Neurology*, 33 (1976), 243–51.

ZEE, D. S., and D. A. ROBINSON, "A Hypothetical Explanation of Saccadic Oscillations," *Annals of Neurology*, 5 (1979a), 405–14.

ZEE, D. S., and D. A. ROBINSON, "Clinical Applications of Oculomotor Models," in *Topics in Neuro-Ophthalmology*, ed. H. S. Thompson. Baltimore: Williams & Wilkins, 1979b, 266–285.

PROBLEMS

3-1 Sinusoids of varying frequencies were applied to a system and the following results were measured. Construct the Bode diagram and estimate the transfer function.

ω (rad/sec)	1.0	2.0	4.0	8.0	10	20	40	80
Magnitude ratio	10	10.4	8.1	5.1	4.6	2.4	1.6	1.2
Phase lag (deg)	12	25	45	73	83	122	173	218

3-2 Construct the magnitude portion of the Bode diagrams for the following transfer functions:

$$\frac{Y(s)}{U(s)} = \frac{0.1(s+10)^2}{s(s+100)} \qquad \frac{Y(s)}{U(s)} = \frac{10(1+s/5)}{s(1+s/50)}$$

3-3 Construct the magnitude and phase Bode diagrams for

$$\frac{Y(s)}{U(s)} = \frac{s}{(1+s/10)(1+s/100)^2}$$

$$\frac{Y(s)}{U(s)} = \frac{5(1+s)(1+s/10)}{(1+s/0.1)(1+s/100)}$$

3-4 The Bode diagram for a system is described as follows; when plotted on log-log coordinates,

(1) The slope for low frequencies is -1.

(2) The slope changes to zero at $\omega = 1$ rad/sec, where the magnitude ratio is $+2$ log units.

(3) Finally, the slope becomes -1, at $\omega = 100$ rad/sec.

(a) Sketch the Bode diagram.

(b) Estimate the transfer function represented by these data.

3-5 The transfer function, $M(s) = V_o/V_{\text{in}}$, for the phase-lead circuit of Fig. 3-7 is most conveniently written as $M(s) = V_o/V_{\text{in}} = K(\alpha Ts + 1)/(Ts + 1)$. Calculate the impulse response and the step response. Find the relationships between α, T, and K, and R_1, R_2, and C. Sketch the Bode diagram.

3-6 Sketch Bode diagrams (both magnitude and phase) for a system that has the pole–zero plot of Fig. P3-6. Comment about such a system.

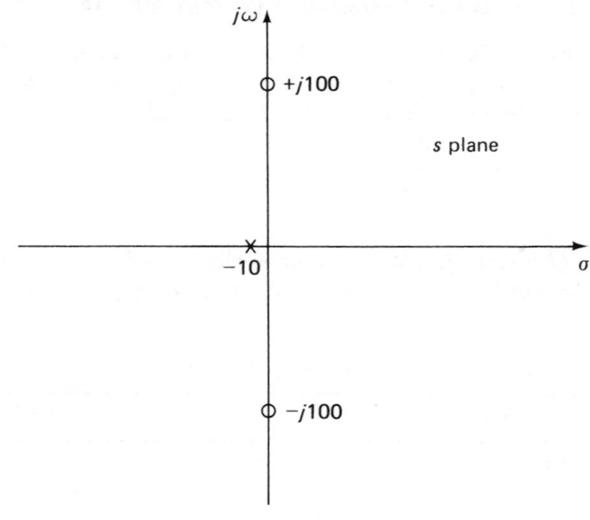

Figure P3-6

3-7 The vestibular system detects translational and rotational movements of the skull and uses this information for postural control and also to control the position of the eyes. The semicircular canals detect rotational accelerations. When the skull and the canal accelerate as shown in Fig. P3-7, the fluid initially remains stationary, so that a relative velocity builds up between the canal and the fluid. This velocity produces a viscous frictional torque which accelerates the fluid in space. Initially, when the fluid is still stationary, the fluid deflects the cupula. This deflection of the cupula produces a bending of hair cells and a proportional neural discharge. The input-output transfer function of this system has been modeled with the following equation:

$$\frac{\phi}{\ddot{\theta}_H} = \frac{k_c}{(10s + 1)(s/200 + 1)}$$

Use this equation to explain how the semicircular canals can fail to provide drive signals for compensatory eye movements under the following conditions. The subject undergoes a constant angular acceleration $\ddot{\theta}_H$(4 rad/sec^2) until 10 rad/sec is reached. Then $\dot{\theta}_H$ remains constant until $t = 40$ sec, after which it is decreased rapidly to zero with constant deceleration of 10 rad/sec^2. Sketch the cupula's time response and the eye's approximate response pattern. The subject is dizzy for a considerable period after such an experience. What happens to the eyes during the dizzy period? Why are the canals said to have failed?

Ballet dancers (and sometimes 1-year-olds) avoid this problem partly by adaptation and partly by using a particular technique of head movement. What is the technique? Analyze corresponding cupula deflections to explain how the problem is overcome. (Based on Milsum 1966.)

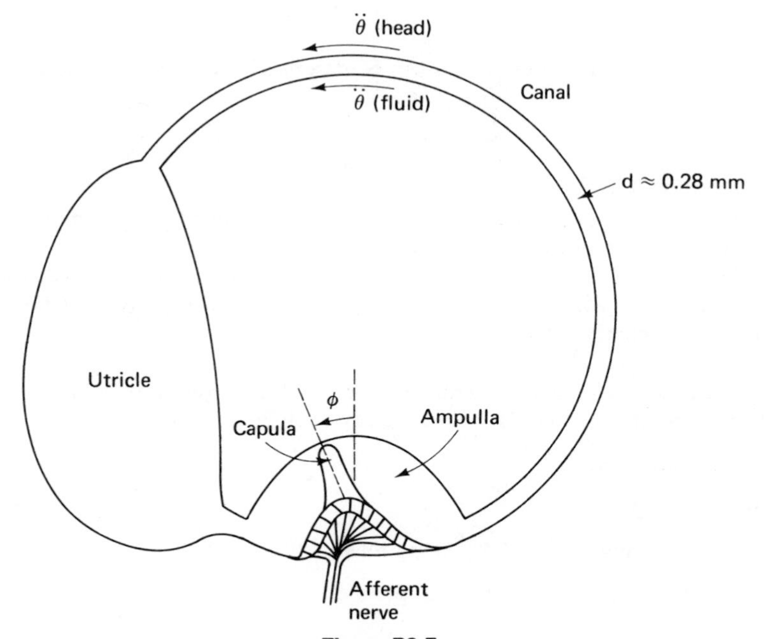

Figure P3-7

3-8 Plot the impulse response for

$$\frac{Y(s)}{U(s)} = \frac{1}{(\tau_1 s + 1)(\tau_2 s + 1)}$$

where $2\tau_1 = \tau_2$, by drawing the two resulting exponentials and subtracting them graphically.

3-9 The period of oscillation of a pendulum clock is given by $P = 2\pi\sqrt{l/g}$, where l is the length of the pendulum and g is the gravitational acceleration, 9.8 m/sec^2. From this equation it can be seen that a 1-m pendulum will have a 2-sec period. The clock keeps good time at a temperature T_0 where the length is l_0. However, the length varies according to $l = l_0(1 + k\,\Delta T)$ where $k = 2 \times 10^{-5}/°\text{C}$ for a brass rod.

(a) Determine the absolute sensitivity function of P with respect to ΔT, S_T^P.

(b) One convenient use of a sensitivity function is to calculate the parameter-induced error of the system, defined as

$$\Delta F = \sum_{j=1}^{r} S_j \Delta\alpha_j$$

This formula will indicate how far the system function will deviate due to the summation of many parameter variations. Use this formula to calculate how many seconds per day the clock will lose if the temperature rises 10°C. (Based on Frank 1978.)

3-10 A linear second-order system can be described with the following equation:

$$M(s) = \frac{\omega_n^2}{s^2 + 2\zeta\omega_n s + \omega_n^2}$$

Use your imagination to decide whether an overdamped or an underdamped system would be more sensitive to variations in the parameter ζ.

Calculate the sensitivity function for $M(s)$ with respect to ζ for several values of s (e.g., $\omega_n/10$, $\omega_n/5$, $\omega_n/2$, ω_n, $2\omega_n$, $5\omega_n$, $10\omega_n$). Plot this sensitivity function as a function of frequency for $\zeta = 1$ and for $\zeta = 0.1$.

3-11 Let F be the force applied and x be the distance from some reference point. Find the transfer function, F/x, and sketch the Bode diagram for:

(a) A spring, K.

(b) A dashpot, B.

(c) Two springs, K_1 and K_2, in parallel.

(d) Two dashpots, B_1 and B_2, in parallel.

(e) Two springs, K_1 and K_2, in series.

(f) Two dashpots, B_1 and B_2, in series.

(g) A spring in parallel with a dashpot.

(h) A spring in series with a dashpot.

(i) The model of unexcited tissue shown in Fig. P3-11.

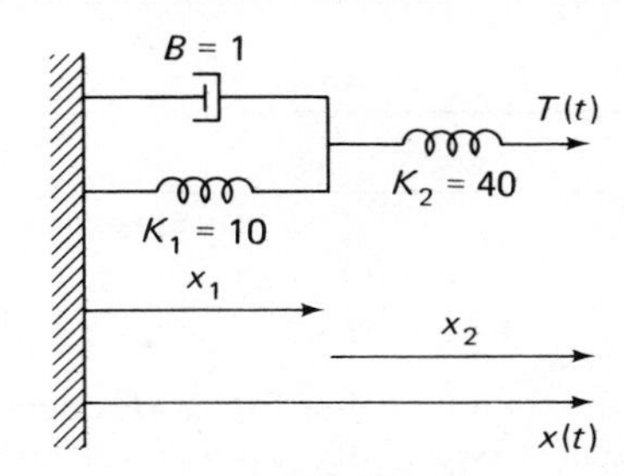

Figure P3-11

3-12 Westheimer's model for the extraocular plant was a linear second-order system

$$\frac{\theta(s)}{F(s)} = \frac{\omega_n^2}{s^2 + 2\zeta\omega_n s + \omega_n^2}$$

with $\omega_n = 120$ rad/sec, $\zeta = 0.707$, and $F(s)$ a step input. For this model sketch the duration and peak velocity as functions of saccadic magnitude. Explain the shapes of your curves. Indicate significant values where possible. Compare these curves to the main-sequence curves of Fig. 3-44.

3-13 Given a linear second-order system with

$$\frac{Y(s)}{R(s)} = \frac{\omega_n^2}{s^2 + 2\zeta\omega_n s + \omega_n^2}$$

with $\omega_n = 100$ rad/sec and $\zeta = 0.707$, calculate and sketch the output for the two inputs shown in Fig. P3-13.

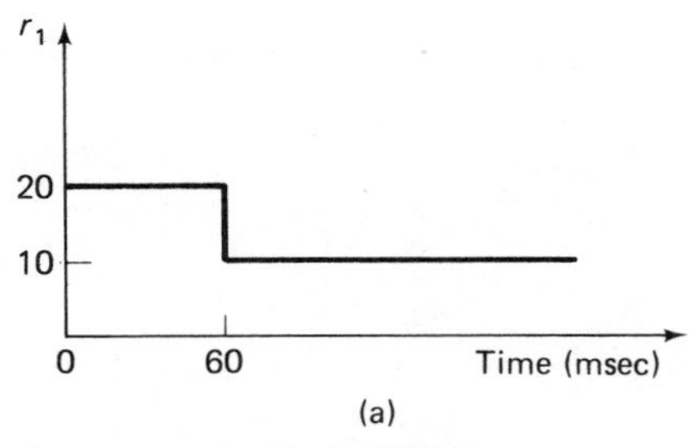

Figure P3-13

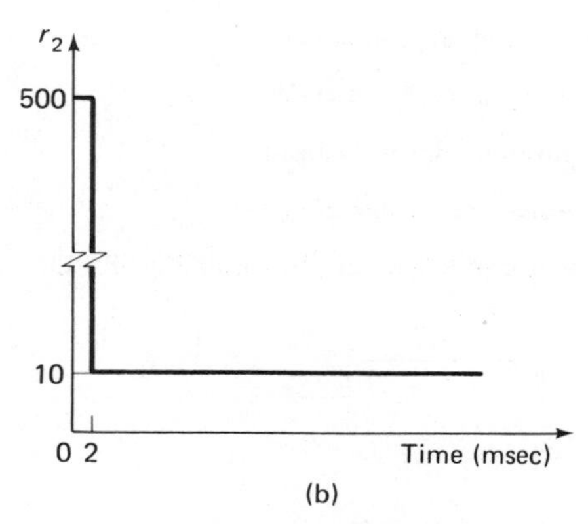

Figure P3-13

3-14 For the reciprocal innervation model of Fig. 3-41, double the value of AG_b, as defined in Fig. 3-38. Sketch a typical 10° saccade for this altered model and explain in detail all changes. How would this change of parameter affect the main-sequence diagrams of Fig. 3-22a?

3-15 The time domain step response for an overdamped linear second-order system is given in Table 3-2. Derive this result.

4

CLOSED-LOOP SYSTEMS

The eye-movement control model that we have just studied is for an open-loop system. This system is just one part of a larger closed-loop system which works in the following way. If a visual target moves 10° and the subject decides to track the target, a command for a 10° saccade is sent to the eye muscles. The eye will then move approximately 10°. The subject again studies the target and if there is an error, another command will be sent, moving the eye closer to the target. This monitoring of eye position and subsequent use of this information to make corrective movements is what makes the system a closed-loop system. Figure 4-1 shows the generalized form of a closed-loop or feedback-control system.

The reference input, $r(t)$, is the target position and the system output, $y(t)$, is the actual eye position. The output is measured and perhaps modified by $h(t)$, which represents the dynamic properties of the sensors and transducers. This signal is subtracted from the input to give the error, $e(t)$. The error is processed by the controller $g_c(t)$, the brain, to produce the controller signal, which is sent to the plant. The plant, $g_p(t)$, is the open-loop model that we studied in Chap. 3.

Manipulating frequency-domain variables is easier than time-domain variables, as we demonstrated in Chap. 3; therefore, the analysis of this

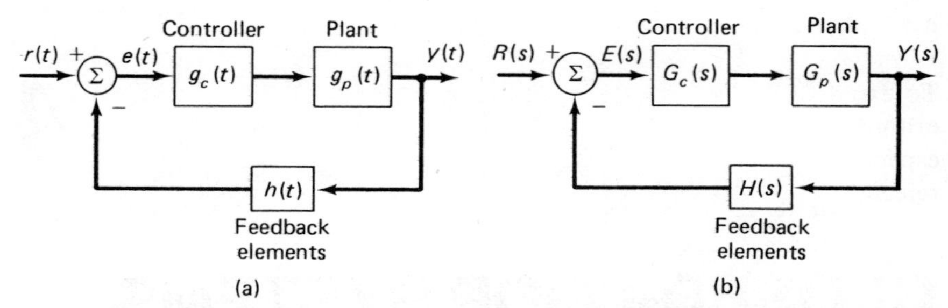

Figure 4-1 (a) Time-domain and (b) frequency-domain representations of a closed-loop or feedback-control system.

closed-loop system will be done with the frequency-domain variables indicated in Fig. 4-1b.

The closed-loop transfer function is defined as the ratio $Y(s)/R(s)$, with all initial conditions zero. To derive this transfer function, we first note that the system output is

$$Y(s) = E(s)G_c(s)G_p(s) \tag{4-1}$$

The system error is given by

$$E(s) = R(s) - Y(s)H(s) \tag{4-2}$$

If we substitute Eq. (4-2) into Eq. (4-1), we can eliminate the error transform and obtain

$$Y(s) = [R(s) - Y(s)H(s)]G_c(s)G_p(s) \tag{4-3}$$

Now, collecting terms, we can obtain the desired closed-loop transfer function $M(s)$:

$$M(s) = \frac{Y(s)}{R(s)} = \frac{G_c(s)G_p(s)}{1 + G_c(s)G_p(s)H(s)} \tag{4-4}$$

Equation (4-4) is the basic expression for the study of linear-control-system behavior. Often, the subscripts and the independent variables are omitted to yield the more terse form

$$\frac{Y}{R} = \frac{G}{1 + GH} \tag{4-5}$$

4-1 WHY USE CLOSED-LOOP SYSTEMS?

To explain why closed-loop systems are so common, let us quantitatively examine the differences between open- and closed-loop control systems, especially noting the sensitivity to plant-parameter variations, disturbance rejection, speed of response, and stability.

4-1-1 Reduction of Sensitivity to Plant-Parameter Variations

In order to compare open- and closed-loop systems, consider the systems shown in Fig. 4-2. The two diagrams in this figure represent a plant or controlled element with negligible frequency dependence, which can be approximated by $G_p(s) = 1$. Assume for the moment that the disturbance $D(s)$ equals zero. Let us design a controller such that the output $Y(s)$ will be 10 times as large as the input. That is, we desire

$$Y(s) = 10R(s)$$

The design of the open-loop controller for this problem is extremely simple. In the absence of disturbance, the transfer function is

$$M(s) = \frac{Y(s)}{R(s)} = G_{co}(s)G_p(s) \tag{4-6}$$

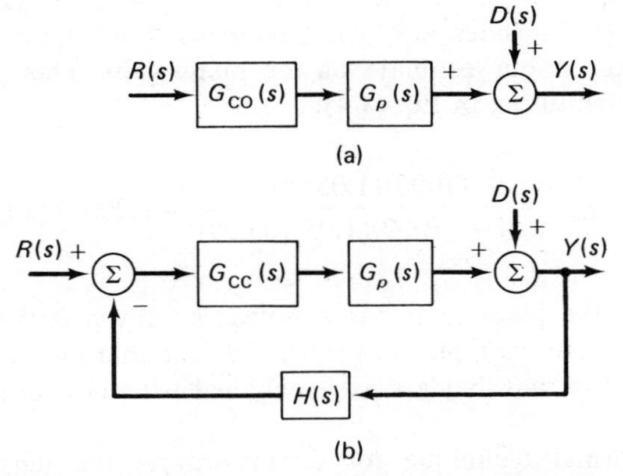

Figure 4-2 (a) Open-loop and (b) closed-loop control systems with the same input-output transfer function.

and the output can be written as

$$Y(s) = G_{co}(s)R(s)$$

Therefore, the controller gain for the open-loop system, G_{co}, must equal 10. Consider now the error in the output $Y(s)$ that would occur if our knowledge of the plant were inaccurate. Let us say that there is a 5% error in our measurement of the plant characteristics, so that, in fact, $G_p(s) = 1.05$. Then the output will be given by

$$Y(s)_{\text{new}} = (10)(1.05)R(s) = 10.5R(s) = 1.05Y(s)_{\text{old}}$$

In other words, the 5 % error in our description of the plant has resulted in a 5% error in the magnitude of the output.

Now let us use the feedback configuration of Fig. 4-2b and design a controller that will make the output 10 times as large as the input. If we again begin by assuming the disturbance to be absent, using Eq. (4-4) and substituting $G_p(s) = 1$, we obtain the requirement

$$\frac{G_{cc}(s)}{1 + G_{cc}(s)H(s)} = 10 \tag{4-7}$$

It can be seen that the feedback configuration has given us an additional degree of freedom, since we can specify both the feedback element and the controller. As a possible solution, let us choose the controller for the closed-loop system, $G_{cc}(s)$, to be 1000 and $H(s) = 0.099$. These values do satisfy Eq. (4-7). Consider now what happens if we again assume a 5% change or error in our estimate of the plant gain. Then the output is obtained by substituting in Eq. (4-4):

$$Y(s)_{\text{new}} = \frac{(1000)(1.05)R(s)}{1 + (1000)(1.05)(0.099)} = 1.0005Y(s)_{\text{old}}$$

A 5% error in the plant gain has resulted in only a 0.05% error in the output. This simple example illustrates the fact that closed-loop systems can be designed so that they are extremely insensitive to variations in plant parameters.

A more formal technique for demonstrating the sensitivity of the system transfer function to a parameter variation is to use the sensitivity function. We are looking for the percentage variation in the transfer function as a result of a percentage variation in some parameter; hence, we want the relative sensitivity function.

$$\bar{S}_{G_p}^{M} = \frac{\%\,\text{change in } M}{\%\,\text{change in } G_p}\bigg|_{G_{po}} = \frac{\partial M/M}{\partial G_p/G_p}\bigg|_{G_{po}} = \frac{G_{po}}{M}\frac{\partial M}{\partial G_p}\bigg|_{G_{po}}$$

Applying this to the open-loop system transfer function equation (4-6) and evaluating at $G_{po} = 1$ yields

$$\bar{S}_{G_p}^{M} = \frac{G_{co}(s)}{G_{co}(s)} = 1$$

Again it confirms that the system transfer function changes by the same percentage as the plant gain.

If we apply the relative sensitivity function to closed-loop system transfer function equation (4-4) and evaluate it at $G_{po} = 1$, we get

$$\bar{S}_{G_p}^{M} = \left|\frac{1 + G_{cc}H}{G_{cc}}\,\frac{(1 + G_{cc}G_p)G_{cc} - G_{cc}^2 G_p H}{(1 + G_{cc}G_p H)^2}\right|_{G_p = 1}$$

$$\bar{S}_{Gp}^{M} = \frac{1}{1 + G_{cc}(s)H(s)} = \frac{1}{100}$$

This shows that if the plant gain changes by 1%, the system transfer function will change by 0.01%.

We might note, however, that we have made the system sensitive to changes in the parameters of the feedback element H.

$$\bar{S}_{H}^{M} = \frac{-GH}{1 + GH}$$

If $GH \gg 1$, this reduces to

$$\bar{S}_{H}^{M} = -1$$

In classical control engineering systems this sensitivity to feedback element parameters was not important, because they usually used unity feedback and a proportional plus integral plus derivative (PID) controller in the forward path. However, in systems using the modern control theory technique of linear state variable feedback the controller resides in the feedback pathways. The increased sensitivity to variations in controller parameters may be the reason why modern state variable feedback controllers have not replaced classical PID controllers in continuous systems.

In biological systems the feedback elements are parts of the nervous system, therefore their parameters are less likely to change than plant parameters; so, controller elements are often included in the feedback pathway.

In this example, only algebraic quantities were used and no frequency dependence was present. Nevertheless, the same type of relationship exists in the presence of dynamic elements. It now becomes clear why biological control systems are generally feedback systems. The closed-loop nature of control systems makes it possible to maintain important physiological variables within very close tolerances, even when there are wide variations in the physiological parameters of the organisms.

4-1-2 Reduction of Sensitivity to Output Disturbances

A second important advantage that closed-loop control systems have over open-loop control systems is their ability to tolerate large disturbances without producing significant changes in the controlled variables. Let us examine Fig. 4-2, and assume that a disturbance or noise is added at the final summing junction.

In the open-loop case, shown in Fig. 4-2a, the output is

$$Y(s) = G_{co}(s)G_p(s)R(s) + D(s)$$

Employing the numerical values of the previous example yields

$$Y(s) = 10R(s) + D(s)$$

which illustrates the fact that the disturbance simply adds linearly to the output produced by the controller–plant combination. In fact, with additive disturbances of the type shown here, there is no way of designing the controller to affect the disturbance of the system output.

Closed-loop systems can compensate for output disturbances. For example, the mammalian thermoregulatory system is a closed-loop control system that maintains a constant temperature in the body core. In terms of Fig. 4-2b, $H(s)$ represents the temperature-monitoring elements, $G_{cc}(s)$ represents the central nervous system, and $G_p(s)$ models the body systems involved in heat production, transportation, and elimination. $D(s)$ represents disturbances in the output temperature which could be caused by, for example, walking from a warm office out into the snow. Our first step in evaluating the effects of this output disturbance will be to assume that the disturbance is applied to the input summing junction. The output then becomes

$$Y(s) = \frac{G_{cc}(s)G_p(s)}{1 + G_{cc}(s)G_p(s)H(s)}[R(s) + D(s)]$$

This system can be transformed into an equivalent system where the disturbance enters a summing junction placed between the controller and the plant. The output of this system is

$$Y(s) = \frac{G_{cc}(s)G_p(s)}{1 + G_{cc}(s)G_p(s)H(s)}R(s) + \frac{G_p(s)}{1 + G_{cc}(s)G_p(s)H(s)}D(s)$$

This summing junction can now be moved again so that the disturbance is added after the plant, as shown in Fig. 4-2b. The resulting output equation is

$$Y(s) = \frac{G_{cc}(s)G_p(s)}{1 + G_{cc}(s)G_p(s)H(s)}R(s) + \frac{D(s)}{1 + G_{cc}(s)G_p(s)H(s)} \tag{4-8}$$

Using the numbers that were employed in the design of the closed-loop system in the preceding section, we obtain

$$Y(s) = 10R(s) + \frac{D(s)}{100}$$

This equation illustrates the fact that the feedback has reduced the effect of the disturbance on the system response by a factor of 100.

To look at the dynamic characteristics of this disturbance rejection, we will let some of our elements have dynamics. Let

$$G_p(s) = \frac{1}{1 + \tau s}$$

with $\tau = 100$ msec. Now from Eq. (4-8),

$$\begin{aligned} Y(s) &= \frac{1000/(1 + \tau s)}{1 + \dfrac{99}{1 + \tau s}}R(s) + \frac{D(s)}{1 + \dfrac{99}{1 + \tau s}} \\ &= \frac{1000}{\tau s + 100}R(s) + \frac{1 + \tau s}{\tau s + 100}D(s) \end{aligned}$$

For the present, we are not interested in the system response to input changes, only its response to disturbances. So, set $R(s) = 0$, let $d(t)$ be a unit impulse, and substitute $\tau = 100$ msec.

$$Y(s) = \frac{s + 10}{s + 1000} = \frac{s}{s + 1000} + \frac{10}{s + 1000}$$

and

$$y(t) = \frac{d}{dt} e^{-1000t} + 10e^{-1000t}$$

$$y(t) = -990e^{-1000t}$$

If the disturbance changes the output, the feedback circuit senses this error, and the controller, G_{cc}, changes its output to restore the output to its desired state. This regulatory action could be improved if the controller did not have to wait for an actual output error to occur before it acted, that is, if it could have some predictor of the future output. One way of approaching this goal is to feed back not only the output but also its derivative. Let us change the feedback element $H(s)$ so that it transmits information about both the output and its derivative.

$$H(s) = 0.099(s + 1)$$

From Eq. (4-8),

$$y(s) = \frac{1000/(1 + \tau s)}{1 + \dfrac{(1000)(0.099)(s + 1)}{\tau s + 1}} R(s) + \frac{D(s)}{1 + \dfrac{99(s + 1)}{\tau s + 1}}$$

$$= \frac{1000}{\tau s + 1 + 99(s + 1)} R(s) + \frac{\tau s + 1}{(\tau s + 1) + 99(s + 1)} D(s)$$

Set $R(s) = 0$, let $d(t)$ be a unit impulse, and substitute $\tau = 100$ msec.

$$Y(s) = \frac{1}{991} \left[\frac{s + 10}{s + \dfrac{1000}{991}} \right]$$

$$y(t) = 9.05 \times 10^{-3} e^{-1.01t}$$

The size of the output variation due to the disturbance has been reduced by a factor of 10^5. (However, the speed of the response is slower. Although we will not discuss this matter, derivative feedback also makes a system more stable.)

Thus closed-loop systems have much more resistance to output disturbances than open-loop systems. The addition of derivative feedback greatly improves the system's ability to reject disturbances.

Once again, it is evident from this point of view why biological systems have in general evolved as closed-loop systems; if this were not the case, living organisms would be unable to maintain any constancy of their internal environment in the presence of large variations in their external environment. It is easy to postulate that, if closed-loop feedback-control systems had not been developed in the evolutionary process, living organisms would only exist in relatively constant environments, such as seawater.

4-2 SPEED OF RESPONSE

A closed-loop system can be designed to have a faster time response than an open-loop system even though they are performing the same task. Let us once again compare the open- and closed-loop systems shown in Fig. 4-2. Let us use the same system components values as before, namely,

$$G_{co} = 10$$
$$G_{cc} = 1000$$
$$H = 0.099$$
$$G_p(s) = \frac{1}{1 + \tau s}$$

and

$$\tau = 100 \text{ msec}$$

Now calculate the system's response to a unit-step input,

$$R(s) = \frac{1}{s}$$

For the open-loop system,

$$Y(s) = \frac{10}{s(1 + \tau s)} = \frac{10/\tau}{s(1/\tau + s)} = \frac{100}{s(s + 10)}$$

and by taking the inverse Laplace transform, we find that

$$y(t) = 10(1 - e^{-10t}) \tag{4-9}$$

Now the closed-loop transfer function, from Eq. (4-4), is

$$\frac{Y(s)}{R(s)} = \frac{1000/(1+\tau s)}{1+\dfrac{(1000)(0.099)}{(1+\tau s)}}$$

$$= \frac{1000}{\tau s + 100}$$

And we can find the step response of the closed-loop system to be

$$Y(s) = \frac{10}{s\left(1+\dfrac{\tau s}{100}\right)} = \frac{10^4}{s(s+1000)}$$

The inverse Laplace transform of this is

$$y(t) = 10(1 - e^{-1000t}) \qquad (4\text{-}10)$$

Clearly, the response of the closed-loop system is 100 times faster than that of the open-loop system. Another way of stating this is that the closed-loop system has a greater bandwidth than the open-loop system.

This speed advantage of the closed-loop system will be diminished if the feedback loop has dynamics. For example, if

$$H(s) = 0.099(1 + as) \qquad (4\text{-}10a)$$

where for convenience we will let $a = 1$, then

$$H(s) = 0.099(1 + s)$$

Then, by Eq. (4-4), the transfer function becomes

$$\frac{Y(s)}{R(s)} = \frac{1000/(1+\tau s)}{1+\dfrac{(1000)(0.099)(1+s)}{(1+\tau s)}}$$

$$= \frac{10.09}{s + 1.009} \approx \frac{10}{1+s}$$

The unit-step response becomes

$$Y(s) = \frac{10}{s(s+1)}$$

Taking the inverse Laplace transform yields

$$y(t) = 10(1 - e^{-t})$$

This is clearly slower than the step response of the closed-loop system without derivative feedback. In fact, this response is even slower than the step response of the open-loop system. Such a closed-loop system will be slower than an open-loop system if the feedback coefficient, a, in Eq. (4-10a) is larger than the plant time constant, τ.

Feedback-control systems are generally faster and are less sensitive to parameter variations and environmental disturbances than open-loop systems are. However, a comparison of the controllers $G_{co}(s)$ and $G_{cc}(s)$ used in our design illustrates the fact that the closed-loop system accomplishes this by requiring a much higher gain controller. In addition, the closed-loop system requires feedback sensors and some form of comparison process for operation. Finally, as an additional price paid for the advantages of closed-loop control, the stability of such systems cannot be taken for granted, even when the plant and controller individually display well-behaved and stable characteristics. The question of stability will be investigated in the next section.

4-3 STABILITY

An open-loop system with no input cannot oscillate unstably; but if such a system is included in a feedback network, it can become unstable. For example, a public address system will not normally oscillate, but if the microphone is pointed toward the speakers and the gain is large enough, the system will oscillate.

A system is stable if the output is bounded for any bounded input. All of the poles of the input-output transfer function are in the left half of the s-plane for a stable system. This fairly strict definition of stability is sometimes called *asymptotic stability*. In such systems all oscillations eventually die out.

Systems with constant-amplitude oscillations (poles on the $j\omega$ axis) are called marginally stable systems. Systems with poles in the right half of the s-plane are unstable.

The closed-loop system of Fig. 4-2 has the transfer function

$$\frac{Y(s)}{R(s)} = \frac{G(s)}{1 + G(s)H(s)}$$

where the plant and controller gains have been lumped into one. To study the stability of the system, we do not have to study the entire transfer function; we must only look at the denominator. The system will be unstable when the denominator is zero: that is, when

$$1 + G(s)H(s) = 0$$

In general, this denominator will be a polynomial in s. All we must do to determine its stability is to factor this polynomial. In general, the factors, or roots, of this polynomial will change as the gain of the system changes. A plot of the location of these roots as a function of gain is called a *root-locus diagram*.

4-3-1 Root-Locus Plots

In Chap. 3 we used pole–zero diagrams to study the behavior of dynamic systems. If we apply this technique to the plant and controller of Fig. 4-3, we will find that there are poles at the origin, at $s = -1$ and at $s = -2$. They are the poles of the open-loop system, and they are shown with x's in Fig. 4-3b. The poles of the closed-loop system will be different and their locations will depend upon the gain K.

In Chap. 3 we stated that a system with poles in the right half of the s-plane would be unstable, but we gave no examples. Let us now show how changing the gain, K, of the system shown in Fig. 4-3 can make the system

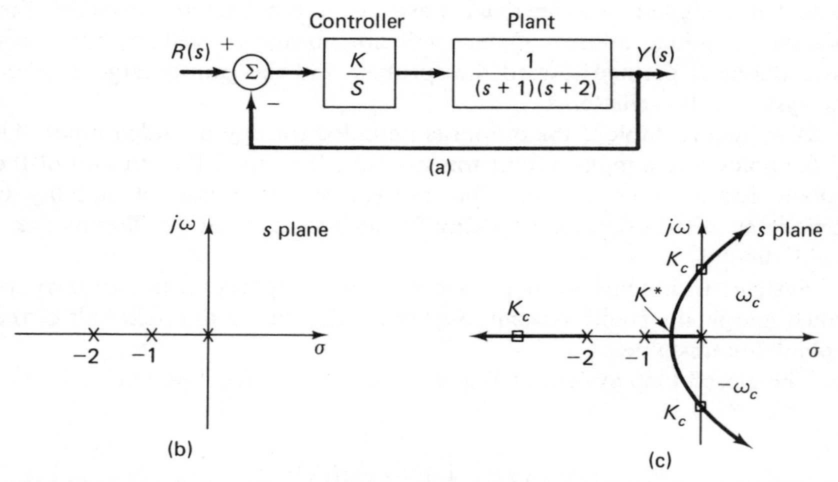

Figure 4-3 (a) A closed-loop control system, (b), its open-loop roots, and (c) its root-locus diagram.

unstable. The transfer function for this system is

$$\frac{Y(s)}{R(s)} = \frac{\dfrac{K}{s(s+1)(s+2)}}{1 + \dfrac{K}{s(s+1)(s+2)}}$$

$$= \frac{K}{s(s+1)(s+2)+K} \tag{4-11}$$

It is only the denominator that is of concern to us now. The roots of the system are the values of s that will make the denominator equal zero. That is,

$$s(s+1)(s+2)+K=0 \tag{4-12}$$

This is a cubic equation in s, and the location of the roots can be found analytically. If they are plotted as a function, K, we obtain the diagram shown in Fig. 4-3c, which shows the location in the s-plane of these roots as a function of K. Note that, at any value of gain, there are three roots. The arrows indicate the movement of the roots as the gain increases. For $K=0$, the closed-loop poles are located precisely at the open-loop poles 0, -1, and -2. For low values of controller gain the closed-loop roots are all real and negative. For example, if $K=0.3$, the poles are at -0.21, -0.66, and -2.13, and the system's response to a disturbance will consist of a sum of three decaying exponentials: the response will be stable. At a certain gain indicated by K^* in the figure, two of the roots of Eq. (4-12) become complex; for example, if $K=1$, the poles are at -2.32 and at $-0.38 \pm j0.56$.

The system response is now made up of a combination of two portions: (1) a decaying exponential that arises from the root located on the negative real axis, and (2) an oscillatory response multiplied by an exponential that arises from the two complex roots. For values of controller gain between K^* and K_c, at which point the root-locus diagram crosses the imaginary axis, this oscillatory response will be damped and will gradually decay toward zero. Hence, a perturbation of the system with this value of controller gain will still result in stable operation. If the controller gain equals 6, the poles are at -3 and $\pm j\sqrt{2}$. If this system is perturbed, the response will contain a rapidly decaying exponential component, given by the root located on the negative real axis, and a constant-amplitude sinusoidal oscillation, at the frequency $\sqrt{2}$. Hence, with this value of controller gain, our closed-loop system is marginally stable. For all values of controller gain that are greater than 6, the closed-loop poles will be located in the right half of the s-plane and the oscillatory response will have an increasing envelope, resulting in an unstable system.

A root-locus plot (like the one shown in Fig. 4-3c) can be obtained, as we have implied, by factoring the denominator polynomial of the transfer function with varying values of K. However, this factorization is usually difficult. The results just presented were derived with the aid of computer programs. The root-locus technique provides a means of obtaining such root-locus plots without algebraic factoring. In effect, it is a graphical technique for factoring the polynomial. Further details about the technique are given in most control theory textbooks. Root-locus techniques are convenient when the algebraic description of the system is already known: that is, when the location of the open-loop poles and zeros are known. Often this description is not available. We may wish to identify the system and also determine its stability characteristics. Bode diagrams, and Nyquist plots are convenient methods of doing this. For both of these techniques sinusoidal inputs are usually applied and the gain and phase of the output is measured. These data are then plotted according to the appropriate set of rules.

We will not discuss any other techniques for studying stability, such as the Routh–Hurwitz criterion, which is an algebraic method that operates on the coefficients of the denominator polynomial; Jury's method of inners, which is ideal for computer analysis of the A matrix; or the direct method of Lyapunov, which is a mathematical technique that is popular in modern control theory studies where the system is modeled with state variables.

4-3-2 Opening the Loop

A very important tool in stability analyses is the concept of opening a loop. Usually, a biological system can be schematically drawn as a closed-loop system, as shown in Fig. 4-4a, or as an open-loop system, as shown in Fig. 4-4b. Let us consider the closed-loop system shown in Fig. 4-4a. One common technique for studying such a system is to "open the loop" and then study the response of the open loop. We define the open-loop transfer function as the total effect encountered by a signal as it travels around the loop. That is, the open-loop transfer function, G_{OL}, is

$$G_{OL} = G(s)H(s)$$

Note that this is not the input-output transfer function of the system with its loop opened, nor is this the transfer function of the equivalent intact open-loop system shown in Fig. 4-4b.

When we open the loop of a closed-loop system the behavior becomes pathological: it is entirely different from the behavior of a properly

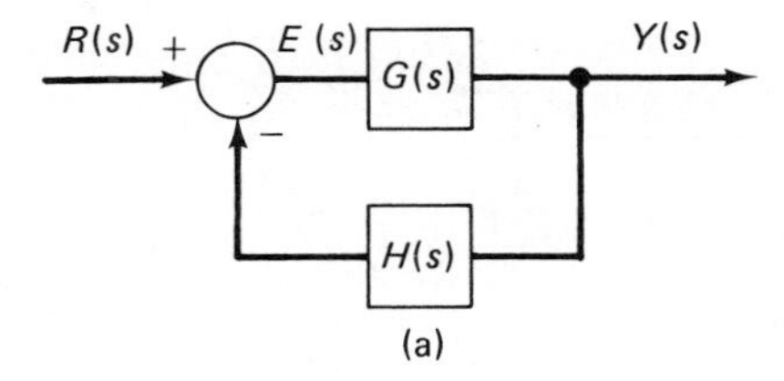

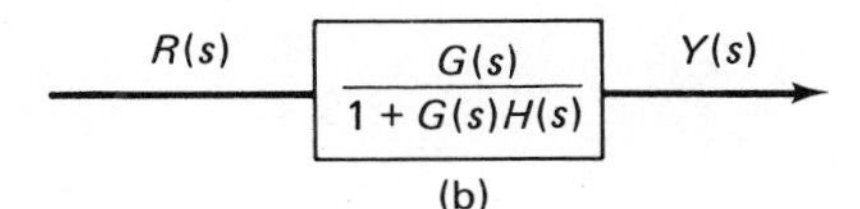

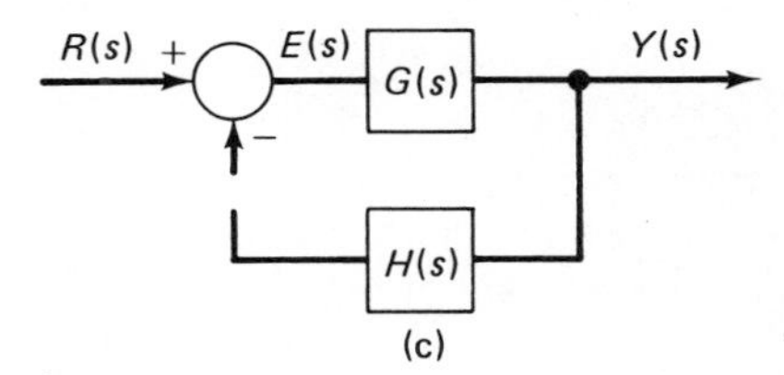

Figure 4-4 (a) A closed-loop control system, (b) an equivalent open-loop representation, and (c) the closed-loop representation with its loop opened. Many analysis techniques require study of the open-looped system of (c).

designed open-loop system. In response to a disturbance, a closed-loop system with its loop opened will vary its output until some nonlinearity limits it. For instance, if $R(s)$ in Fig. 4-4c is a step, and $G(s) = 1/s$, the error will be constant and the output will increase until it reaches saturation.

The open-loop transfer function is simpler than the closed-loop transfer function. It is used for Bode diagrams, Nyquist plots, and other methods of studying the stability of closed-loop systems.

Often the success of a bioengineering study is due to finding some clever way of opening the loop of the system so that these engineering analysis tools can be applied.

4-3-3 Bode Diagrams

Bode diagrams are plots of the input-output amplitude ratio and phase as functions of the logarithm of the frequency. In order to study stability, we plot the Bode diagram for the product of the controller, plant, and feedback transfer functions, $G(s)H(s)$: the open-loop transfer function. Consider, for example, the system shown in Fig. 4-5 and its corresponding Bode diagram. The controller is indicated as having an adjustable gain K. If we examine the Bode diagram, we can see that when $\omega = \sqrt{2}$ the system phase characteristic equals $-180°$. This implies that, at this frequency, the feedback signal (which in this case is equal to the system output) now adds

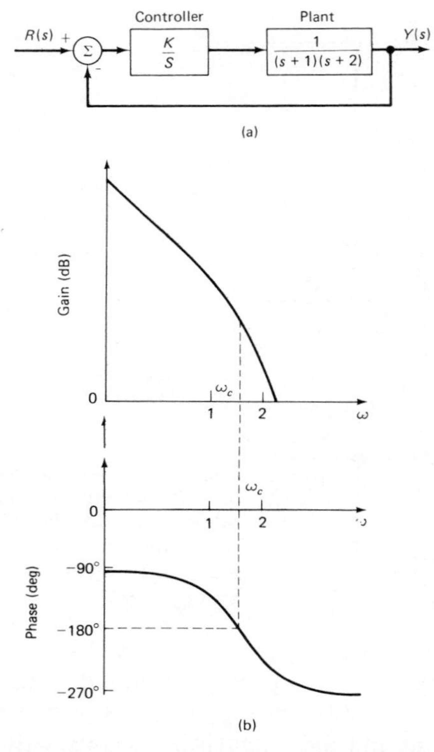

Figure 4-5 (a) A third-order closed-loop control system and (b) its Bode diagram: $KGH = 6/s(s+1)(s+2)$.

to the system input rather than subtracting from it. Hence, at this frequency, our system behaves as a positive feedback system rather than as a negative feedback system. It can be seen that, if the system is now perturbed and if the perturbation contains any energy at the frequency $\sqrt{2}$, the resulting feedback signal at that frequency will add to the input. Whether such positive feedback leads to instability will now depend on whether the combined gain of the controller and plant at that frequency exceeds unity or not. To find the value of K that will make this gain equal unity, substitute $j\omega$ for s and let ω assume the value of $\sqrt{2}$. The product $G(s)H(s)$ becomes

$$\frac{K}{j\omega(j\omega+1)(j\omega+2)} = \frac{K}{6}\angle 180^\circ$$

Therefore, for any value of K greater than or equal to 6, the system is unstable. This is the same result that we derived from the root-locus plot.

4-3-4 Nyquist Plots

A very popular way to study these Bode diagrams is to transform the magnitude and phase information into a polar plot called a *Nyquist plot*. The resulting plot is then examined to see if it circles the -1 point in the s-plane. If it does, the system is unstable.[1]

Because Nyquist plots are similar to Bode diagrams, they, too, are studies of the behavior of the closed-loop system with its loop opened, that is, of $KG(s)H(s)$. Nyquist plots are best drawn from the actual behavior rather than the asymptotic approximation.

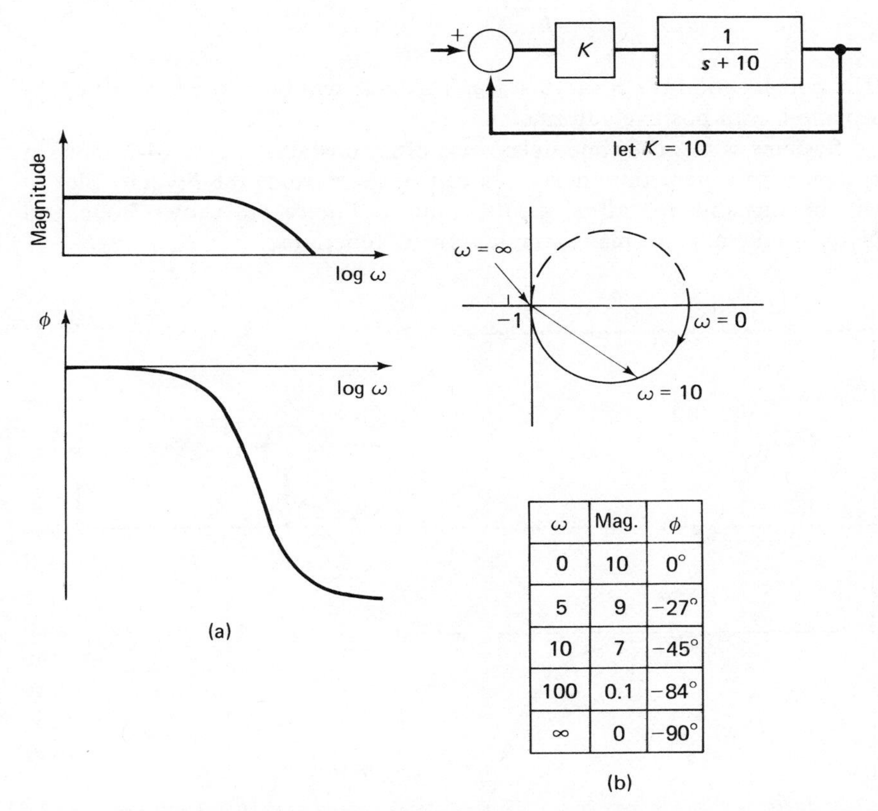

ω	Mag.	ϕ
0	10	0°
5	9	−27°
10	7	−45°
100	0.1	−84°
∞	0	−90°

Figure 4-6 (a) Bode diagram and (b) Nyquist plot for a first-order control system: $KGH = K/(s+10)$.

[1] The rigorous statement of the Nyquist stability criterion is that the number of clockwise encirclements minus the number of counterclockwise encirclements of the point $s = -1 + j0$ by the Nyquist plot of $KG(s)H(s)$ is equal to the number of poles of $Y(s)/R(s)$ minus the number of poles of $KG(s)H(s)$ in the right half-plane.

Figure 4-6 shows a simple first-order system, its Bode diagram, and its Nyquist plot. The Nyquist plot is obtained by picking a value of ω (e.g., $\omega = 10$), finding the corresponding values of the magnitude and phase, 7.07 and $-45°$, and in this case, drawing this vector on the s-plane plot. The procedure is repeated for all values of ω, then the mirror image of this plot is drawn, as shown with the dashed line. This plot does not circle the -1 point, so the system is stable.

For this simple case we should be able to check this by solving for the pole of the closed-loop transfer function.

$$\frac{Y(s)}{R(s)} = \frac{K}{s + 10 + K}$$

The pole location is $s = -(10 + K)$. This pole will be in the left half of the s-plane for all positive values of K.

Systems with pure time delays are often unstable. Figure 4-7 shows a system with a pure time delay. As can be seen from the Nyquist plot, it will be unstable for all K greater than 1. Figure 4-8 shows Bode and Nyquist diagrams for many simple transfer functions.

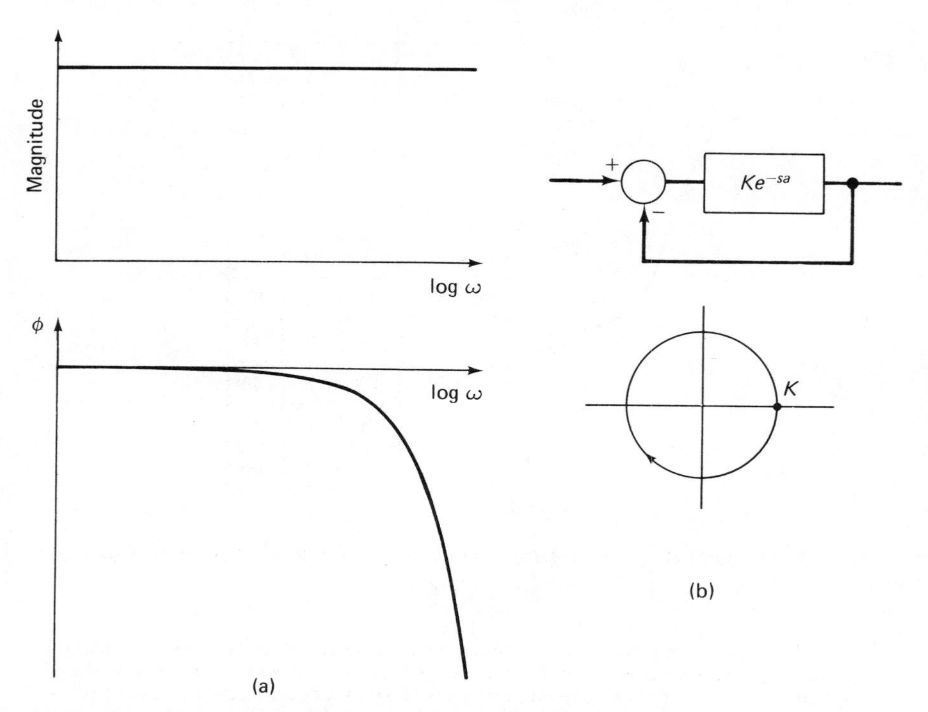

Figure 4-7 (a) Bode diagram and (b) Nyquist plot for a pure time delay: $KGH = K\,e^{-j\omega a}$

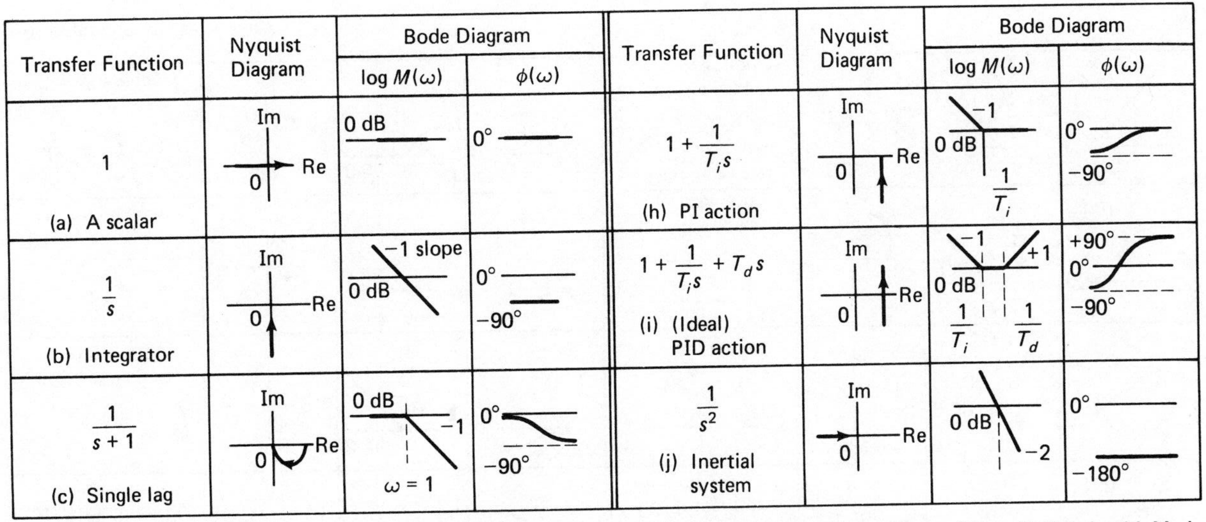

Transfer Function	Nyquist Diagram	Bode Diagram		Transfer Function	Nyquist Diagram	Bode Diagram	
		log $M(\omega)$	$\phi(\omega)$			log $M(\omega)$	$\phi(\omega)$
1 (a) A scalar	Im, Re, 0	0 dB	0°	$1+\frac{1}{T_i s}$ (h) PI action	Im, Re, 0	−1, 0 dB, $\frac{1}{T_i}$	0°, −90°
$\frac{1}{s}$ (b) Integrator	Im, Re, 0	−1 slope, 0 dB	0°, −90°	$1+\frac{1}{T_i s}+T_d s$ (i) (Ideal) PID action	Im, Re, 0	−1, +1, 0 dB, $\frac{1}{T_i}$, $\frac{1}{T_d}$	+90°, 0°, −90°
$\frac{1}{s+1}$ (c) Single lag	Im, Re, 0	0 dB, −1, $\omega = 1$	0°, −90°	$\frac{1}{s^2}$ (j) Inertial system	Im, Re, 0	0 dB, −2	0°, −180°

Figure 4-8 Nyquist plots and Bode diagrams, on next page, for some common transfer functions. (From Y. Takahashi, M. J. Rabins, and D. M. Auslander, *Control and Dynamic Systems*, Addison-Wesley Publishing Co., Inc. Reading, Mass.; 1970, p. 368.)

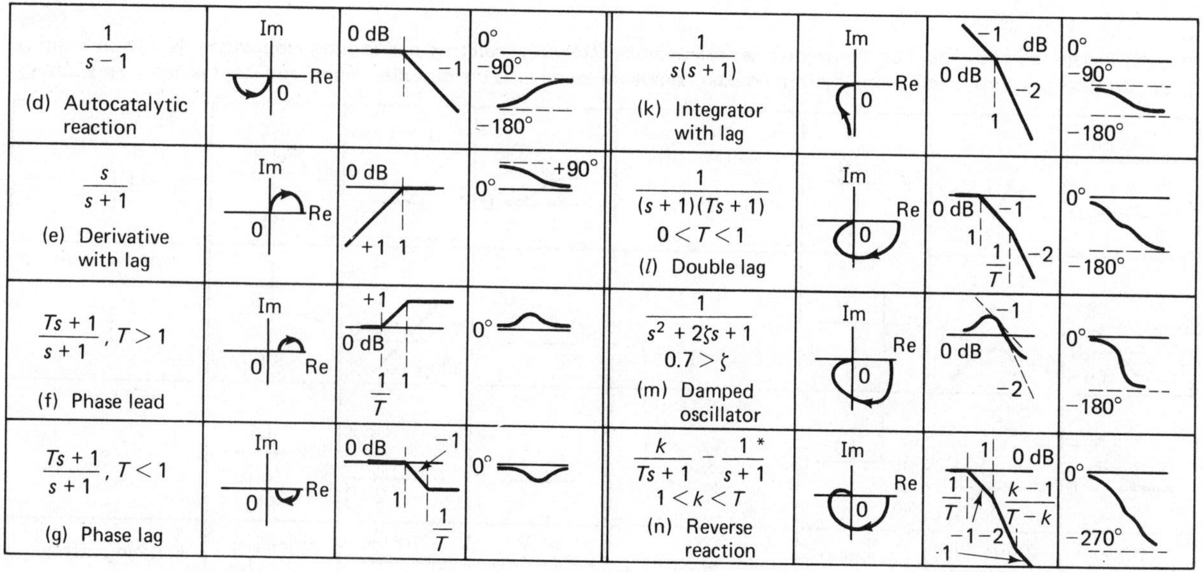

Transfer function	Polar plot	Magnitude	Phase	Transfer function	Polar plot	Magnitude	Phase
$\frac{1}{s-1}$ (d) Autocatalytic reaction	Im, Re, 0	0 dB, −1	0°, −90°, −180°	$\frac{1}{s(s+1)}$ (k) Integrator with lag	Im, Re, 0	−1, dB, 0 dB, −2, 1	0°, −90°, −180°
$\frac{s}{s+1}$ (e) Derivative with lag	Im, Re, 0	0 dB, +1, 1	0°, +90°	$\frac{1}{(s+1)(Ts+1)}$ $0<T<1$ (*l*) Double lag	Im, Re, 0	0 dB, −1, 1, $\frac{1}{T}$, −2	0°, −180°
$\frac{Ts+1}{s+1}$, $T>1$ (f) Phase lead	Im, Re, 0	+1, 0 dB, $\frac{1}{T}$, 1	0°	$\frac{1}{s^2+2\zeta s+1}$ $0.7>\zeta$ (m) Damped oscillator	Im, Re, 0	−1, 0 dB, −2	0°, −180°
$\frac{Ts+1}{s+1}$, $T<1$ (g) Phase lag	Im, Re, 0	0 dB, −1, 1, $\frac{1}{T}$	0°	$\frac{k}{Ts+1}-\frac{1}{s+1}$ * $1<k<T$ (n) Reverse reaction	Im, Re, 0	1, 0 dB, $\frac{1}{T}$, $\frac{k-1}{T-k}$, −1, −2, −1	0°, −270°

*Nonminimum phase system.

Figure 4-8 Continued.

4-3-5 Instability in Nonlinear Systems

The preceding sections have shown the nature of the stability problem for linear control systems. However, most biological systems are nonlinear systems because no variable can increase arbitrarily in magnitude without bounds. In all real systems, all variables are constrained in magnitude by physical limitations. If an amplitude constraint or saturation characteristic is placed on a linear system, the system becomes nonlinear. Because the amplitude will no longer grow without bounds. nonlinear systems are capable of oscillating with a constant amplitude determined by the constraint. The resulting oscillation will not be sinusoidal in shape because of the distortions that arise as a result of saturation. Such oscillations in nonlinear systems are known as limit cycles and are known to occur under certain conditions in biological systems. However, nonlinear systems are hard to study, so most biological systems are approximated with linear models and linear control theory is used to study them.

4-4 THE NEUROMUSCULAR SYSTEM

Most biological control systems are closed-loop systems. To gain an appreciation for such systems, we investigate a small part of the neuromuscular system.

4-4-1 The Stretch Reflex

Neuromuscular systems seem unnecessarily complex for the task they perform. For example, there are 30 major muscles in each leg, even though there are only three major joints to be controlled. We do not wish to study such a complicated, poorly understood system, so our first task is to choose a simplified model to analyze. We first study the biceps muscle, which bends the elbow.

This muscle, with its associated sensory fibers and motoneurons, acts as a feedback-control system. In 1924, Liddel and Sherrington showed that stretching a muscle results in a prompt contraction of that muscle, which tends to restore it to its former length. They showed that this response depended upon afferent signals from the muscle to the spinal cord (the feedback signals) and upon efferent signals that were sent from the spinal cord to the muscle. This reflex is called the *stretch reflex*. This is the reflex physicians check when they tap the knee with a rubber hammer.

For simplicity, Fig. 4-9 shows only a single spindle receptor (in the belly of the muscle), a single motor neuron, mn, and single afferent and

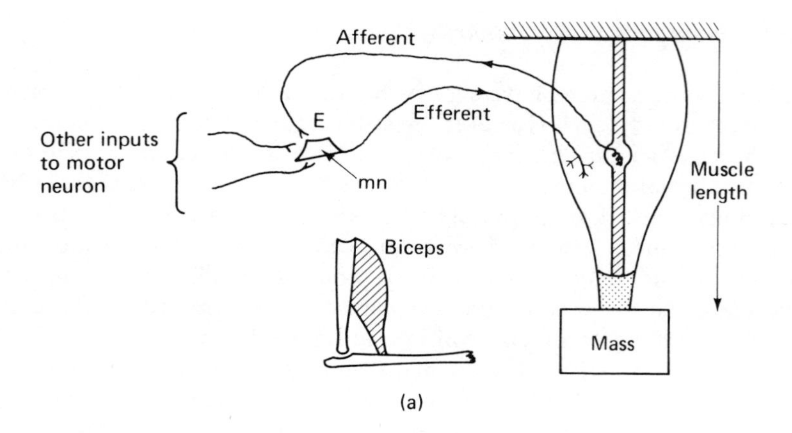

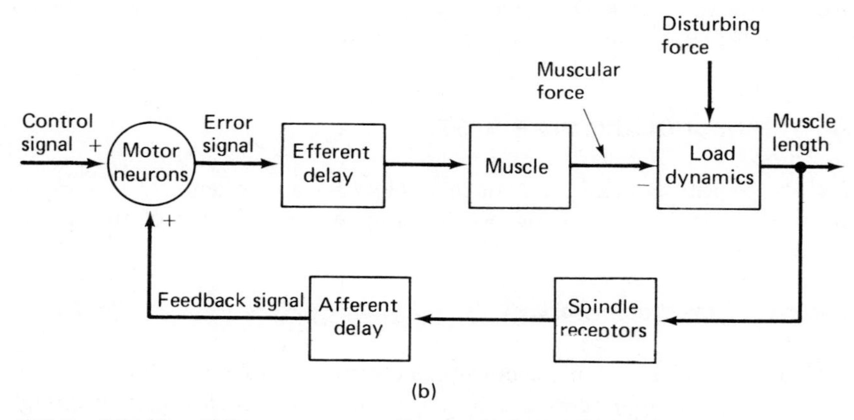

Figure 4-9 Closed-loop neuromuscular control system, showing (a) anatomic connections between physiologic components that participate in stretch reflex and (b) Block diagram of the stretch reflex. (Modified from J. Houk and E. Henneman, "Feedback Control of Muscle: Introductory Concepts," in *Medical Physiology*, 13th ed., V. B. Mountcastle, The C. V. Mosby Company, St Louis, 1974.)

efferent nerve fibers are shown. E stands for excitatory synaptic connection. Inset shows the similarity between the human biceps muscle supporting the forearm and the isolated muscle of the stretch reflex circuit.

The controlled output of Fig. 4-9 is the position of the load. This can be related directly to the length of muscle. It appears that muscle shortening might be a better choice for the output, because the only action a muscle can produce is a contraction. However, it is more convenient to put a negative sign into the muscle dynamics and call the controlled output muscle length. This is the one negative sign that is required in the loop for a negative feedback-control system.

The transducers that measure muscle length are called *muscle spindle organs*. These sensory organs are located within the belly of the muscle and are connected in parallel with the other muscle fibers (Fulton and Pi-Suner 1928), so that they are stretched whenever the muscle is stretched (see Figs. 4-9 and 4-10). These nerve fibers increase their rate of firing when they are stretched. There are two types of nerve fibers, called *primary* and *secondary*; the primary endings are sensitive to velocity as well as length.

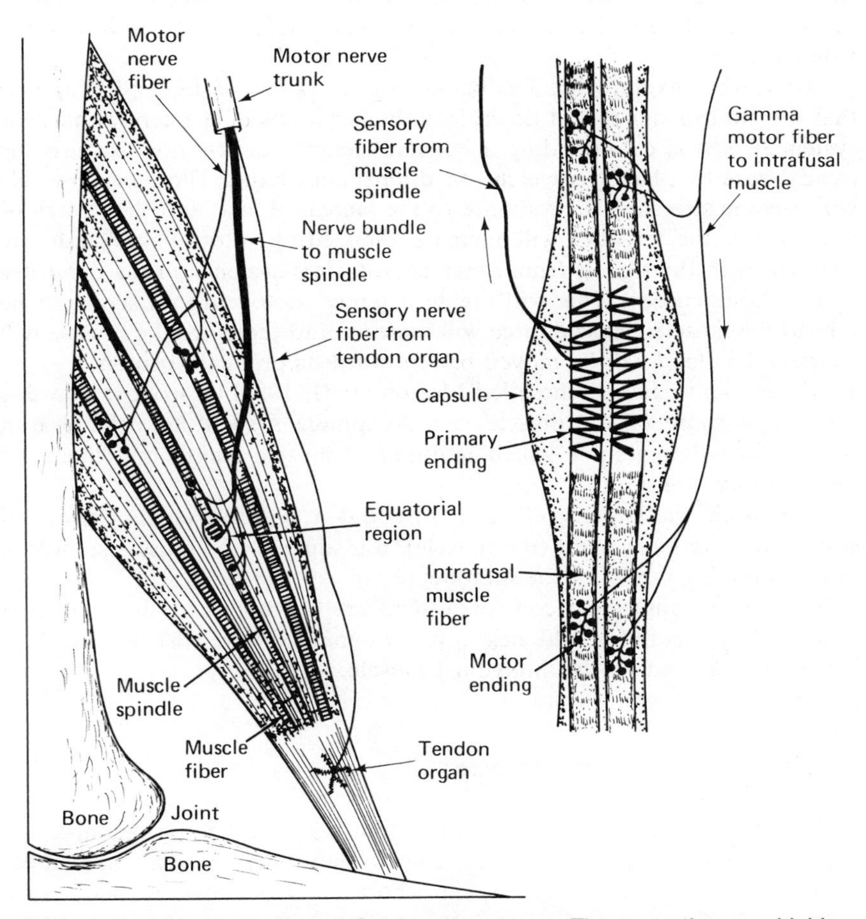

Figure 4-10 Muscle spindle and Golgi tendon organs. The proportions are highly distorted: a real muscle fiber would be many centimeters long but less than a millimeter in diameter. The muscle spindle organ contains primary sensory nerve endings which terminate in an annulospiral configuration around the center of the receptor, and a muscular portion which contains muscle fibers that are innervated by gamma motor fibers. [From P. A. Merton, "How We Control the Contraction of Our Muscles," *Scientific American*, 226(5) (1972), 30–37.]

From Fig. 4-10 we can also see that the ends of the muscle spindles are composed of small motor fibers. They fall into the gamma range of fiber sizes (1 to 8 μm) and are therefore called *gamma motor fibers*. When the gamma motor fibers are fired, the muscular portions of the spindle receptor contract, stretching the central portion of the receptor. This response has the same effect as a pull on the muscle, elongating the receptor, and results in a discharge of the receptor. With an increased excitation of gamma discharge, a smaller stretch of the spindle is required to elicit an afferent discharge. We will return later to the role of these gamma fibers.

To further explore the feedback-control system of Fig. 4-9, suppose that the control signal sent down from higher levels of the central nervous system (CNS) is commanding a constant length. Let us now disturb the steady state by placing a weight on the person's hand. The biceps muscle will become stretched. In response to the increased load and the stretch of the muscle, the spindles will increase their firing rates. The signals are summed with the signals from other neurological systems on the dendrites of the motoneurons. This will create a larger motoneuronal signal to be sent to the muscle. Muscle force will increase and therefore the muscle will shorten. The load will be moved back toward its original position.

Experiments such as these led Merton (1951, 1953) to propose that this was the purpose of the muscle–muscle spindle system; to behave as a constant-length regulator which maintained muscle length and therefore joint position constant.

Therefore, the stretch reflex of muscle is a negative feedback-control circuit. It contains a plant (the muscle), transducers (the muscle spindles), and a summing junction (the motoneurons).

Actual neuromuscular control systems are far more complex than that described by this reflex. The next level of complexity we wish to introduce is the level created by the antagonist muscle.

4-4-2 The Antagonist Muscle

The biceps muscle contracts the elbow joint, and its antagonist, the triceps muscle, extends the joint angle. These two muscles are direct antagonists, for when one muscle shortens, the other is necessarily lengthened. These muscles are reciprocally innervated to prevent them from working against each other. These innervation pathways are shown in Fig. 4-11a. The direct, or monosynaptic, pathways that the muscle spindle afferent fibers make with the motor neurons of their own muscle are excitatory pathways (E). In addition, inhibitory pathways (I) are shown from the same muscle spindle to the antagonist muscle. These pathways include interneurons.

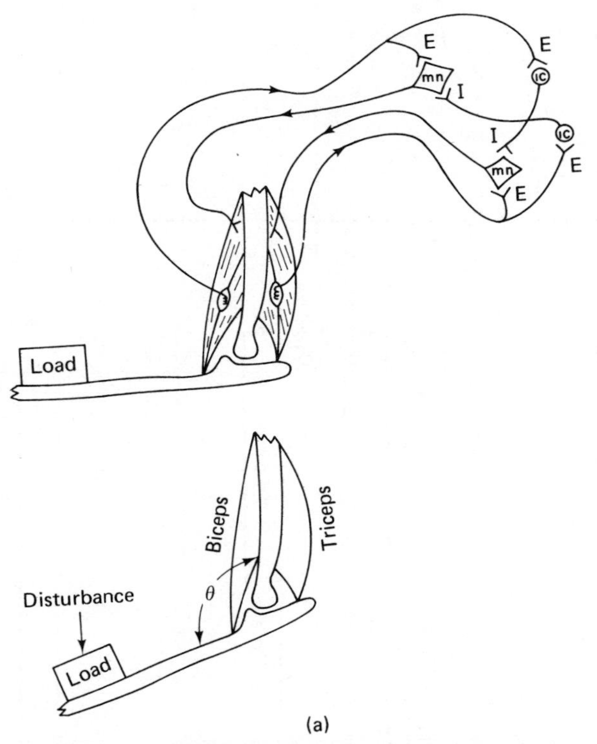

Figure 4-11 (a) Anatomical sketch and (b), continued on next page, block diagram of the previous neuromuscular control system (Fig. 4-9), with the addition of an antagonist muscle. I stands for inhibitory synaptic connection. (Modified from J. Houk and E. Henneman, "Feedback Control of Muscle: Introductory Concepts," in *Medical Physiology*, 13th ed., ed. V. B. Mountcastle, The C. V. Mosby Company, Saint Louis, 1974.

4-4-3 Two Control Mechanisms

From the schematic of Fig. 4-12, which now includes the role of the gamma fibers, we can see that the higher levels of its CNS can control the output position with either of two strategies. The open-loop control strategy uses the alpha motoneurons and holds the innervation to the muscle spindles, via the gamma fibers, constant. Thus, if it were desired to decrease the elbow angle θ, the CNS innervation to the biceps motoneurons would increase and the innervation to the triceps motoneurons would decrease. This strategy is fast and it is used for large or fast movements (Stark 1968; Houk 1974).

The second strategy, the closed-loop control strategy, uses the gamma fibers. To decrease the elbow angle, θ, the innervation to the gamma fibers

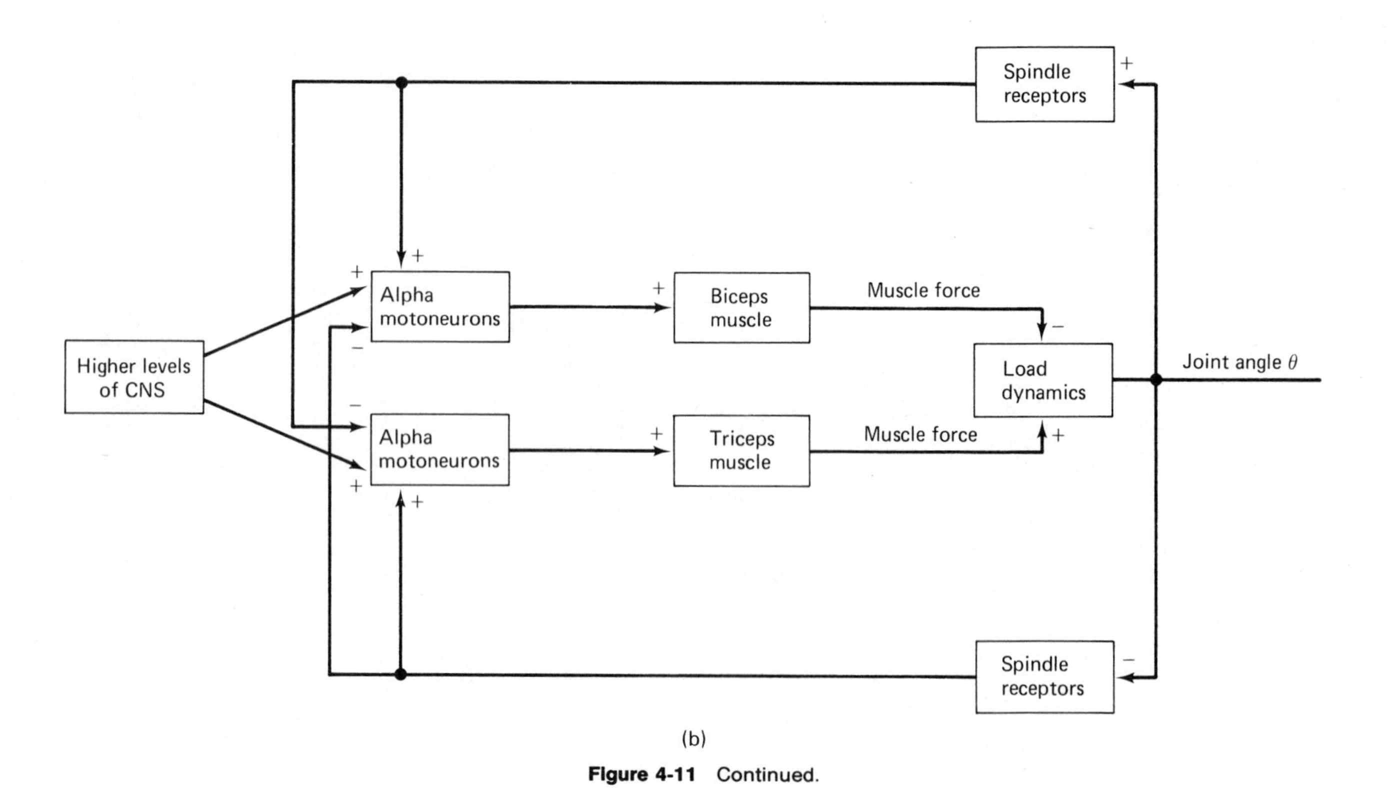

(b)

Figure 4-11 Continued.

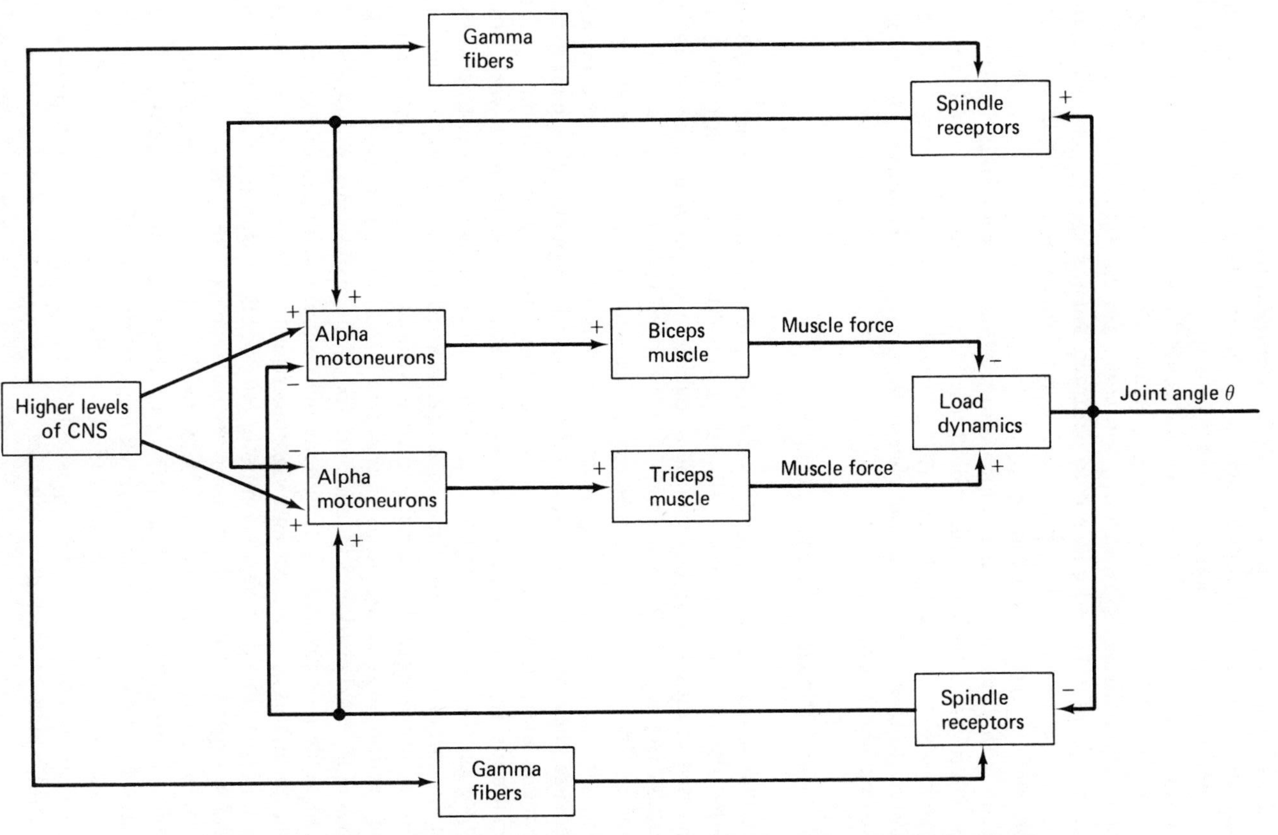

Figure 4-12 Neuromuscular control system with the addition of the gamma fiber inputs.

of the biceps muscle is increased, and concomitantly the innervation to the gamma fibers of the triceps muscle is decreased. Now considering only the biceps muscle, this increased firing will cause the biceps spindle fibers to contract. This, in turn, will stretch the central region of the spindle, where the sensory fibers are. Therefore, the biceps sensory fibers will increase their rate of firing. This, in turn, will stimulate the alpha motoneurons of the biceps muscle and cause it to contract. Analogous changes will allow the triceps to relax. Gamma fiber control may be used for very fine movements and for postural control, but most movements are probably made with a combination of alpha and gamma control (Houk 1974).

There are other theories about the role of the alpha and gamma systems. For example, Inbar (1975) proposed that the purpose of the spindle organs is to allow the system to implement adaptive control and thus change its dynamics.

4-4-4 The Golgi Tendon Organs

The neuromuscular control system also contains another sensory organ, called the *Golgi tendon organ* (Binder, Kroin, Moore, and Stuart 1977). In contrast to muscle spindles that lie in parallel with the main muscle fibers, tendon organs are located at the junction between the tendon and the muscle fibers and are therefore in series with the contractile elements. Tendon organs are relatively insensitive to passive stretch of the muscle because they are in series with the contractile element. The contractile element absorbs most of the stretch and prevents elongation of the tendonous region. Therefore, changes in muscle length have very little effect on the tendon organs. However, when a muscle contracts, the tendon organs in it discharge in proportion to the developed tension. If contraction merely shortens the muscle without developing much tension, the tendon organs are weakly excited. If contraction occurs when the muscle is lengthened and its ends are fixed, the shortening of the muscle's contractile part necessarily lengthens the noncontractile regions where the tendon organs are located, and vigorous firing results. A particular tension may be developed at various muscle lengths. By virtue of their location, tendon organs measure this tension regardless of length. The outputs of the sensory tendon organs are fed back to the alpha motoneurons through interneurons. They inhibit the firing of the alpha motoneurons and are therefore part of a negative feedback loop, as shown in Fig. 4-13. However, paradoxically, this is a constant-force feedback-control system, and thus this system should work in opposition to the constant-position feedback-control system of the muscle spindles.

For example, if the load in Fig. 4-9 is suddenly increased, the muscle will lengthen. As the muscle is stretched, its length–tension curve (Figs.

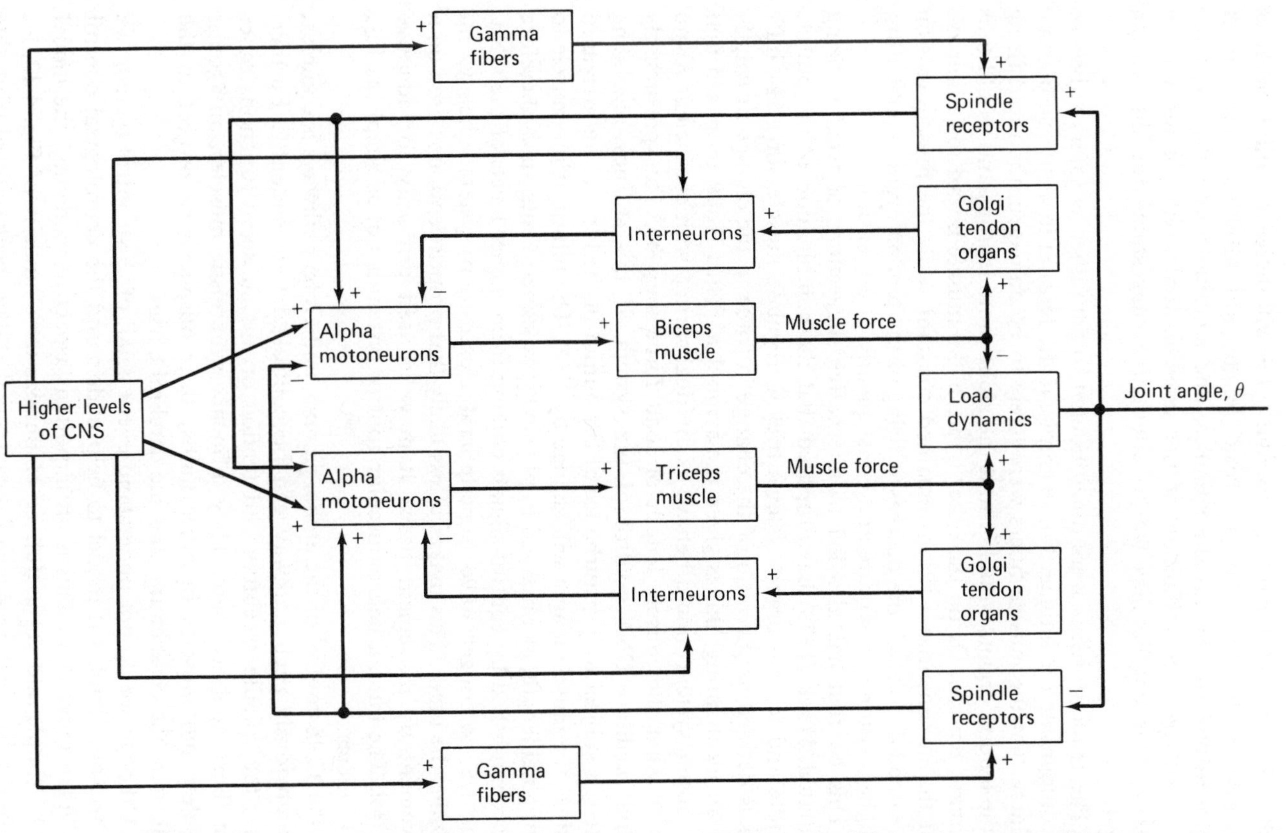

Figure 4-13 The neuromuscular control system with the addition of the Golgi tendon organ system.

3-28, 3-31, and 3-32) and the muscular force will increase until it balances the new load force. The increased length will produce excitation of motoneurons via the muscle spindles, and simultaneously the increased force will produce inhibition of the same motoneurons via the tendon organs. This conflict has led to alternative suggestions for the tendon organs.

The tendon organs were once thought to act either as safety valves or to compensate for fatigue. For example, if the tension in a muscle became so large that the muscle fibers were likely to be damaged, the rapid firing of the tendon organs would inhibit the alpha motoneurons and would thus decrease tension. Or on the other hand, if the muscle began to fatigue, so that the motoneuronal firing remained constant while the muscle tension developed decreased, the decreased firing of the tendon organ would allow a higher motoneuronal rate and consequently more tension.

This conflict has also led to alternative suggestions of what is being regulated. Houk (1978) has proposed that the combination of the muscle spindle and tendon organ systems tries to regulate muscle stiffness. Stiffness is defined as the ratio of the change in force to the change in length. For an ideal spring, this is given the symbol K. Thus Houk proposed that the muscle system should behave as an ideal spring, with coefficient K, no matter what the muscle length or load force may be. This presumably would facilitate CNS control of the system. If this stiffness-regulating system is subject to a disturbance at the output in the form of an increased load, F, the muscle length will change by F/K. The return of the muscle to its original length, as observed in physiological experiments, must therefore be accomplished by control signals coming from higher levels of the CNS. Thus we no longer have a peripheral closed-loop system: the system includes the brain. This makes it less likely that proprioception is used on a movement by movement basis: it may be used primarily for adaptive control. The human neuromuscular control system is not as simple as Fig. 4-13 implies.

The CNS control of the alpha, gamma, and Golgi pathways is complex, and obviously each pathway can be controlled independently. Furthermore, the spindle receptors and tendon organs subserve functions other than those analyzed here. They influence synergistic muscles, antagonist muscles, and muscles in other limbs; their outputs also project to the brainstem, the cerebellum, and the cerebral cortex.

Although we do not understand every detail of this control system, we can use our simplified model to help understand the neurological control of human movement. This is an important aspect of modeling; the model can be simple and yet useful. To emphasize the simplicity of Fig. 4-13, let us just briefly mention some of the effects that have been omitted from this system: the muscle dynamics that we studied in Chap. 3, such as the series elasticity, and the length–tension and force–velocity diagrams; CNS inputs from joint proprioceptors and from the visual system; the very important effects of temperature; statistical properties of the neural spike trains; hysteresis, habituation, and accommodation; corollary discharge; the role of Renshaw cells; the nonlinearities of the system; and saturation. Finally, a very simple geometry has been assumed where all fibers are in series or parallel. Most of these omissions have been previously elaborated upon by many others, as summarized by Partridge (1978). But in spite of, or perhaps because of, all these simplifications, the model is useful in studying the control of human movement in normal human beings and in patients.

4-4-5 Experimental Validation of the Model

Lawrence Stark (1968) reported a series of investigations on this neurological control system. The following discussion is based upon his work. A subject was instructed to rotate a handle back and forth as rapidly as possible. A record of handle angle as a function of time was obtained, shown in the top of Fig. 4-14. During the high-velocity portion of the movement, the antagonists were quite relaxed and limp. Because of this inactivity, despite marked stretching of the antagonists by the agonists, he concluded that the stretch reflex was inoperative. Either the afferent feedback from the muscle spindle was markedly reduced, or more likely, it was functionally ineffective in exciting the alpha motor neurons. The block diagram of Fig. 4-11b would be appropriate for this situation if the effects of the muscle spindles were removed, so that the system behaved in an open-loop manner, solely under control of the higher levels of the CNS.

The subject was then instructed to perform successive movements again, but was told that his primary object was to prevent deflection of his hand by random input disturbances. Only secondarily was he to oscillate his hand at as high a frequency as possible. Figure 4-11b is now the appropriate model for this mode of operation. The operation of the muscle spindle system as a high-gain length regulator causes an increased stiffness of both the agonists and the antagonists. As a result, the frequency of the oscillations is decreased, as shown in the middle trace of Fig. 4-14.

As we stated previously, the primary endings of the muscle spindle organ are sensitive to both position and velocity. That means the feedback pathway has dynamics. This allows the possibility that the response of the closed-loop system may be slower than the response of the open-loop system. This is exactly the response that Stark and his group found.

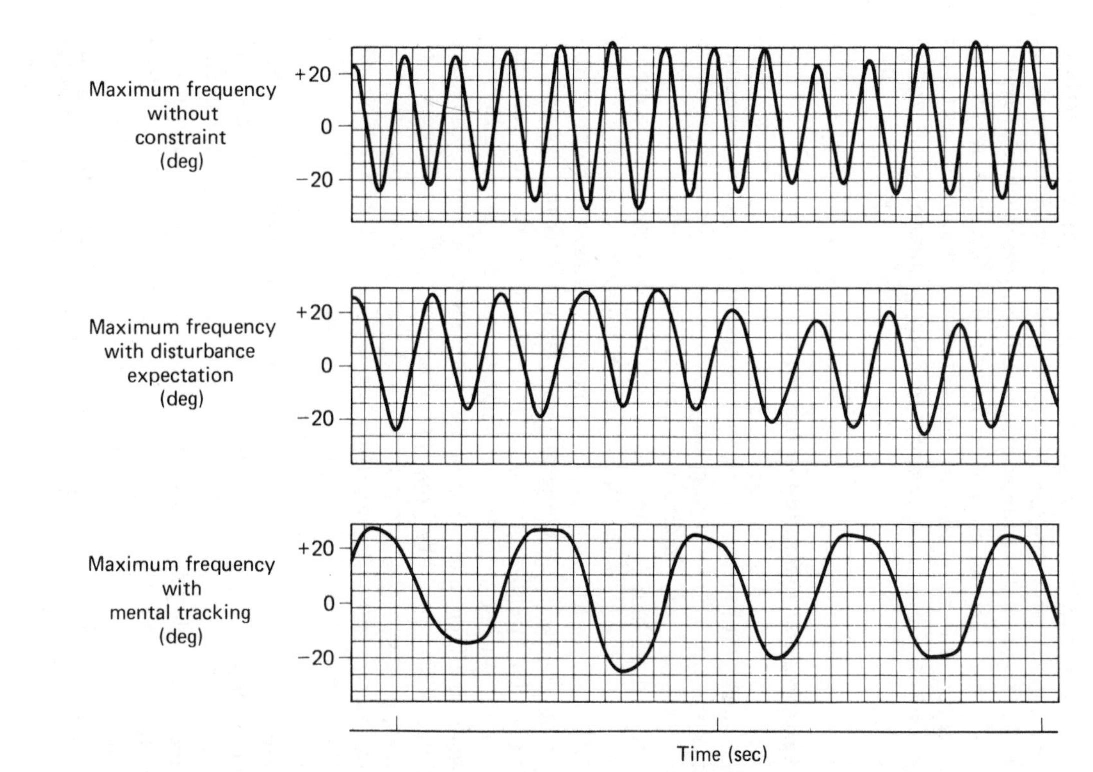

Figure 4-14 Arm position as a function of time for three sets of instructions: (top) rotate the arm as fast as possible; (middle) rotate the arm but be prepared to resist disturbances, (bottom) move the arm to track an imaginary visual target. From L. Stark *Neurological Control Systems Studies in Bioengineering*, New York: Plenum Press, 1968.

In the final set of instructions given to the subject, he was asked to imagine a pointer moving back and forth, to track this imaginary pointer, and then attempt to oscillate his hand as fast as possible, as in the first mode. The postulated control model is now the model of Fig. 4-12 with the muscle spindle feedback replaced with a feedback loop from limb position to visual–motor cortical processes, and finally back to the muscles. Perhaps because of the necessity of transmitting and processing all control signals through the imagery of the mental tracking process, the oscillation is markedly slowed. This is shown in the bottom recording of Fig. 4-14.

4-4-6 Parkinson's Syndrome

Parkinson's syndrome is a central nervous system disease afflicting elderly persons and is well known in the neurological literature. The muscles and peripheral nerves of Parkinson patients are generally not affected by the disease (until late secondary changes occur). Evidence of this comes from observations both of patients able to walk normally when sleepwalking and of the modification of this syndrome by certain brain operations. The interruption of the stretch reflex markedly reduces the rigidity that is a prominent sign of this syndrome.

Stark (1968) and his colleagues performed the experiments described in the last section on a group of 20 Parkinson patients. When asked to rotate the handle back and forth as rapidly as possible, the patients could not attain frequencies as high as the normals, implying that they could not open the loop on the muscle spindle system. This elucidates the essential nature of the rigidity of Parkinson's syndrome. In these patients the spindle-length regulator is always (except in sleep) on full gain, and thus opposes and weakens corticospinal inputs.

Stark and his colleagues then had normals and patients track unpredictable target movements. They recorded the responses and constructed Bode diagrams, and from these they constructed Nyquist plots. As can be seen in Fig. 4-15, the patients had much lower gains at high frequencies. These results can once again be explained by postulating that the Parkinson patients could not turn off their muscle spindle systems and allow unimpeded fast control by the cortical–motoneuronal pathways.

The Nyquist plot of Fig. 4-15 implies that the feedback control system of the patients should be more stable than in the normals. (They are farther away from the -1 point.)

This brings up a most perplexing symptom of this disease. Parkinson patients often have a tremor of the limbs. In Stark's patients the oscillations were about 10° in amplitude and ranged from 3 to 10 Hz. These oscillations cannot be explained on the basis of the feedback-control model of Fig. 4-15. The cause of the oscillations, therefore, probably

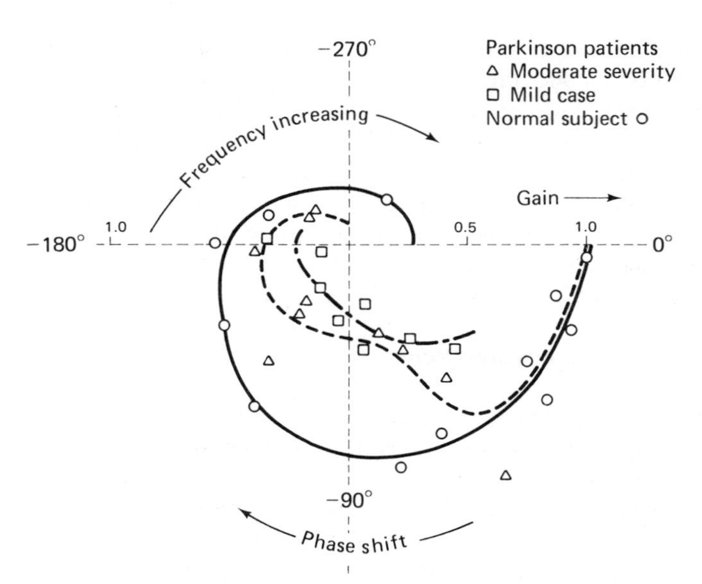

Figure 4-15 Nyquist diagram for arm position target tracking of normal humans and Parkinson patients. The patients could not track rapidly moving targets. From L. Stark *Neurological Control Systems Studies in Bioengineering*, New York: Plenum Press, 1968.

resides in higher levels of the CNS. Thus both the inability to turn off the muscle spindle system and also the Parkinsonian tremor are caused by disturbances in the CNS rather than in the peripheral nervous system. This viewpoint is supported by Toll (1974) in his biochemical explanation of Parkinson's syndrome.

4-5 THERMOREGULATION SYSTEMS

Mammals, man in particular, regulate their temperatures very precisely. This precision is made possible by closed-loop feedback control. Figure 4-16 shows a simple model of this system.

Thermoregulation is necessary for survival; species with good thermoregulation systems have a competitive advantage over other species. The biochemical reactions of living tissues are temperature-dependent. For example, muscles are more efficient at higher temperatures. Some of the faster-swimming and more successful predatory fish maintain elevated temperatures in their swimming muscles (Carey 1973). However, at high temperatures proteins are broken down faster than they can be constructed. The upper temperature limit for survival is about 45°C (113°F), when proteins denature and lose their biological properties. The lower limit for animal survival is slightly below 0°C (32°F), when intracellular water forms ice crystals that rupture and kill cells.

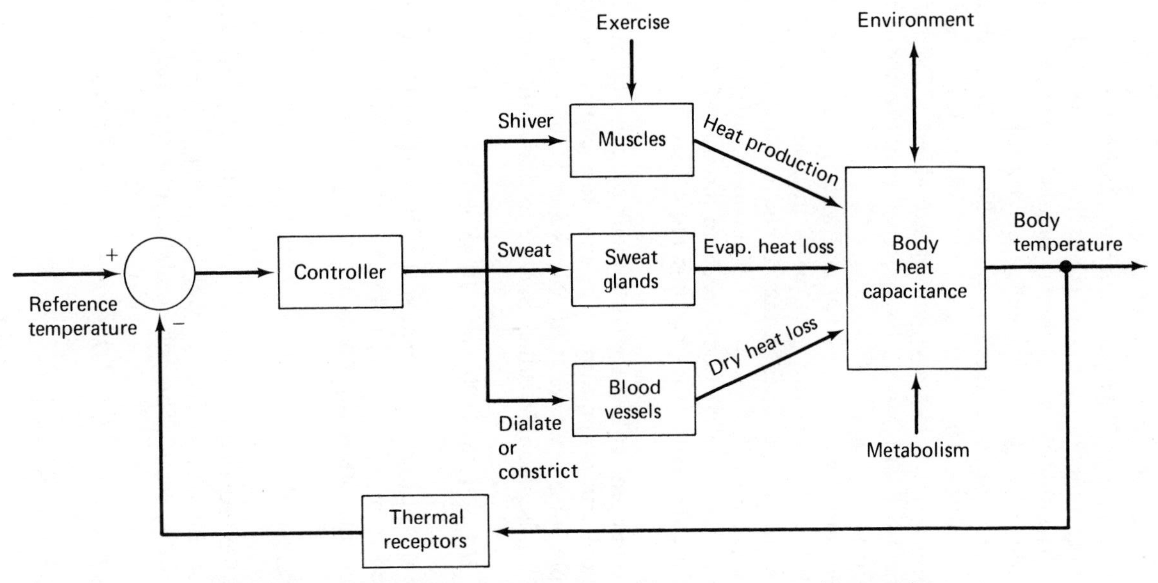

Figure 4-16 Simplified block diagram of human thermoregulatory system.

The temperature sensors for the thermoregulatory system are located primarily in the skin and the hypothalamus, although there are receptors in the spinal cord, the medulla, abdominal viscera, and perhaps in muscles and the respiratory tract. There are separate sensors for warm and cold.

The summing junction and the controller are located in particular regions of the hypothalamus. There may be one region for warm and one for cold or perhaps one for each mechanism controlled.

There are three principal mechanisms of heat production: metabolism, exercise, and shivering. The basic body metabolism produces about 86 W. This power would raise the body temperature about 1°C/hr if there were no mechanisms for heat removal. Exercise can produce about 1000 W and shivering can produce roughly 100 W.

Heat is exchanged within the body and with the environment. Convection of heat by the bloodstream is a major mechanism for thermoregulation. In the cold, the arteries going to the surface of the extremities constrict, and when it is hot these vessels dilate. In the cold, differences of 7°C are common between the body core and the skin of the extremities. In cases of frostbite the extremities become more than 37°C colder than the core.

Heat is exchanged with the environment by convection of air, radiation, conduction, and evaporation. The heat loss caused by convection is quite noticeable when a winter wind whips across the body. The equation used by the U.S. National Weather Service to derive the wind-chill factor is

$$H_{\text{con}} = 1.16A(100\sqrt{w} + 10.45 - w)(33 - T_a)$$

where H_{con} is the power loss in watts, A is the exposed surface area, w is the wind speed in m/sec, and T_a is the air temperature in °C. The body radiates energy into the environment according to the equation

$$H_{\text{rad}} = k'A(T_s^4 - T_a^4)$$

where H_{rad} is the radiated power in watts, A is the effective area, T_s is the temperature of the skin in degrees Kelvin, and T_a is the ambient radiant temperature in degrees K. The polynomial can be factored into

$$H_{\text{rad}} = k'A(T_s - T_a)(T_s^3 + T_s^2 T_a + T_s T_a^2 + T_a^3)$$

which is approximately equal to

$$KA(T_s - T_a)$$

where K is approximately 8 W/m^2°C. The body also loses heat by conduction, when in contact with chairs or walls.

Heat loss by convection, radiation, and conduction are not under CNS control, whereas heat loss by evaporation can be controlled. Sweat glands, stimulated by the cholinergic fibers of the autonomous nervous system, can secrete up to 1 liter of perspiration per hour during exercise. The latent heat of vaporization of a liter of water is 2.3 MJ. However, not all of this water is evaporated; only about 0.5 MJ is liberated by this process.

Evaporative heat exchange through the lungs and respiratory tracts is also important. Usually, inspired air has a lower water vapor pressure than air in the lungs. This is particularly true of cold, dry, winter air. Expired air is saturated with water vapor. Therefore, this heat-loss term is proportional to the ventilation rate, which in turn is proportional to oxygen consumption and the metabolic rate. There is less heat loss by breathing through the nose than by breathing through the mouth.

4-5-1 Model of the Plant

The model of Fig. 4-17 for the thermoregulatory plant is a simplified version of Stolwijk and Hardy's model (1966, 1977). Resting state values of metabolic heat production (M_0), thermal conductance between compartments (resistors between units, in the right half of the figure), thermal conductance between skin segments and environment (at temperature T_A) and insensible evaporative heat loss (E_V) are given. Metabolic heat production and evaporative heat loss are in watts, thermal conductances are in watts/deg C. The left half of the figure shows the blood flow in liters per hour. The brain is perfused at 45 and the core of the trunk at 210 liter/hr. All other blood flows change according to demand. The extra arrows to the core of the head, muscles of the trunk and extremities indicate that extra heat inputs occur in shivering and in exercise. The body is modeled as a series of compartments. The head and trunk are modeled by cylinders, and the aggregate of the extremities is modeled by another cylinder. These cylinders are subdivided into coaxial shells representing the skin, fat, muscle, and viscera. The central blood volume, contained by the heart and the large arteries and veins, is a separate compartment. The dimensions of the cylinders were chosen so that their mass/surface area ratio corresponds to that of an average man (74 kg and 1.9 m^2 of total skin area). Heat exchange by conduction occurs only radially within each cylinder. This conductive heat must flow through the thermal resistance indicated by the long resistor symbols in Fig. 4-17. It is assumed that there is no conduction between cylinders.

Heat is transferred between cylinders by convection by blood flow. This flow is under the control of the vasomotor centers and varies significantly. These pathways are represented by the short resistor symbols of Fig. 4-17. Only the flow to the brain and the viscera is not variable.

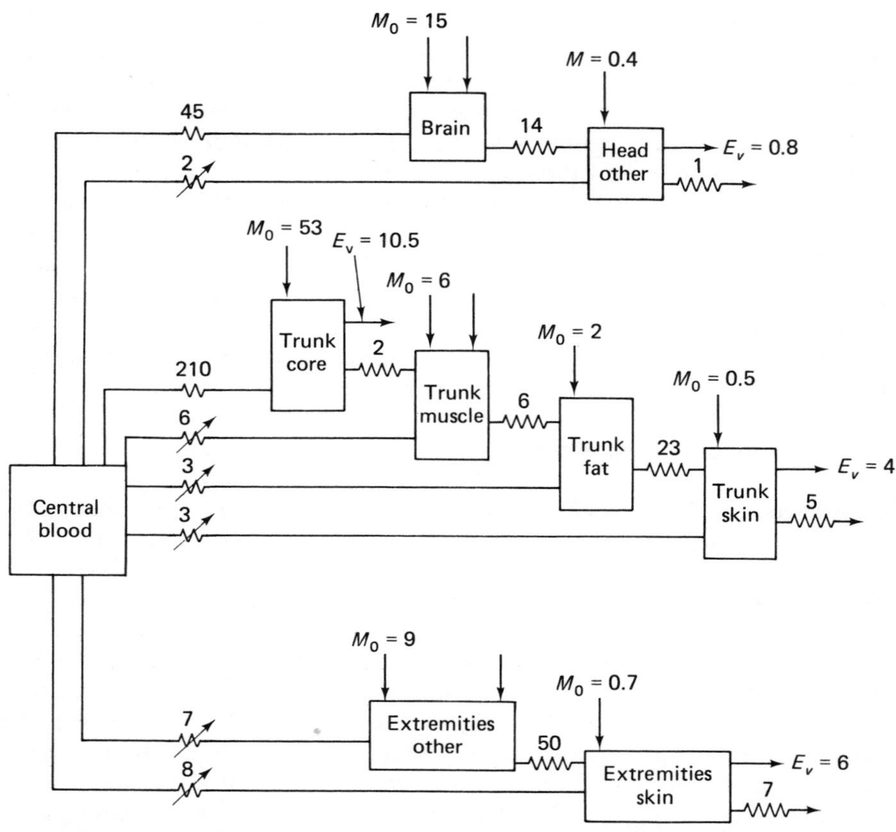

Figure 4-17 Diagram for thermoregulatory plant. After Stolwijk and Hardy (1966 and 1977).

Basal metabolic heat sources and heat sources caused by muscular activity or shivering are shown with arrows into the appropriate compartments. Heat losses resulting from evaporation are shown with arrows out of the compartments.

It is of passing interest to note that resting blood flow is not proportional to metabolic rate (which is proportional to oxygen consumption). For example, the ratio of basal blood flow to basal metabolic rate for the brain is 3 liters/W-hr. This ratio is 4 liters/W-hr for the organs of the trunk and 10 liters/W-hr for the skin. This means that the brain is more efficient at extracting oxygen from blood than is the viscera or the skin.

4-5-2 Controller Model

Our primary defense against overheating is perspiration (unless we are an elephant and we use our ears as giant radiators, or we are a dog and we pant).

In Stolwijk and Hardy's model (1966) there are two reference set points and two temperatures that are measured: the temperature of the core of the head, T_{HC}, and the average skin temperature, $\overline{T}_s$. Both of these set points must be exceeded to produce perspiration. The resulting evaporative power loss becomes

$$E_V = K_2(T_{HC} - 36.6°C)(\overline{T}_s - 34.1°C)$$

if

$$(T_{HC} - 36.6) > 0$$

and

$$(\overline{T}_s - 34.1) > 0$$

K_2 is equal to 80 W/(°C)^2. This equation is shown diagrammatically in Fig. 4-18b.

This is unfortunately a nonlinear control law. A summation of the two terms was tried as the control law, but it was found that the transient response was not as good. This model also allowed high muscle temperature and high core temperatures to induce sweating, and allowed high temperatures to increase the flow of blood to the skin. Their 1977 model achieved a linear controller.

One defense against body cooling is shivering. The power produced by shivering is

$$\Delta M = K_3(T_{HC} - 36.6°C)(\overline{T}_s - 34.1°C)$$

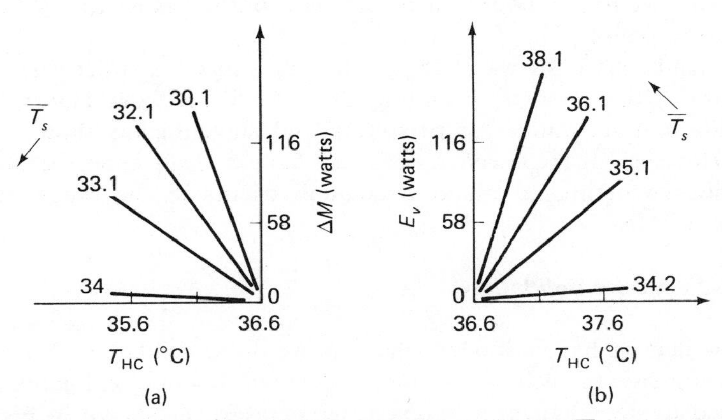

Figure 4-18 Controller action for high average skin ($\overline{T}_s$) and brain core (T_{HC}) temperatures (a) and for low average skin and brain temperature (b). From S. A. Talbot and U. Gessner, *Systems Physiology*, John Wiley & Sons, 1973. Reprinted by permission of John Wiley & Sons, Inc.

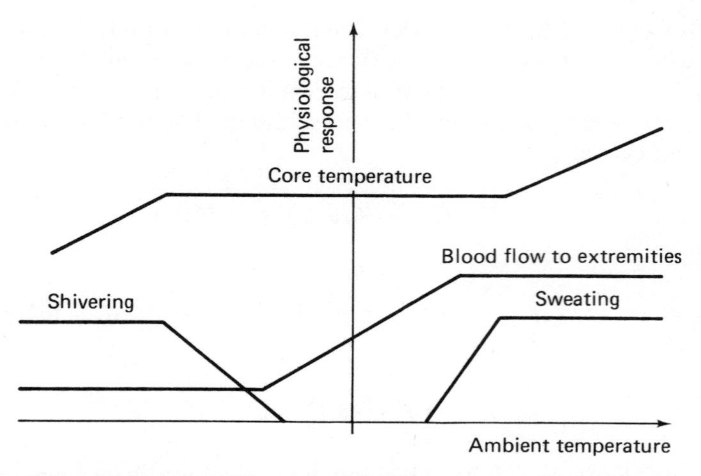

Figure 4-19 Controller action if set points are different for shivering, sweating and blood flow control.

if

$$(T_{HC} - 36.6) < 0$$

and

$$\left(\overline{T}_s - 34.1\right) < 0$$

The constant K_3 equals 70 W/(°C)^2. The function is shown in Fig. 4-18a. Once again we have a nonlinear control law, but it was necessary for good transient response.

No dead zone is shown in the model; the subject is either shivering or perspiring. Other models, including that of Stolwijk and Hardy (1977), have different set points for perspiring and shivering (as shown in Fig. 4-19). However, the controller does not have a dead zone, because the regulation of peripheral blood circulation occurs in the range between these set points.

4-5-3 Model Validation

Now that we have a model, what can we do to validate it? For a start, we can compare the behavior of the model with that of a real human being subjected to step changes in ambient temperature, as shown in Fig. 4-20. The temperature of the tympanic membrane in the ear was measured for an approximation of T_{HC}. The subject's weight loss was used to calculate the amount of perspiration that evaporated. The results of the model

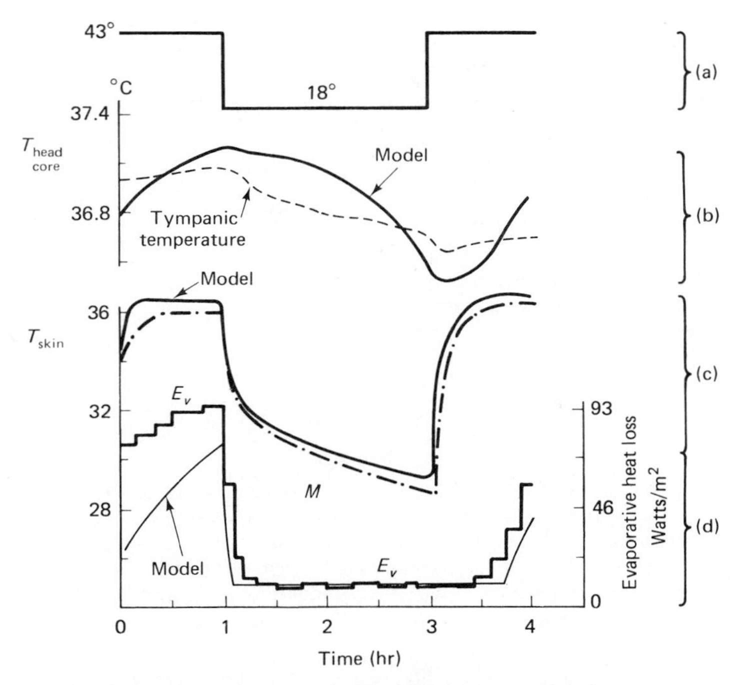

Figure 4-20 Response of the closed-loop model, in comparison with experimental data obtained from a young nude man, to external transients. (a) Ambient temperature showing stimulus to system; (b) computed head core temperature together with temperature measured at the tympanic membrane; (c) skin temperatures, computed and measured; (d) evaporative heat loss E_v computed in the model and calculated from loss of weight of the experimental subject. After Stolwijk and Hardy (1966).

compare well with the physiological data. The "ice cream experiment," shown in Fig. 4-21, was the next test of the model. In this test an abrupt change in trunk core temperature was produced by rapidly eating 420 g of ice cream. Again the model results adequately represent the physiological data.

Stolwijk and Hardy's original model was implemented on an analog computer. One of the most important validations of their model had to wait until the model was rewritten in FORTRAN for general-purpose digital computers. In this new form, other experimenters in independent laboratories could study and use the model. Konz et al. (1977) ran a series of dry-ice cooling experiments on this new version of the model and found a good qualitative match between their implementation of this model and their new physiological data.

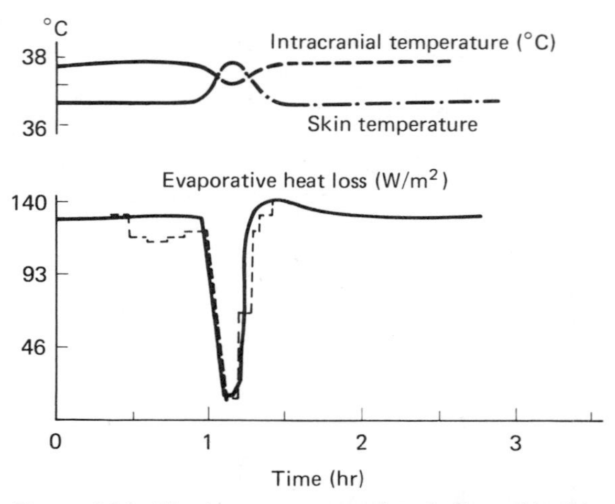

Figure 4-21 The ice cream experiment: the subject is seated in a room at 42° C; at the end of the first hour he eats 420 g of ice cream. The model (solid lines) predicts experimental data (dashed lines) satisfactorily for head core temperature T_{HC}, skin temperature $\overline{T}_s$ and the evaporative heat losses E_v. After Stolwijk and Hardy (1966).

4-5-4 Model Variations

So far we have considered the reference input to be a constant. There are a variety of factors that can cause it to change. Stolwijk and Hardy found that their model could match an artificial fever, created by injecting immune typhoid vaccine, if the reference temperatures, or set points, were changed. Exercise (Talbot and Gessner 1973) and the onset of hibernation behavior (Heller, Crawshaw, and Hammel 1978) also change the reference temperatures. Studies showing that a tropical animal such as the tenrec could be made to hibernate led to the use of hypothermia (body cooling) on patients undergoing heart surgery.

This thermoregulation model is valid for time periods of minutes to hours. It does not model diurnal variations, long-term changes, or adaptation. For example, to adapt for a cold environment the human adrenergic autonomous nervous system can boost the metabolic rate throughout the body. To adapt for a cold climate the thyroid gland can produce more thyroxin, which will stimulate metabolic rate (Talbot and Gessner 1973; Bligh 1973). Aging reduces the animal's ability to control its temperature. When thermoregulatory competence was tested in rats by measuring the speed of recovery of body temperature following a 3-min immersion in ice water, the drop in body temperature was significantly greater in older animals than in adults, their recovery rate was significantly slower, and their body temperature remained lower throughout the test period (Timiras 1978).

In this simplified model for thermoregulation we have only used one summing junction; its presumed location was the hypothalamus. Our controller had no dead zone, which is in concert with the physiological fact that there is no dead zone in normal human beings. In studies where the hypothalamus has been destroyed, it has been found that there is still thermoregulation, but there is now a dead zone. Satinoff (1978) has proposed an evolutionary explanation for this. He suggests that primitive animals had thermoregulatory systems involving only local feedback loops. These systems had large dead zones. As the species advanced, higher and higher levels of the central nervous system became involved and produced regulation with smaller and smaller dead zones. A model that fits this description is shown in Fig. 4-22. In some cases there will be multiple nested levels of feedback (not shown in Fig. 4-22). In some cases the

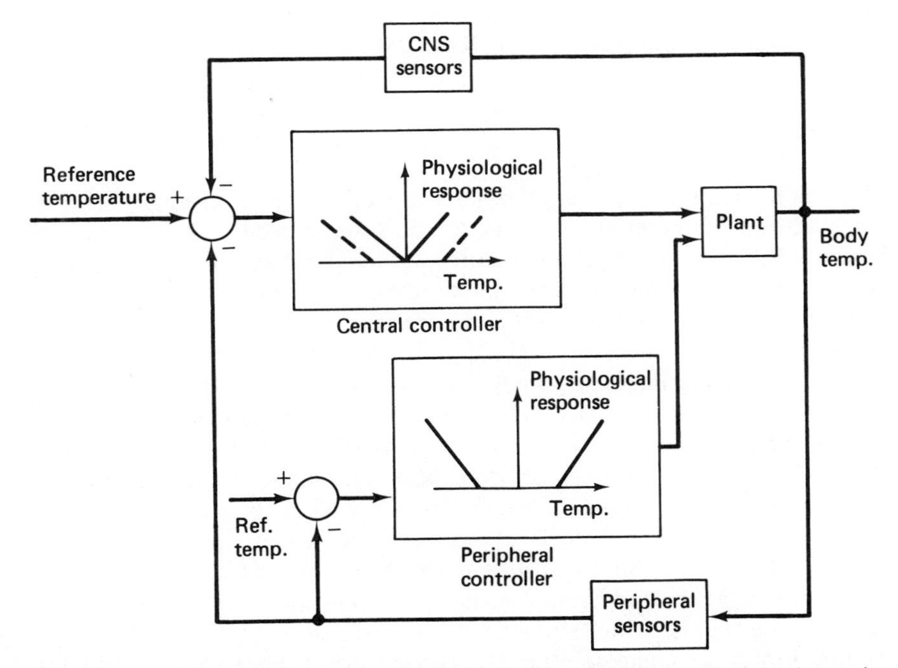

Figure 4-22 Expanded model showing peripheral feedback loop with a dead zone in the peripheral controller. This loop would only be used if the CNS controller were damaged or if the pathway back to the CNS were never established. The central controller normally operates without a dead zone (solid lines). Disease or lesions can cause it to operate with a dead zone (dashed lines).

higher-level feedback loops may not develop. For example, the panting reaction in a dog is apparently not controlled by the hypothalamus, but rather is controlled only by the peripheral sensor loop.

4-5-5 Industrial Applications

A person in a space suit cannot use perspiration as a defense against overheating because the water vapor cannot escape. Cooling-water coils have been installed in space suits to remove the heat generated by exercise and work. However, there is then a problem of controlling the cooling. Manual control by the subject does not work, because the subject often reacts too late or overreacts, possibly indicating that there are different physiological and psychological temperature sensors. Thermoregulation models were used to help design the automatic control systems for these space suits. Two methods that worked involved measuring the oxygen consumption, which is proportional to metabolic activity, and measuring the skin temperature at several locations. Although these controllers were not as sophisticated as those of the human body, they were adequate for use in the space suits (Hwang and Konz 1977).

4-6 SUMMARY

Most biological systems are impossible to study in their entirety, but they are made up of hierarchies of smaller subsystems which can be studied. Simon (1962) discusses the necessity for such hierarchies in complex systems. He shows that most complex systems are decomposable, enabling subsystems to be studied outside the entire hierarchy. For example, in studying the motion of a baseball, it is sufficient to apply Newtonian mechanics considering only gravity, air, the ball, and the bat. One need not worry about electron orbits or the motions of the sun and the moon. Forces that are important when studying objects of one order of magnitude seldom have an effect on objects of another order of magnitude. Bioengineers therefore study the small subsets of larger physiological systems. For example, there is a great deal of interest in the control of human locomotion. But because it is much too complicated a system to study, bioengineers study small parts of it: some try to determine what is being optimized in a normal step cycle (e.g., time, mechanical energy, torque, forces, or chemical energy); some study the step cycles and the control of foot-fall patterns in cats, spiders, or crayfish; some study the subsystems involved, such as the stretch reflex or the golgi tendon organ system; some study the dynamic properties of muscle; some study deficits caused by pathology; and some study the interactions of various sensory systems and their influence on the control of locomotion. All of these bioengineering studies are studies of small subsystems of the locomotory control system. However, even studies of these small subsystems cannot take into account all the physiological factors that influence the overall control strategy. For example, a 10°C change in temperature can drive a spindle receptor through its entire output range, yet few studies even consider temperature. Successful bioengineering studies have ignored the unimportant feedback loops and have found ways of opening the loop on the important feedback loops. Good control theory studies of small subsystems seem to provide the best promise for eventual understanding of how the brain controls movement. If we can understand how the brain controls movements, we may be able to understand how the brain performs its other tasks, and eventually we may be able to help the brain to compensate for defects caused by pathology or trauma.

REFERENCES

BINDER, M. D., J. S. KROIN, G. P. MOORE, and D. G. STUART, "The Response of Golgi Tendon Organs to Single Motor Unit Contractions," *Journal of Physiology*, 271 (1977), 337–50.

BLIGH, J., *Temperature Regulation in Mammals and Other Vertebrates*. Amsterdam: North-Holland Publishing Co., 1973.

CAREY, F. G., "Fishes with Warm Bodies," *Scientific American*, 228 (February 1973), 36–44.

FULTON, J. R., and J. A. PI-SUNER, "A Note Concerning the Probable Function of Various Afferent End-Organs in Skeletal Muscle," *American Journal of Physiology*, 83 (1928), 554.

HELLER, H. C., L. I. CRAWSHAW, and H. T. HAMMEL, "The Thermostat of Vertebrate Animals," *Scientific American*, 239 (August 1978), 102-13.

HENNEMAN, E., "Peripheral Mechanisms Involved in the Control of Muscle," in *Medical Physiology*, ed. V. Mountcastle. St. Louis: The C. V. Mosby Company, 1974, 617–35.

HOUK, J., "Feedback Control of Muscle: A Synthesis of the Peripheral Mechanisms," in *Medical Physiology*, ed. V. Mountcastle. St. Louis: The C. V. Mosby Company, 1974, pp. 668–77.

HOUK, J. C., "Participation of Reflex Mechanisms and Reaction Time Processes in the Compensatory Adjustments to Medical Disturbances," in *Cerebral Motor Control in Man: Long Loop Mechanisms*. Progress in

Clinical Neurophysiology, Vol. 4, ed. J. E. Desmedt. Basel: Karger, 1978.

HWANG, C., and S. KONZ, "Engineering Models of Human Thermoregulatory System—A Review," *IEEE Transactions on Biomedical Engineering, BME*-24 (1977), 309–25.

INBAR, G. F., "Modulation of Dynamic Parameters of Muscle by Selective Activation of Its Gamma System," *Biological Cybernetics*, 19 (1975), 169–80.

KONZ, S., C. HWANG, B. BHIMAN, J. DUNCAN and A. HASUD, "An Experimental Validation of Mathematical Simulation of Human Thermoregulation," *Computers in Biology and Medicine*, 7 (1977), 71–82.

LEWIS, E. R., *Network Models in Population Biology*. New York: Springer-Verlag, 1977.

LIDDELL, E. G. T., and C. S. SHERRINGTON, "Reflexes in Response to Stretch (Myotatic Reflexes)," *Proceedings of the Royal Society, London* (*B*), 96 (1924), 212.

MELSA, J. L., and S. K. JONES, *Computer Programs for Computational Assistance in the Study of Linear Control Theory*. New York: McGraw-Hill Book Company, 1970.

MERTON, P. A., "The Silent Period in a Muscle of the Human Hand," *Journal of Physiology*, 114 (1951), 183–98.

MERTON, P. A., "Spectulations on the Servo Control of Movement," in *Spinal Cord*. Boston: Little, Brown and Company, 1953, pp. 183–98.

PARTRIDGE, L. D., "Muscle Properties: A Problem for the Motor Physiologist," in *Posture and Movement: Prospective for Integrating Sensory and Motor Research on the Mammalian Nervous System*, ed. R. E. Talbott and D. R. Humphrey. New York: Raven Press, 1978, pp. 189–229.

SATINOFF, E., "Neural Organization and Evolution of Thermal Regulation in Mammals," *Science*, 201 (July 1978), 16–21.

SIMON, H. A., "The Architecture of Complexity," *Proceedings of the American Philosophical Society*, 106 (1962), 467–82.

STARK, L., *Neurological Control Systems, Studies in Bioengineering*. New York: Plenum Press, 1968.

STOLWIJK, J. A. J., and J. D. HARDY, "Temperature Regulation in Man—A Theoretical Study", *Pflugers Archiv fuer die Gesamte Physiologie*, 291 (1966), 129–62.

STOLWIJK, J. A. J., and J. D. HARDY, "Control of Body Temperature," in *Handbook of Physiology*. Baltimore: The Williams & Wilkins Co., 1977, pp. 45–68.

STROJNIK, P., A. KRALJ, and I. URSIC, "Programmed Six-Channel Electrical Stimulator for Complex Stimulation of Leg Muscles during Walking," *IEEE Transactions on Biomedical Engineering*, BME-26 (1979), 112–116.

TAKAHASHI, Y., M. J. RABINS, and D. M. AUSLANDER, *Control and Dynamic Systems*. Reading, Mass.: Addison-Wesley Publishing Co., Inc., 1970.

TALBOT, S. A., and U. GESSNER, "Body Temperature and Its Control," in *Systems Physiology*. New York: John Wiley & Sons, Inc., 1973.

TIMIRAS, P. S., "Biological Perspectives on Aging," *American Scientist*, 66 (1978), 605–13.

TOLL, K. V. S., "Basal Ganglia and Biogenic Amines," *Medical Physiology*, ed. V. Mountcastle, St. Louis: The C. V. Mosby Company, 1974, pp. 700–702.

PROBLEMS

4-1 In the example of Sec. 4-1-2 (rejection of output disturbance), let

$$G_p = \frac{1}{1+\tau s} \qquad \text{where } \tau = 100 \text{ ms}$$

$$H(s) = as + 0.1$$

$$G_{cc}(s) = 1000$$

Find the output as a function of time in response to an impulse disturbance of the output when a takes on the values 0, 0.001, 0.01, 0.1, and 1.0. Why is $a = 0.001$ so unique?

4-2 Construct Bode diagrams for the following transfer functions:

$$G_1(s) = \frac{1}{s-a} \qquad \text{a right-half-plane pole}$$

$$G_2(s) = s - a \qquad \text{a right-half-plane zero}$$

$$G_3(s) = a - s$$

4-3 The following transfer function is used on an analog computer to approximate a pure time delay:

$$G(s) = \frac{a-s}{a+s}$$

(a) Determine the Bode and Nyquist plots and compare them to the Bode and Nyquist plots of a pure time delay.

(b) Find the step response of this pseudo time delay for initial conditions of zero.

4-4 The time course of an epidemic can be modeled using the law of mass action, which states that the rate of growth in the number of infected people is proportional to the number of contacts between susceptible people and infected people (Takahashi, Rabins, and Auslander 1970; Lewis 1977). Let

$x_1(t)$ = number of people susceptible to the disease

$x_2(t)$ = number of unisolated infected people

$x_3(t)$ = number of people removed from group two because of recovery (with immunity), isolation (hospital), and death

Our model requires that

$$\frac{dx_1}{dt} = -k_1 x_1 x_2$$

$$\frac{dx_3}{dt} = k_2 x_2$$

For a constant-size population,

$$N = x_1 + x_2 + x_3$$

and therefore

$$0 = \frac{dx_1}{dt} + \frac{dx_2}{dt} + \frac{dx_3}{dt}$$

yielding

$$\frac{dx_2}{dt} = K_1 x_1 x_2 - K_2 x_2$$

Sketch x_1, x_2, and x_3 as a function of time.

What is the critical size of the population, N, required for an epidemic to grow?

4-5 The open-loop transfer function for the system of Fig. P4-5 is

$$G(s) = \frac{K(2s+1)e^{-0.2s}}{s^2 - 1}$$

Sketch Bode diagrams and a Nyquist plot for this system. Is the system stable?

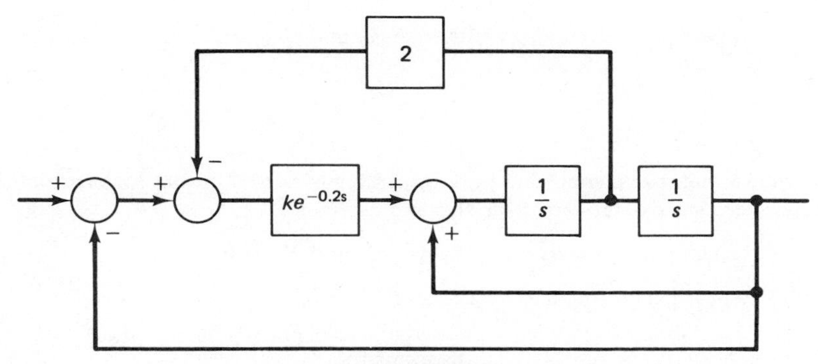

Figure P4-5

4-6 Sketch a Nyquist plot and comment upon the stability of a feedback control system with the following open-loop transfer function:

$$\mathrm{KGH} = \frac{K(s+10)(s+20)}{s(s-10)(s+40)}$$

4-7 Draw a Nyquist plot and comment upon the stability of a feedback control system with the following open-loop transfer function:

$$\mathrm{KGH} = \frac{2(s+0.5)(s+10)}{(s+1)(s+2)(s^2+2s+5)}$$

4-8 The controller, H_1, of the closed-loop system shown in Fig. P4-8 is called a proportional plus integral controller.

(a) Sketch the step response of this controller.

(b) Find the closed-loop transfer function $Y(s)/R(s)$ and also $E(s)/R(s)$ for this system.

(c) Use the final value theorem, namely,

$$\lim_{t\to\infty} f(t) = \lim_{s\to 0} sF(s)$$

to find the steady-state error for a step input.

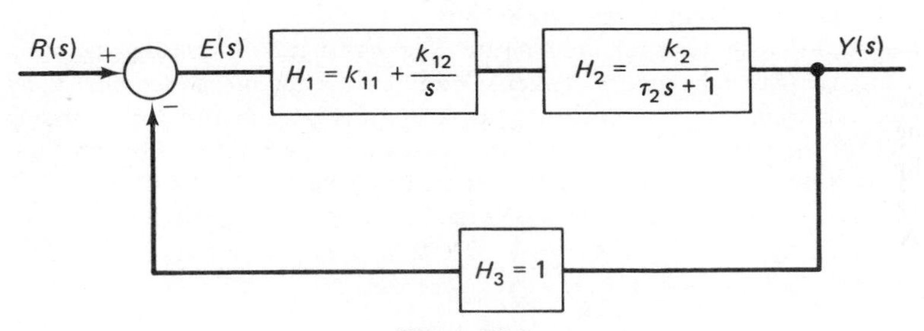

Figure P4-8

(d) Repeat parts (b) and (c) with $k_{12}=0$, and

$$H_2 = \frac{k_{21}}{s(\tau_2 s + 1)}$$

This shows that, in general, one pure integrator is needed in the forward path to produce zero steady-state error for step inputs.

(e) Repeat part (c) with $H_3 = 1/(\tau_3 s + 1)$ and $K_{12} = 0$.

(f) Repeat part (c) with $H_3 = 1/s$ and $K_{12} = 0$.

4-9 Sketch the Bode and Nyquist diagrams for

$$H_1(s) = \frac{6}{(s+1)[(s/2)+1][(s/10)+1]}$$

and

$$H_2(s) = \frac{(s/5)+1}{(s/50)+1}$$

Sketch the Bode and Nyquist diagrams when $H_1(s)$ and $H_2(s)$ are connected in series. Compare the resulting system to $H_1(s)$ acting alone. Is it more stable? Is it faster? That is, would a step response be completed earlier?

4-10 Often the success of an engineering analysis depends upon being able to "open the loop" on a system. If it is an electrical circuit, we can merely cut a wire. If it is a human physiological system, such an approach is not feasible. Stark (1968) developed two ingenious techniques for opening the loop in the human pupil system. The first was an electronic technique where the measured pupil area was used to modulate the light intensity. When the light intensity was increased, the pupil area decreased. The decrease in measured pupil area was used to again increase the light intensity; thus the effects of a change in pupil area were negated and the loop was opened. The controlled light falling on the retina became independent of the negative feedback of the pupil system. The light was then sinusoidally modulated and the pupil area was measured. The ratio of the input light intensity to the output pupil area, the open-loop gain, was calculated and Bode diagrams and Nyquist plots were constructed.

The second technique for opening the loop consisted of focusing the light in a beam smaller than the smallest possible pupil size. Thus no matter how the pupil diameter changed, the light falling on the retina was dependent upon system input, the light intensity, not system output, the pupil area (see Fig. P4-10b). The data for this method of opening the loop are shown in Fig. P4-10d.

Assume that the dynamics of the retina are fast; for example,

$$H(s) = \frac{1}{1+0.01s}$$

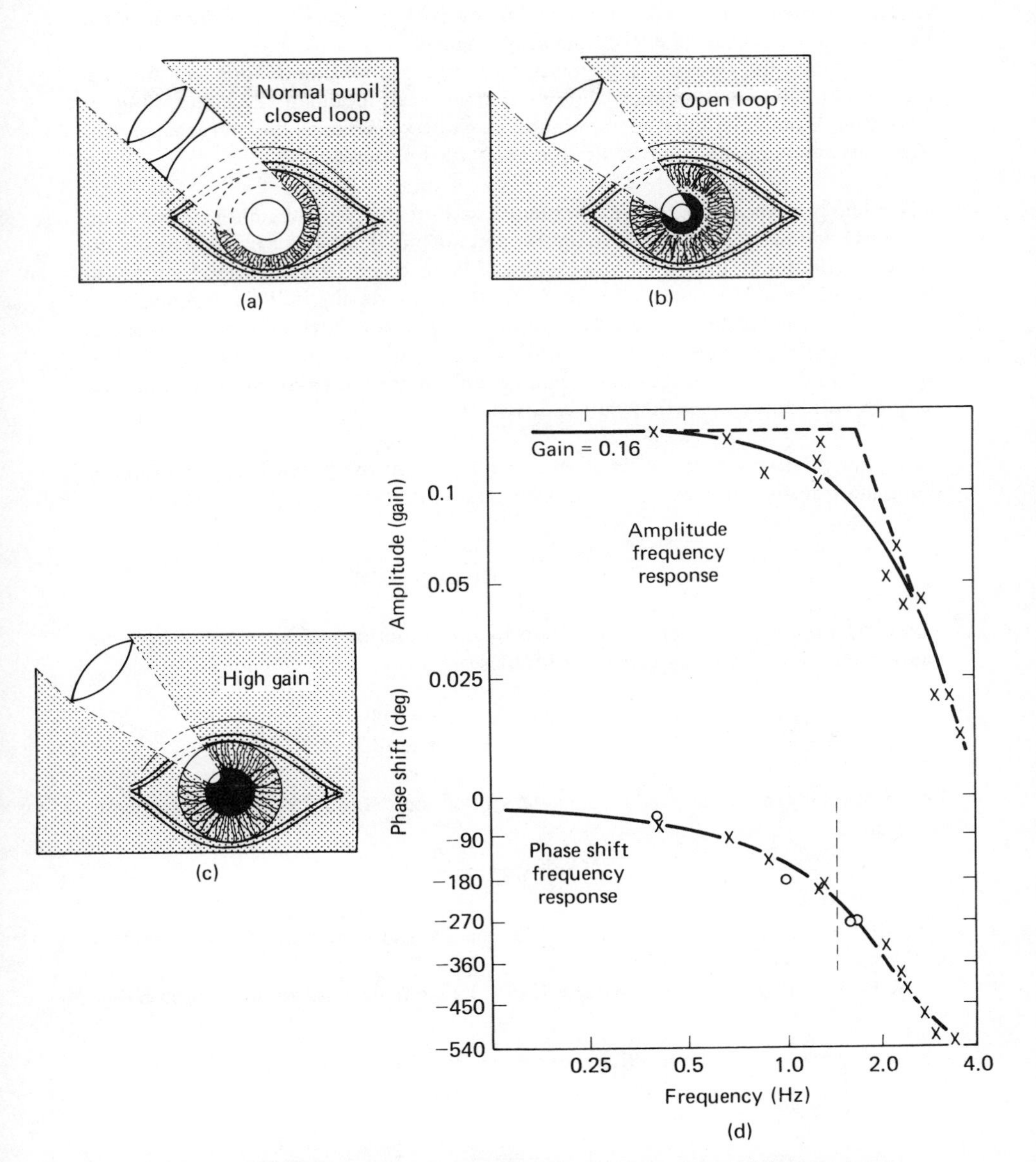

Figure P4-10 Light stimulus configuration to create normal (a), open-loop (b), high gain (c) pupil systems and Bode diagrams for the open-loop system (d). From L. Stark *Neurological Control Systems Studies in Bioengineering*, New York: Plenum Press, 1968.

Find the closed-loop transfer function for the data of Fig. P4-10d. Draw a block diagram showing where the loop has been opened.

The gain of the system can be made very large by focusing the light on the edge of the pupil. Now small changes in pupil diameter can modulate the light falling on the retina from maximum to minimum brightness. What is the purpose of this high-gain experiment? What is the frequency of the oscillation?

4-11 The primary ingredient in ointments used for the relief of minor muscle pain is methyl salicylate. It has an analgesic action, but it also causes (perhaps indirectly) a dilation of the blood vessels and thus an increase in the blood supply to the affected area. The muscle relaxes while the body repairs it. This chemical may affect the psychological temperature sensors, the physiological temperature sensors, the pain sensors, the blood vessels, and perhaps even the spindle organs. Explain which elements of the thermoregulatory and neuromuscular systems may be affected and explain how they can be affected.

4-12 The relative sensitivity function is the most convenient sensitivity function to use if the transfer function is written as a quotient.

$$M(s,\alpha) = \frac{N(s,\alpha)}{D(s,\alpha)}$$

where $N(s,\alpha)$ and $D(s,\alpha)$ are real continuous functions of s and α. Then the relative sensitivity function can be written as

$$\bar{S}_\alpha^M = \frac{\partial \ln(N/D)}{\partial \alpha}\bigg|_{\alpha_0} = \frac{\partial \ln N}{\partial \ln \alpha}\bigg|_{\alpha_0} - \frac{\partial \ln D}{\partial \ln \alpha}\bigg|_{\alpha_0} = \bar{S}_\alpha^N - \bar{S}_\alpha^D$$

For example, the system of Fig. P4-12 (based on Frank 1978) has a transfer function of

$$M = \frac{HP + RLP}{1 + LP}$$

Find the relative sensitivity function $\bar{S}_p^M$ of the transfer function $M(s)$ with respect to the plant $P(s)$.

Why is it significant that this sensitivity does not depend upon $R(s)$ or $H(s)$?

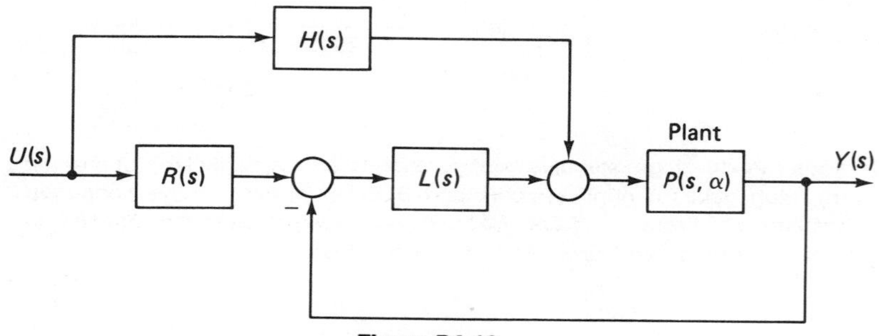

Figure P4-12

4-13 Some stroke patients become hemiplegic and cannot walk unassisted, although there is nothing wrong with their muscles. Electronic stimulators are available that monitor the ground contact of the heel and toe and provide electronic pulses to excite specific muscles (Strojnik, Kralj, and Ursic 1979). The system is a feedback-control circuit and is modeled in Fig. 4-1b. Assume that the electronic circuits, G_c, the leg, G_p, and the feedback elements, H, can be modeled as

$$G_c(s) = A$$

$$G_p(s) = \frac{e^{-0.1s}}{s(s+20)}$$

$$H(s) = 1$$

Sketch Bode and Nyquist diagrams for amplifier gains $A = 2$ and $A = 100$. At what frequency will the system oscillate if the gain is sufficiently increased? What value of gain, A, would be necessary to produce such oscillations?

4-14 The behavior of a squid giant axon membrane during a voltage spike can be modeled with the simple closed-loop system shown in Fig. P4-14. Assume that a pseudo impulse of a transmitter which opens sodium channels is applied to the membrane. Sketch the membrane voltage and the Na^+ and K^+ conductances as functions of time.

Marmont and Cole developed, and Hodgkin, Huxley, and Katz used, a voltage-clamp technique for the squid giant axon as a method of opening the loop of this normally closed-loop system. See section 1-9. Indicate on the block diagram where the loop was opened. Sketch the K^+ and Na^+ conductances as functions of time after the membrane voltage undergoes a step change from resting potential to zero. Indicate numerical values where possible.

Sketch the Na^+ and K^+ currents as functions of time for this open-loop situation. On the same axis sketch these currents for a normal closed-loop voltage spike. Label each curve. Explain differences between open- and closed-loop responses.

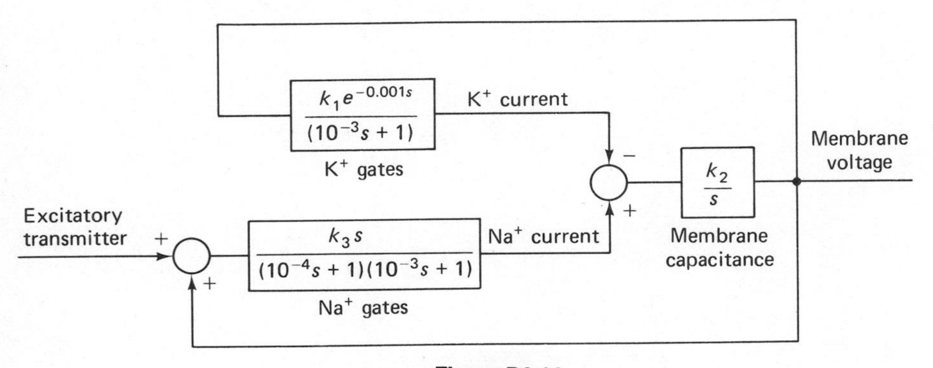

Figure P4-14

4-15 Would sensitivity functions for the transfer function, for the impulse response, and the step response provide the same information about a system? For the closed-loop feedback-control system of Fig. 4-1, let

$$G_c = H = 1$$

$$G_p = \frac{a}{\tau s + 1} = \frac{a\omega}{s + \omega}$$

Find the relative sensitivity of the transfer function with respect to the gain a and cutoff frequency ω. Derive the impulse response for the system [i.e., find $y(t)$ if $r(t)$ is a unit impulse]. Find the relative sensitivity of the impulse response with respect to the parameters a and ω. How does this differ from the results derived for the transfer function? Without solving the equations, estimate whether the steady-state step response will be more sensitive to a or to ω.

5

ELECTRICAL SAFETY

Electrical safety is a very important topic. However, it is usually not covered in traditional electrical and biomedical engineering courses. The principles are simple, but not necessarily obvious. They are equally applicable to the hospital or the home. More attention is paid to electrical safety in the hospital because patients, who are already ill, are more likely to be connected to a myriad of electrical devices. Furthermore, very low levels of current, which would be imperceptible to a normal individual, may be sufficient to electrocute a critical care patient. These dangers can be prevented by good circuit design and continual safety checks.

5-1 TYPES OF HAZARDS

5-1-1 Physiological Harm

Electric shock can cause pain, injury, or death. The amount of current flowing through the body determines the severity of the shock. This current level depends upon the impedances of the body and the contact interfaces. The impedance of the human body can be modeled as a core of low resistance (around 500 Ω) composed of electrolytes and tissues, and the skin with a higher resistance (1 to 100 KΩ). When current passes through the heart, it can cause a desynchronization of the muscle fibers, which prevents the heart from pumping the blood. This effect is called *ventricular fibrillation*. These and other effects of electrical shock on human beings are summarized in Table 5-1. The values given are only approximate; there is a great deal of intersubject variation.

Microshock is caused by small electric currents well below the threshold of perception. When such imperceptible currents pass through the heart they can produce ventricular fibrillation. As little as 80 μA may constitute a hazard. Currents of this magnitude are not regulated by ordinary electrical codes because they are not dangerous except in certain unusual environments, such as the cardiac care unit and the catheterization laboratory. They are commonly encountered in laboratories, homes, and workshops by normal people who feel nothing and suffer no ill effects. Yet the same current flowing through a nurse and into a pacing catheter may electrocute a patient.

Animal studies were used to set the initial current limits for the electrically susceptible patient. These limits were usually set at about 10 μA. More recently, in three separate studies, measurement of the amount of electrical current necessary to induce ventricular fibrillation in the human heart during open heart surgery showed a minimum of 80 μA (Rowley 1976), 100 μA, and 180 μA (Grass 1978). Accordingly, safety standards have been revised to allow currents as large as 100 μA.

In studies of the functional dependence of ventricular fibrillation threshold on frequency in animals, it has been found that 60 Hz is the optimum frequency for producing ventricular fibrillation. As the frequency increases or decreases, it requires more current to cause fibrillation. The current pathway also has an affect on the amount of current necessary to cause fibrillation. If the electrical path does not cross the chest, fibrillation is unlikely even if the skeletal muscles are stimulated into uncontrolled contraction. If a continuous electrical current is applied to the heart below

TABLE 5-1 Effect of 60-Hz ac currents through the body of adult human beings.

Less than 1 mA	Microshock: imperceptible when applied externally; as little as 80 μA applied through myocardium may induce ventricular fibrillations
1 mA	Threshold of perception
5 mA	Maximum harmless current intensity
10 mA	Let-go current: if passed through the hand or arm, may cause contraction of flexor muscles and cause inability to release grip
100–300 mA	Causes ventricular fibrillation, the desynchronization of activity in ventricles of the heart, and cessation of pumping
5 A	Causes burning of tissues.

the ventricular fibrillation threshold, it is easier for spurious electrical pulses or successive premature ventricular beats to induce fibrillation. Human beings are more susceptible to electrical current during the recovery phase of the heart beat. Some factors that can affect the patient's susceptibility are:

1. Voltage.
2. Frequency.
3. Current path through the body.
4. Electrical resistance of the body.
5. Occurrence in the cardiac cycle.
6. Length of time the current is applied.

The interrelationship of time and current is seen in the following equation from Dalziel and Lee (1968), which expresses the predicted current necessary to fibrillate one-half of 1% of the adult human population.

$$I = \frac{116 \text{ mA}}{\sqrt{\text{time}}}$$

There are also many psychological and clinical factors which affect electrical sensitivity, such as body weight, wounds, myocardial eschemia, anoxia, hypoxia, skin moisture, and recent use of drugs. These and other factors affecting ventricular fibrillation thresholds are discussed in Olson (1978), and the AAMI Standards (1978).

Patients are generally more susceptible to harm from small electric currents than other adults. The impedance of the skin is often deliberately reduced by scrubbing in order to give a good contact for skin electrodes. The current path may bypass the skin if intravenous fluid units or catheters are in use. Therefore, the electrically susceptible patient may have the lowest possible path of resistance to the heart.

5-1-2 Static Electricity

Static electricity may be dangerous to people and sensitive equipment (e.g., anything containing CMOS integrated circuits). Static electricity shocks could startle a person and cause an accident. Sparks from static electricity could ignite flammable gases, causing an explosion. It has also been suggested that shocks from static electricity could cause cardiac arrest if applied to a pacing catheter.

Floor carpeting is a very common source of static electricity charge buildup. Therefore, carpeting is not recommended for patient care areas where pacing catheters or similar instruments are in use. Methods intended to abolish static electricity are imperfect and necessitate grounding of the carpeting, thus creating an additional grounded surface, which further complicates the problem of making the electrical environment safe.

5-1-3 Explosion Hazard

Gases used for anesthesia, particularly the older varieties, are explosive. Therefore, operating rooms and the equipment used in them require special design to reduce the risk of anesthetic explosions resulting from electric sparks. To minimize the occurrence of static sparks, the floor is made conductive and is grounded. An isolation transformer is then used to reduce the shock hazard to personnel. The explosive gases and vapors can also be ignited by sparks from electronic and electric equipment, so power connectors and equipment used in operating rooms must be designed to prevent sparks.

5-1-4 Interruption of Power

Interruption of electrical power to life-support equipment can also be hazardous. If a delay occurs before emergency power is brought into operation, the failure of a respirator, monitor, defibrillator, pacemaker, or other life-support equipment can be fatal. The possibility of a power failure must be considered in the planning of a power distribution system. Electrical service to life-support equipment should be, as nearly as possible, uninterruptable. Emergency power-generation equipment within the hospital is necessary to provide limited power in case of power failure. The National Electrical code allows 10 sec for the transfer of power from normal to emergency sources (Summers, 1978). This may necessitate constantly running turbines operating in a no-load mode, quick starting motor-generator sets or battery systems. The emergency power system must share the same common ground with the regular power system.

5-2 WAYS OF AMELIORATING THE SITUATION

5-2-1 The Three-Wire Electrical Distribution System

Most modern buildings have a three-wire power distribution system. The three wires are the hot conductor (black insulation), the neutral conductor (white insulation), and the ground conductor (green insulation or bare wire). Note that this color coding differs from the coding used inside electronic devices, where black is usually the ground. The ground conductor is sometimes called the *U-ground* because of its U-shaped hole in the receptacle. At the service panel (Fig. 5-1), both the neutral and ground conductor are tied to an earth ground. Earth ground should be a combination of the cold-water inlet pipe to the building, a ground wire provided by the electric utility company, and some other grounding device. Although both neutral and ground are tied to the same point at the service panel, they have different purposes and are not interchangeable. In a properly wired system, electrical current flows down the hot conductor and returns via the neutral conductor. Little or no current flows in the ground conductor. In the absence of the U-ground conductor, frayed insulation on the hot wire could place the instrument case at a high voltage potential. The current could then return to ground through the user (see Fig. 5-2). If the insulation on the hot wire frays when the U-ground is connected to the instrument case, a large current immediately flows through the U-ground connector. This should blow the protective fuse or circuit breaker, thus interrupting power to the device.

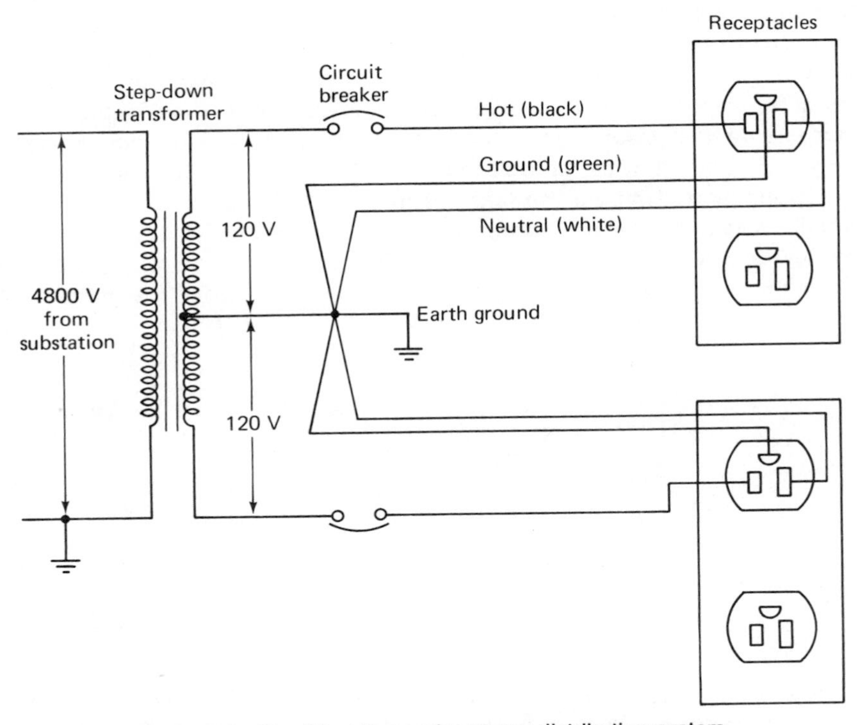

Figure 5-1 Simplified three-wire power distribution system.

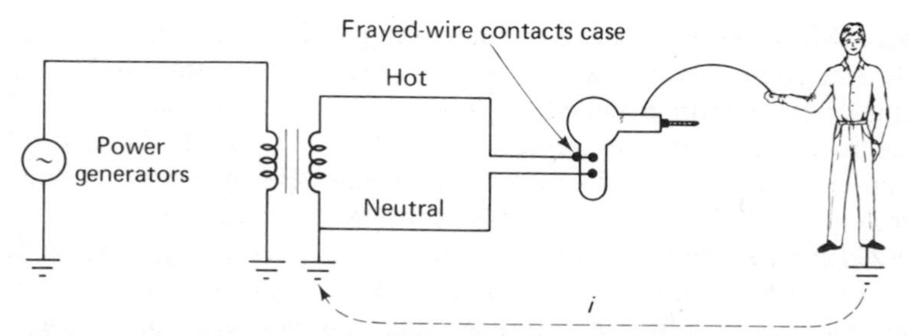

Figure 5-2 Current pathway for electrical shock. Current can pass through the user to the ground if the instrument does not have a ground conductor and if the insulation of the hot conductor becomes frayed.

5-2-2 Ground Integrity

Most electrical power sources have a common reference point, earth ground; therefore, it is possible to be shocked if you touch the hot side of the system while any part of your body is in contact with a ground path. The purpose of the ground conductor is to provide a low-resistance path to ground for any fault currents that could be produced by leakage, improper wiring, or misuse. For example, the cases of most instruments are grounded in normal use. If, because of a fault in the instrument, the hot conductor should come in contact with the case, the ground conductor provides a path for current flow. If a two-wire adapter plug is used, negating the purpose of the ground line, the fault condition would not be remedied (Fig. 5-3), and all of the fault current would flow through the user.

Equipment grounding should not be confused with patient grounding, which is used to reduce noise. A patient ground should offer a current limited pathway to building ground, unlike an equipment ground, which is intended to divert large hazardous fault currents from the patient. A

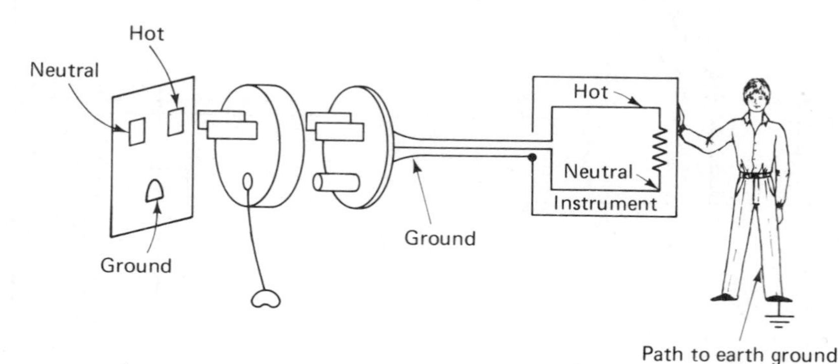

Figure 5-3 Hazard created by bypassing the U-ground circuit. If a three prong-to-two prong adaptor is used a potential current pathway through the person develops.

low-resistance ground connected to the patient would be dangerous, if he or she were to touch a second device which acted as a current source, such as a monitor or electric bed. This low-resistance ground could also provide an unintended current pathway during electrosurgery or defibrillation.

In older buildings, the hot and neutral conductors were often run in a metal conduit. The metal conduit was then used as the ground conductor. This system is not optimal. When new, a metal conduit is a good conductor; however, with time, joints open or become oxidized, increasing electrical resistance. When 60-Hz noise interferes with the desired physiological signal, the instruments are sometimes directly grounded to the nearest coldwater pipe. However, sections of the water line may have been built with plastic pipe and, although the water may provide a conductive path, the integrity of the pathway is compromised. Of the three grounding methods (grounding conductor, metal conduit, or water pipe), the grounding conductor is the best choice for safe operation and is the method required in health care facilities.

5-2-3 Single-Point Grounding

Single-point grounding requires that all equipment in a room be connected to ground at a single point. This discussion of grounding is based on Roth, Teltscher, and Kane (1975). The dangers of using two ground conductors are illustrated in Fig. 5-4, which shows a patient's room and an adjoining room. If a fault occurs in a piece of equipment located in the adjoining room, a large current may flow in the ground wire to earth ground. A part of this current may flow through the patient, despite the fact that the patient is located far from the source of trouble. In Fig. 5-4a, the two instruments connected to the patient are grounded via the U-ground conductor in the power cord. However, each receptacle is grounded to a different house ground. The current from the fault divides so that a portion of the fault current travels to the house ground through the instruments and the patient. These hazardous currents may be due to faults or to leakage currents. The motor windings of a vacuum cleaner are often exposed to dust (often damp), which provides a good current path from the hot line to the case. This ground wire current is called *leakage current*.

Single-point grounding means not only that one ground conductor should be used, but also that only one point on that conductor should be used for ground connections. Figure 5-4b shows two receptacles tied to different points on a ground conductor. Even though both receptacles are connected to the same conductor, their attachment at different locations permits a portion of the fault current to pass through the patient. This current through the patient may be dangerous. If the resistance between the two ground points is 0.05 Ω and the patient is modeled as a 500-Ω

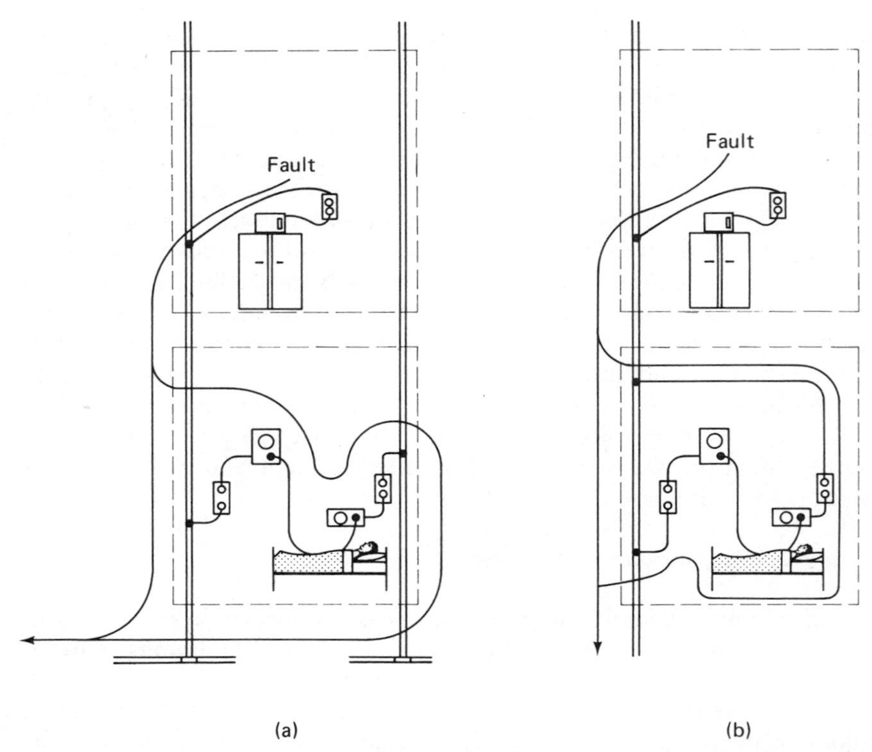

Figure 5-4 A fault outside the patients's room can cause current to flow through the patient if the equipment in the patient's room is (a) grounded to different building grounds or (b) grounded at different locations along a single building ground. Based on Roth, Teltscher and Kane, 1975.

resistor, one ten-thousandth of the fault current will pass through the patient. For a 20-A fault, this current would be 2 mA.

Interestingly, single-point grounding is also an important principle in noise-reduction design (Ott 1976). If the patient's room in Fig. 5-4 measures 10 ft by 10 ft and single-point grounding were not used, there would be a large conductive pathway surrounding the room called a ground loop. Its area is about 100 ft^2. A voltage will be induced in this loop in accordance with Faraday's law.

$$e = N\frac{d\phi}{dt} = A\frac{dB}{dt} = 100\ \text{ft}^2\frac{dB}{dt} = 9\text{m}^2\frac{dB}{dt}$$

Assuming a 60-Hz magnetic field of 10^{-4} weber per square meter (the same order of magnitude as the earth's magnetic field or the magnetic field 1 cm away from a long, straight conductor carrying a 5-A current), the

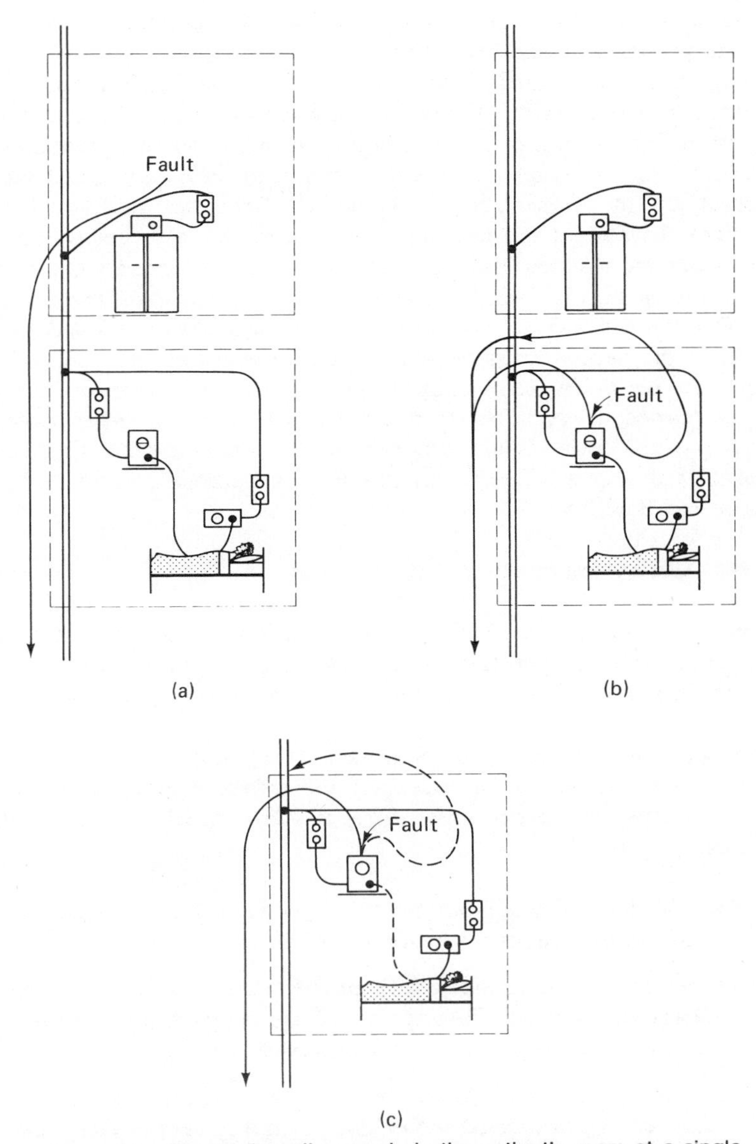

Figure 5-5 Connecting all grounds in the patient's room at a single point (a) will protect the patient from fault currents originating outside the room, but (b) not from those originating in the room. To protect against faults in the room, isolated patient leads (c) can be used. Based on Roth, Teltscher and Kane, 1975.

induced voltage is 0.34 V. This induced voltage will produce noisy signals as well as present a hazard to the patient.

Dangers caused by faults originating inside the patient's room are not eliminated by single-point grounding. If a fault occurs in one instrument, as shown in Fig. 5-5b, a possible fault current will enter the ground of the instrument and divide into two paths. One path leads to ground via the U-ground of the power plug; the other flows to ground through the patient. To correct this problem, instruments are designed so that the patient leads are isolated from ground. With such equipment, the current flow, as shown in Fig. 5-5c, is altered so that virtually no fault current passes through the patient. The actual current depends upon the degree of isolation of the patient leads. With proper equipment design, this isolation can usually be made sufficiently large so that fault currents are, for practical purposes, precluded from entering the patient. When possible, isolated patient leads should be used in conjunction with single-point grounding techniques. (Design techniques for isolated patient leads are discussed in Chap. 2.)

5-2-4 Grounding in Critical Care Areas

Special grounding techniques are required for critical care areas, such as emergency rooms, operating rooms, intensive care units, coronary care units, catheterization rooms, recovery rooms, delivery rooms, and dialysis units.

The National Electric Code (Summers 1978) describes the grounding requirements for anesthetizing locations and critical care areas. Although the requirements for the two areas differ in some respects, both use these principles:

1. Establishing a single-point ground by installing a single ground connection from points inside the room to points outside.
2. Limiting wire size and length in the room in order to keep the voltage drops small. Testing the voltage drops to ensure that they are below the maximum acceptable level.
3. Designing the surroundings so that the patient is certain to be confined and protected within the ground system of one room.

In critical care areas there are three separate ground points: the patient equipment grounding point, the room reference grounding point, and the building ground. The following example, based on Roth, Teltscher, and Kane (1975), shows how these grounds are interconnected.

All grounds in the patient's area are connected to the patient equipment grounding point. These connections may be made with a patch cord, with the ground wire of the power cord, or with a separate fixed wire. Each of these techniques is shown in Fig. 5-6. In Fig. 5-6a, design criteria are noted on each path. The patient equipment grounding bus (A) is the central point for all grounding for a given patient. Another patient may share this bus or may be tied to another patient equipment grounding bus. However, no patient will be connected to more than one patient equipment grounding bus. The power receptacles in the patient's area (B) are connected to the bus, permitting electrical instruments to be grounded by way of the U-ground connection in the receptacles. Special devices may be connected to the bus by use of a patch cable (C), and fixed points such as the bed may be connected by a fixed wire (D). This patient equipment ground bus is connected to the room reference grounding bus (E) by a single wire. The room reference grounding bus is in turn connected to the building ground.

The design limits for wire length and size were translated into resistance values in Fig. 5-6b. No allowance was made for contact resistance, except for the patch cord, which included an allowance of 0.005 Ω for each connection.

In this example no ground was permitted to have more than 0.05 Ω resistance from the room reference grounding bus. Therefore, it was important to minimize the resistance between E and A. To hold the total resistance below 0.05 Ω required that the length of wire between points E and A be less than 26 ft, for AWG No. 10 wire. For a 26-ft length, the resistance between points E and A was $26 \times 0.000999 = 0.026$ Ω, and that, added to the resistance between A and B, just reached the limit of 0.05 Ω.

If a longer wire were needed to connect A and E, a larger size wire would have to be used. For example, using AWG No. 6 wire from E to A permits that length to be increased to 66 ft. The wire resistance of AWG No. 6 is 0.000395 Ω/ft, which at 66 ft yields $66 \times 0.000395 = 0.026$ Ω. It is also possible to gain some advantage by using a larger wire size for run AB, but this improvement is limited because the patch cords between A and C quickly become the limiting resistance values for the installation. Patch cords are usually AWG No. 10, so lower resistance values can only be obtained by shortening the cords. A length of 6 ft is about as short as is reasonable, and such a cord has a resistance of 0.012 Ω. Adding the contact resistance of 2×0.005 Ω yields a total patch-cord resistance of 0.017 Ω.

With this grounding configuration the patient is protected against faults, as shown in Fig. 5-7a. Assume that an instrument plugged into a receptacle (B) develops a fault condition in which some of the power-line

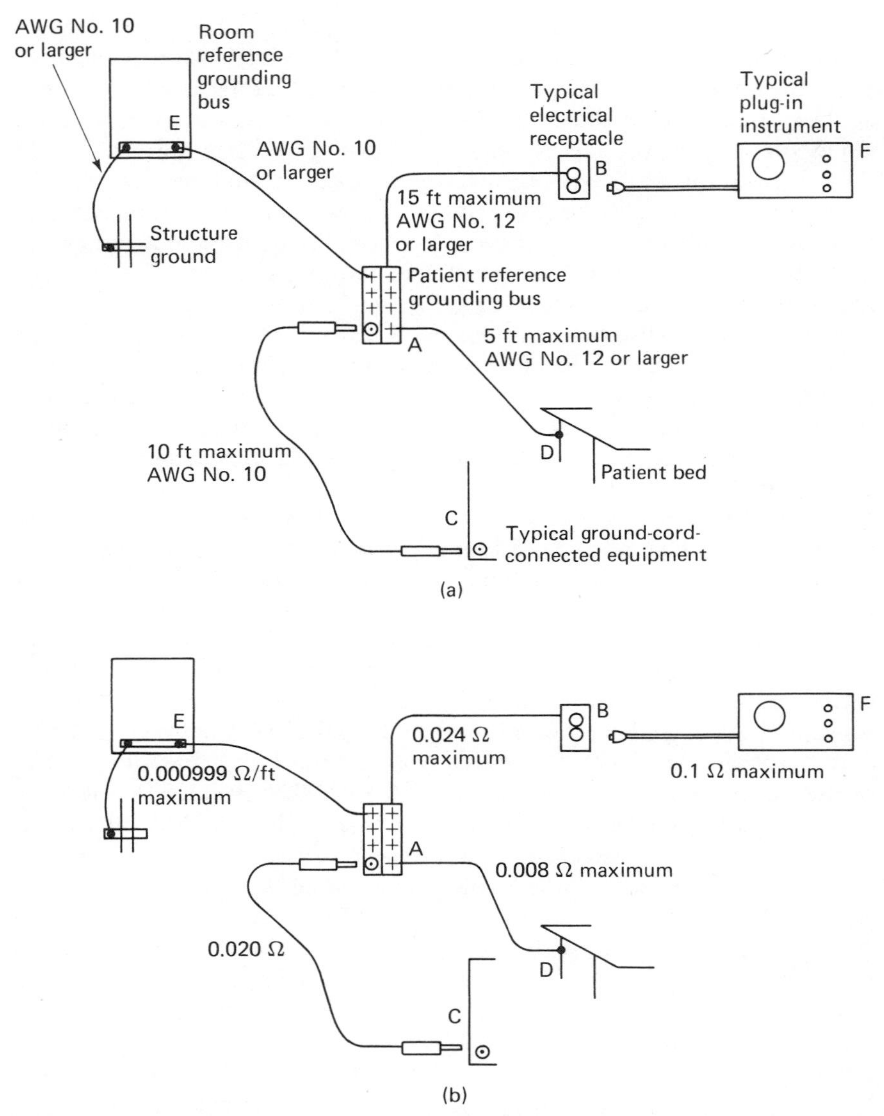

Figure 5-6 (a) Wire length and size limitations for an example of ground resistance control. (b) Nominal resistance between controlled points A, B, C, and D in the patient area and point E outside the patient area. (From H. H. Roth, E. S. Teltscher, and I. M. Kane, *Electrical Safety in Health Care Facilities*, Academic Press, Inc., New York, 1975.)

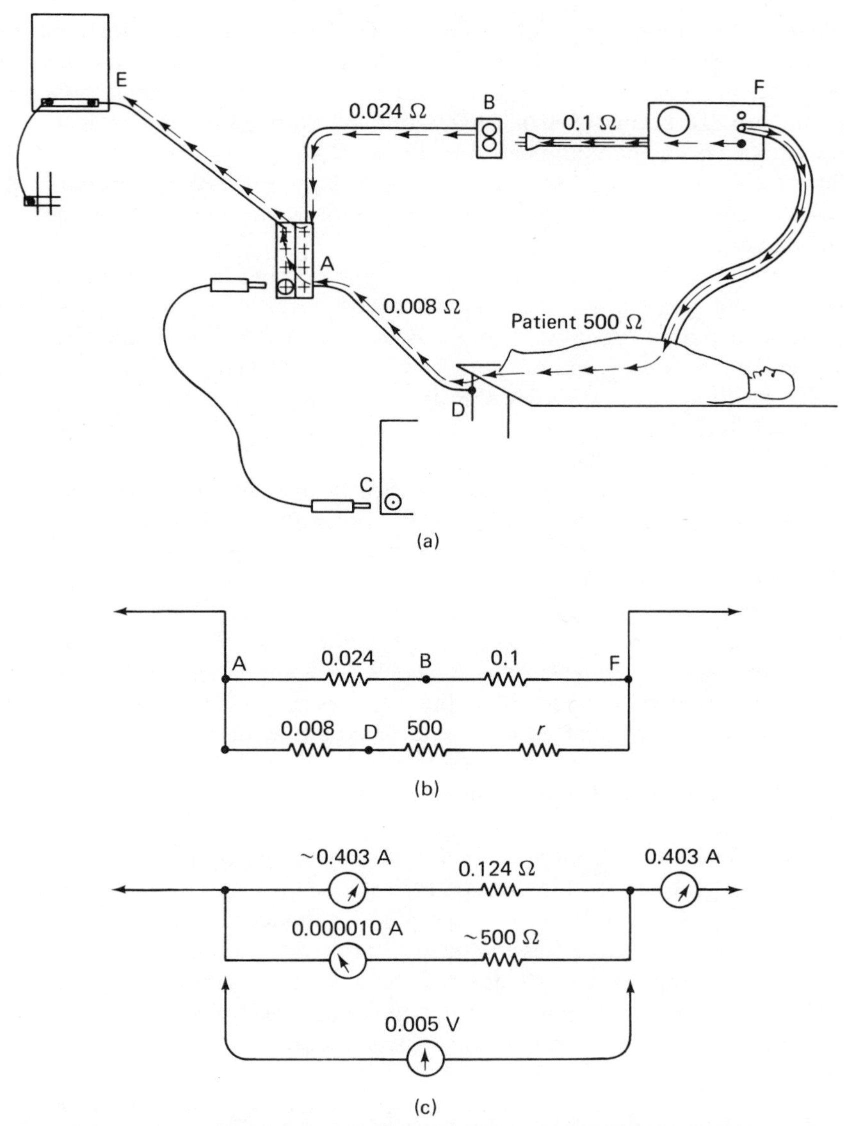

Figure 5-7 Effects of 40-mA fault current. (a) Dual path taken by current. (b) Branch through patient has much larger resistance, thus diverting most of the current around patient. (c) Current in each branch and voltage on patient. (From H. H. Roth, E. S. Teltscher, and I. M. Kane, *Electrical Safety in Health Care Facilities*, Academic Press, Inc. New York, 1975.)

current flows into the ground wire of the instrument. The fault current returns to the power source via two pathways, one of which leads through the patient. The path via the ground wires has a very small resistance value, not exceeding 0.1 + 0.024 Ω, while the body path has a resistance of about 500 Ω. The two paths are shown, in terms of resistances, in Fig. 5-7b. The value of r is not known, but since the conductor is metallic, the resistance is not likely to be significant compared to the body resistance of the patient.

Roth, Teltscher, and Kane tried to keep voltage drops within grounds to less than 5 mV in the example. The fault current required to reach this limit is shown in Fig. 5-7c. When 0.403 A emanates from the instrument, the voltage between points A and F reaches 5 mV. Of the 0.403 A, only a small fraction, 10 μA, flows through the patient; the remaining current bypasses him or her by way of the ground wires. The 10-μA current in the patient is safe for human beings.

The 1978 version of the National Electrical Code is more lenient than the 1971 code used in this example. The newer code only prohibits any two exposed conductive surfaces in the patient area (except tables, chairs, trays, pitchers) from having potential differences greater than 500 mV in general care areas or greater than 100 mV in critical care areas. Furthermore, each receptacle must be grounded with an insulated copper grounding conductor, and at least one receptacle for each patient bed in a critical care area must be connected to the emergency power system. The 1981 Code limits potential differences in critical care areas to 40 mV.

5-2-5 Isolation Transformers

As previously mentioned, many anesthetic gases are explosive. To minimize the likelihood of ignition from static electricity, the floors of operating rooms are made of a conductive material and are grounded. This, however, creates a possible hazard to the people using the rooms, for if there is a malfunction, there will be a path to ground through their bodies. To minimize this hazard, the entire electrical service for the operating room is provided by an isolation transformer, as shown in Fig. 5-8.

There are, however, many serious drawbacks to using isolation transformers for electrical distribution systems. They are expensive and hard to maintain. They increase the source impedance and put restrictions on the maximum power capability. Devices (ground-fault monitors) must be provided to automatically check the integrity of the system, and these devices actually provide a path to ground and introduce noise on the power lines.

It is, however, practical and desirable to have the power supply in each device incorporate an isolation transformer so that there is no direct

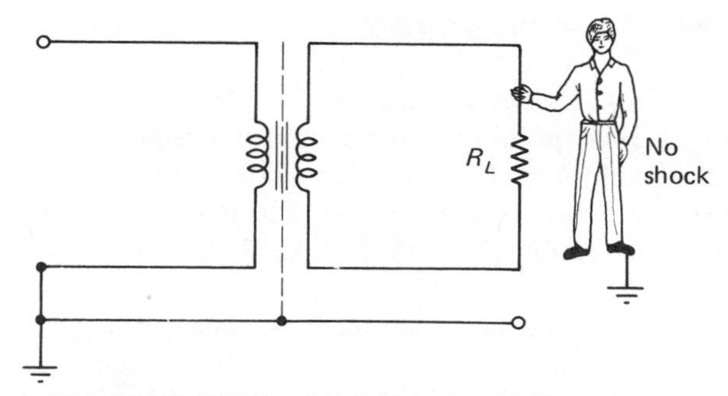

Figure 5-8 Schematic diagram illustrating how an isolation transformer can protect against shock even if the subject touches a hot conductor.

connection between the power line and the patient circuitry. At least one electrostatic shield should be incorporated in the transformer between the primary and secondary coils, and this should be connected to the power ground. This shield will improve the isolation by reducing the capacitive coupling between the primary and secondary windings.

5-2-6 Double Insulation

Most new small hand tools, such as electric drills, are now double-insulated. This concept in electrical safety is also used in medical instrumentation. The housing associated with electrical circuitry may be grounded as usual via the U-ground; however, this housing should not be in electrical contact with any metal part to which the patient or the operator might be exposed (see Fig. 5-9). To double-insulate an instrument you use a completely functional electronic device, wrap it in insulation, and then put it in its final enclosure. All the shafts penetrating the insulation barrier must be nonconductive. For use in critical care areas the outer case should not have any exposed conductive surfaces. Double-insulated equipment is not always grounded through the power supply cord: it sometimes has only a two-pronged plug.

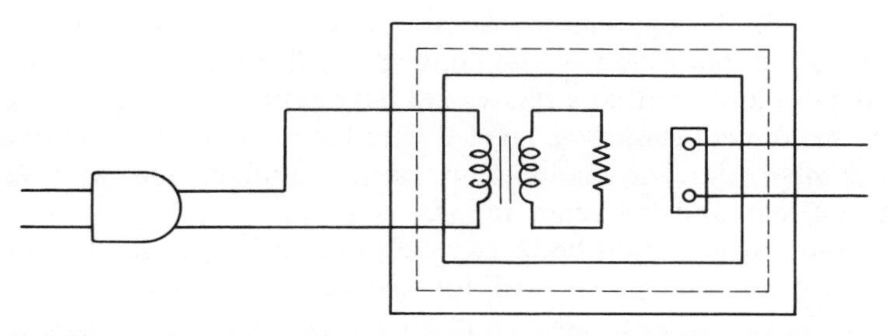

Figure 5-9 Double insulation ensures that the outer case cannot come in contact with the hot line.

5-2-7 Good Engineering Design

Electrical safety is greatly enhanced by good engineering design. The following paragraphs point out some useful design concepts. They are based on I. S. Wright, and D. T. Frederickson, "Inter-Society Commission for Heart Disease Resources, Report of Electronic Equipment in Critical Care Areas," *Circulation*, 44 (1971), A237–61, by permission of the American Heart Association, Inc.

Equipment should not become unsafe when exposed to any conditions likely to be encountered in normal use, such as high or low humidity, extreme temperatures, prolonged storage, or mechanical stress. It should be possible to disinfect a device and, if reasonable, sterilize it. The markings on the case and the dials should not be affected by cleaning agents. The instrument's center of gravity should be low enough to prevent its tipping under any but the most extraordinary circumstances. Knobs and levers should not protrude excessively. Surfaces that are accessible to patients should be nontoxic, even to children. The device should be designed without top openings, so that spilled liquid will be unlikely to flow inside. In some cases, the device should be designed to prevent entry even of splashed or sprayed liquids and should perhaps be immersible. Ventilation and other openings into the case should be small enough to make it impossible for a child to insert a finger or small metal object, and should be placed so that if a conductor were inserted, it would not contact a high voltage. The instrument case should prevent dispersion of molten metal or flaming particles in the event of a fault. Cathode ray tubes should be mechanically secure and protected against damage; implosion of the tube should not be a hazard to the user. The power cord's point of entry into the device requires protection by a high-quality strain-relief device to prevent the transmission of force to the cord's internal connections. The power-cord insulation should be protected from damage by the edge of the entrance opening.

To test the adequacy of this strain relief device, a 35-lb weight is suspended on the cord and supported by the appliance so that the strain-relief device is stressed from any angle that is permitted by the construction of the appliance. The strain relief is not acceptable if, at the point of disconnection of the conductors, there is such movement of the cord as to indicate that stress would have resulted on the connections.

Each instrument should be labeled with the name of the manufacturer, model number, date of manufacture, serial number, and with voltage, wattage, current, and frequency ratings.

The front panel should be clearly designed with controls that are easy for the user to understand and to operate. They should not require simultaneous reference to written instructions. Dials should be calibrated in standard units of measurement, and clockwise rotations should increase the labeled function; for example, turning a gain knob clockwise should increase gain, turning an attenuator knob clockwise should increase the attenuation. Rear-panel jacks should be clearly identified. Equipment should be labeled to indicate whether it is suitable for use in a critical care area or in the presence of flammable gases. Emergency equipment may need to have simplified instructions affixed to it.

The method of lead attachment should be, as far as possible, noncritical. Controls should not interact with one another, unless this is desirable for the operation of the device. The shafts of switches or other controls should not be electrically alive.

Overload protecting devices (fuses, circuit breakers) should not open during normal use. They should be labeled as to current rating. Routine replacement of fuses, pilot lamps, batteries, or other user changeable components should not be a shock hazard. Such parts should be readily accessible without the use of tools. Access to live parts should be possible only with the aid of a tool, unless automatic disconnection from the power supply is incorporated.

5-3 LEAKAGE CURRENT

A *leakage current* is an extraneous current flowing along paths other than those intended. It could be due to resistive, inductive, or capacitive effects, but in most instruments with ac power, the leakage current is capacitive. There are three primary sources of leakage: the power cord, the power-line filter, and power transformer, as shown in Fig. 5-10.

The capacitor between the hot wire and ground, C_3, models the distributed capacitance in the power cord. It is normally 1 to 2 μA per foot. Special low-leakage power cords are available with maximum leakage current to the green ground wire of 0.5 μA/ft. Leakage into the ground wire is dangerous because if the ground were broken at the plug of a 10-ft cord, 10 to 20 μA of current could flow from the instrument case through the patient to a ground.

Some instruments have capacitors, or some combination of capacitors and inductors, connected between the power lines and ground to filter out spurious signals. These filters are usually bilateral: they will prevent noise from entering the instrument, and they will prevent noise from leaving the instrument and being introduced onto the power lines. These power-line filters produce leakage currents in the grounds and are therefore undesirable in biomedical instruments.

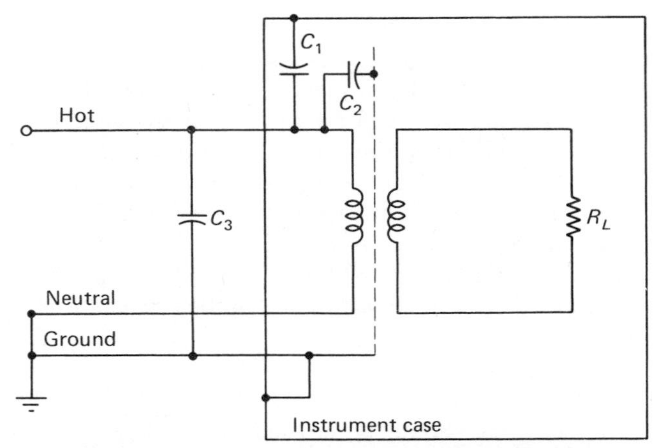

Figure 5-10 Model of a biomedical instrument attributes most of the leakage current to capacitive coupling between the hot wire and the case C_1, the transformer shield C_2, and the ground wire C_3.

Most instruments have a power transformer that transforms the power-line voltage into various secondary voltages. The input is applied to the primary windings and the outputs are taken from the secondary windings. Both of these windings are usually placed on a ferromagnetic core. Thus there are two conductors separated by an insulator, which provides capacitive coupling between the primary windings and the core. Most of the leakage current is a result of the capacitance between the primary winding and ground. Leakage currents vary considerably among instruments, depending upon the physical size of the transformer, how it is constructed, and which side of the primary winding is connected to the neutral conductor. Regulatory limits for power-line leakage currents are around 100 μA.

5-3-1 Dangers of Leakage Currents

The following hypothetical example from Rowley (1976) illustrates the dangers of leakage current. A patient in an electrically operated bed, as shown in Fig. 5-11, has a pacemaker with a bipolar catheter going to the right ventricle of the heart via the right jugular vein. The pacemaker instrument case is connected to the ground conductor of the power cord. The three-wire, 8-ft power cord is connected to a two-wire, 10-ft extension cord which is plugged into the three-wire power outlet. The bed frame is properly grounded to the power system. The patient's left hand is resting on the bed frame. There is a total of 180 μA of leakage current in the

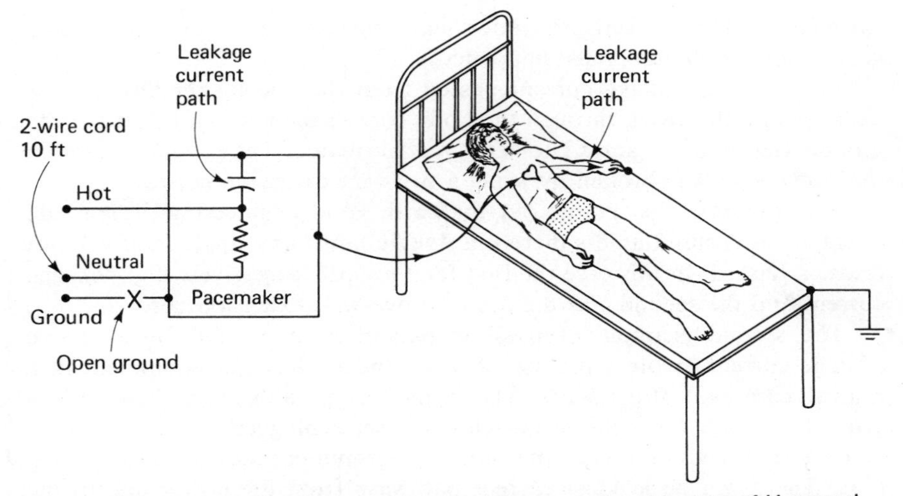

Figure 5-11 Safety hazard created by leakage current and bypass of U-ground.

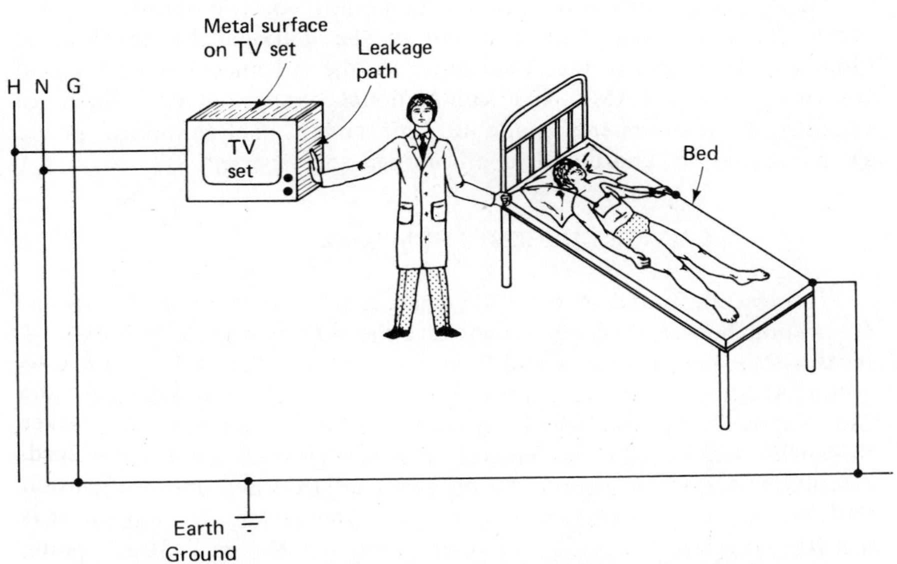

Figure 5-12 Safety hazard created by leakage current and lack of a U-ground.

pacemaker. The patient's monitor has triggered an alarm indicating ventricular fibrillation. What happened?

In this case, leakage current passed from the pacemaker through the catheter into the heart, through the body core to the left hand, and then to ground via the bed frame. The heart was fibrillated. The ground system at the pacemaker was broken by using a two-wire extension cord.

This example also points out that with good engineering design safe operation is ensured unless there is a double fault: any single fault will not create a hazard. In this case the first fault was the excessively high leakage current and the second was the use of a two-wire extension cord.

The second example of possible patient harm caused by excessive leakage current involves the use of ungrounded electrical equipment in a critical care unit (Fig. 5-12). The patient is in a hospital bed with a grounded frame. Near the bed a television set is plugged into an electrical receptacle. Television sets are not normally grounded: they use a two-prong plug. There will be leakage current pathways from the power line to the chassis. An attendant touches a metal point on the television and also touches the patient, or an intermediary such as a urine spill. Current passes through the patient and causes fibrillation.

5-3-2 Leakage-Current Testing

Three leakage-current tests should be performed on biomedical instruments. They are described in detail in the National Electrical Code (Summers 1978) and in the Association for the Advancement of Medical Instrumentation's (AAMI) American National Standard of 1978. Each test consists of measurements made in four circuit configurations: power switch on and off, and polarity both normal and reversed.

5-3-2.1 LEAKAGE CURRENT FROM CASE

The test illustrated in Fig. 5-13 ensures safe operation if the ground connection between the instrument and the wall receptacle is broken. It measures the current that would flow through the patient if he or she were grounded and touched the instrument case. It measures the leakage current from exposed metal surfaces to ground when the grounding prong of the three-wire power cord is disconnected from the grounding connector of the receptacle. A current meter and a dummy load are wired in series between the exposed metal surfaces and ground. The leakage current is then measured for the following four configurations: the instrument's power switch on and off, and the polarity reversing switch in alternative positions.

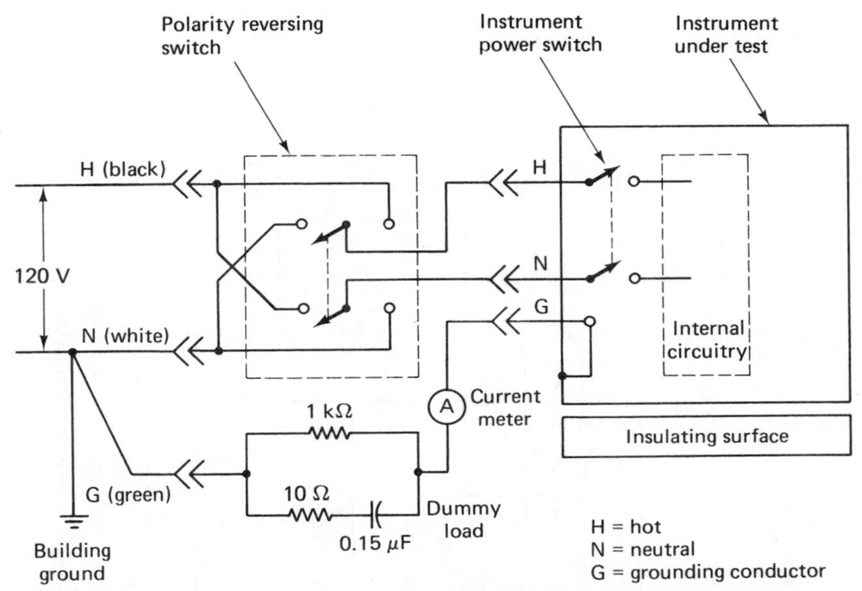

Figure 5-13 Test circuit for measuring leakage current from the case.

An alternative, but simpler way to perform this test is to insert the three-prong plug into a three prong-to-two prong adapter plug and measure the current flowing through the dummy load from the green wire to ground. The polarity can be reversed by reversing the two-prong plug. The maximum allowable value for this leakage from the chassis to ground is 100 μA. The dummy load is frequency-dependent. The standards allow more leakage current of frequencies about 1 kHz because they pose less of a physiological risk.

5-3-2.2 LEAKAGE CURRENT OUT OF PATIENT LEADS

The second test measures the leakage current that the instrument supplies to the patient through the patient leads if the power-cord ground pathway were interrupted. The test is made with patient leads active: for example, in the case of a cardiograph, the lead selector switch is in an operating position, not in the standardizing position, and an appropriate dummy load is connected to the leads. This test consists of four measurements: the appliance power switch on and off, and the polarity reversing switch in alternative positions (see Fig. 5-14). The maximum source current that can emanate from the patient end of the cables is 20 μA in critical care areas and 50 μA in other areas.

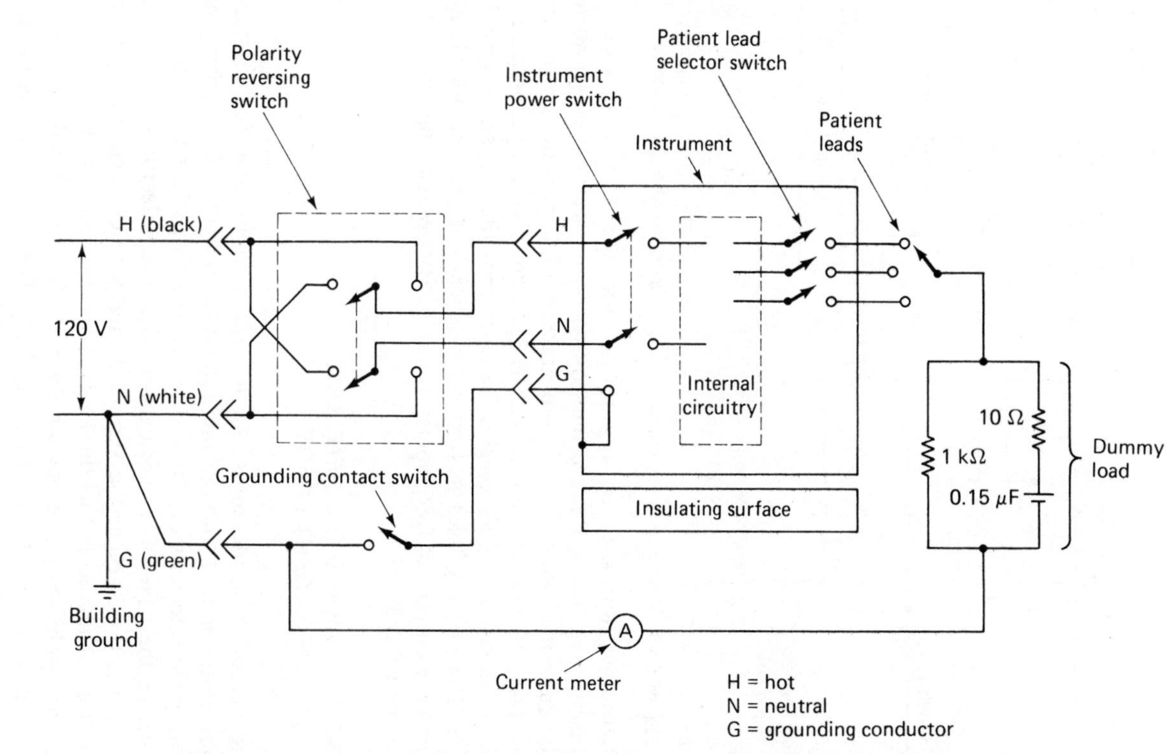

Figure 5-14 Test circuit for measuring leakage currents sourced from patient leads.

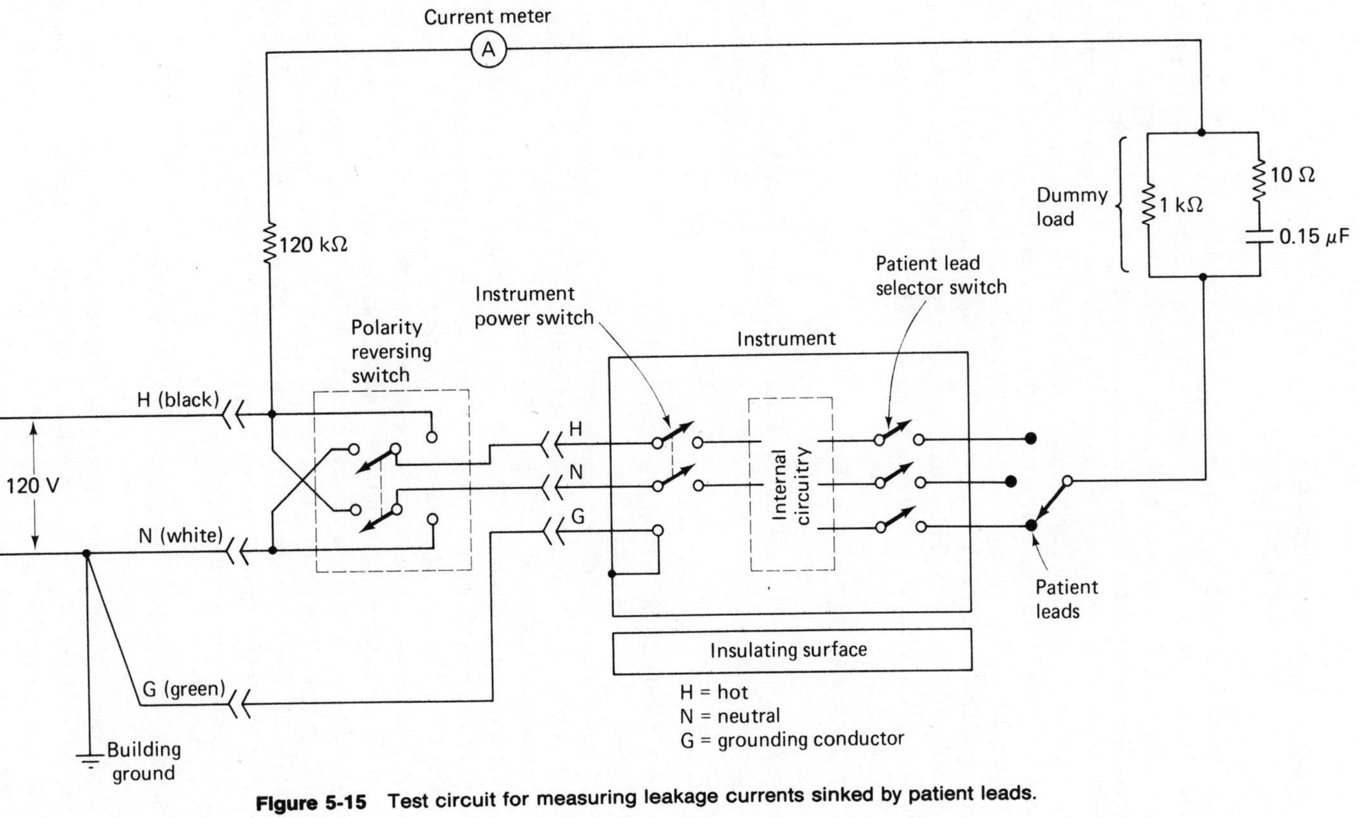

Figure 5-15 Test circuit for measuring leakage currents sinked by patient leads.

5-3-2.3 LEAKAGE CURRENT INTO PATIENT LEADS

If the patient contacts a hot wire, current will flow through his or her body only if there is a body-to-ground path. The instrument and its patient leads may offer the pathway to sink this current. The internal circuitry can protect the patient from these dangers by limiting the possible leakage to small values. The third test measures the leakage from external sources. The current in each isolated lead is measured by applying an external source of power line frequency and voltage between the lead and ground. A large resistor, for example 120 kΩ, should be placed in series with the meter and all conductors carrying 120 V should be insulated to protect the operator (see Fig. 5-15). The maximum current that any patient lead can sink is 20 μA in critical care area devices and 50 μA in other devices.

5-4 PLUGS AND RECEPTACLES

The study of electrical plugs and receptacles may seem mundane; however, their proper selection and installation can add considerably to electrical safety.

5-4-1 Receptacle Wiring

Ac power receptacles are sometimes miswired in both the original construction and replacement. Often the neutral and hot lines are reversed. Occasionally, the ground terminal becomes the neutral or hot. Proper receptacle wiring is shown in Fig. 5-1. The ground can be located at either the top or the bottom of the receptacle. When looking at an outlet with the ground on the top, the right terminal is the neutral and the left terminal is the hot. The neutral terminal may be larger than the hot terminal. On the back of the receptacle, the black hot line should be connected to the brass or dark terminal. The white neutral line should connect to a nickel-plated or light terminal. The ground line, which is either a bare wire or has green insulation, should be connected to the green terminal. The ground prong should be the first to make contact and the last to break contact when the plug is inserted into a receptacle.

5-4-1.1 RECEPTACLE POLARIZATION TESTING

In a correctly connected three-wire power distribution system, it requires two mistakes before a shock can occur. (For example, an open ground and a shorted hot lead to the case.) The purpose of polarization testing is to ensure that the first mistake does not occur.

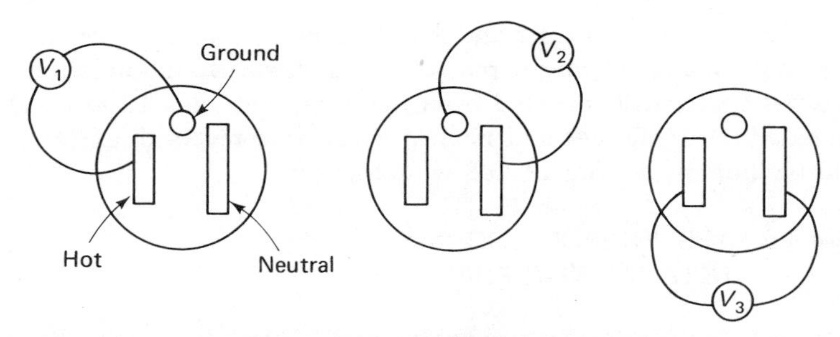

Figure 5-16 Detecting faults in receptacles. Possible faults are H, N, or G not connected, and H–N, H–G, or N–G reverse-connected. Five of these faults can be detected from the three voltmeter measurements shown. Reversal of N and G cannot be easily detected.

Measurements can be made between the three terminals of a receptacle to determine whether proper polarization exists. Six fault conditions are possible: any one of the three leads might be open, and there are three ways of interchanging the connections, assuming that only one fault occurs at a time. The fault may be uniquely deduced from measurements of V_1, V_2, and V_3 shown in Fig. 5-16. The reversal of neutral with the U-ground may not be obvious.

If a large number of receptacles are to be tested, it may be more convenient to build or buy a simple circuit that will allow faster checking of the receptacle wiring. Such a commercially available device is shown in Fig. 5-17.

The tester in Fig. 5-17 uses three light-emitting diodes (LEDs) to indicate the wiring status. The wiring is correct only if LEDs 1 and 3 are

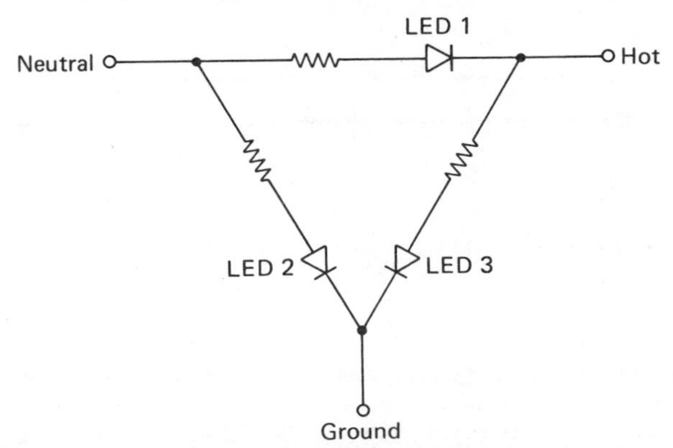

Figure 5-17 Receptacle fault analyzer using light-emitting diodes. In a correctly wired receptacle, only LEDs 1 and 3 will be illuminated. This circuit cannot detect neutral-ground reversals.

on and LED 2 is off. LEDs 1 and 3 will actually be blinking on and off at 60 Hz, but a human being perceives this as a steady illumination. If the ground wire were disconnected from the U-ground, only LED 1 would be illuminated. Or if the hot and neutral wires were reversed, LEDs 1 and 2 would be illuminated and LED 3 would be off.

5-4-1.2 TESTING FOR GROUND NEUTRAL REVERSAL

The receptacle tester of Fig. 5-17 will not differentiate between properly wired receptacles and those which have reversals of the U-ground and the neutral. To test for such reversals, a load is applied to a nearby receptacle on the same circuit, and measurements are made with an oscilloscope of the voltages on the ground and the neutral as shown in Fig. 5-18. When the receptacles are properly connected, the power current flows through only the hot and neutral conductors. The U-ground conductor carries only a small leakage current. The power current in the neutral is larger and has greater noise variations than the leakage current in the U-ground. Therefore, the neutral voltage V_1 will be larger and noisier than the ground voltage V_2. A device is commercially available for making this measurement in the same receptacle.

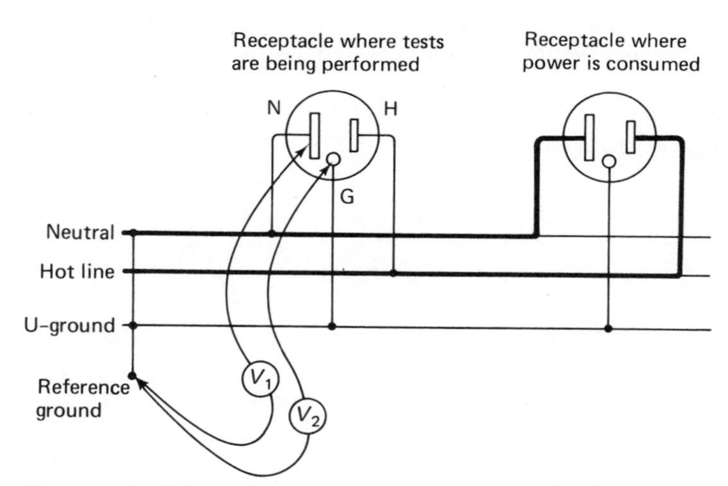

Figure 5-18 Circuit for detection neutral-ground reversals. V_1 and V_2 will be different, owing to the large current flowing in the neutral conductor. (From H. H. Roth, E. S. Teltscher, and I. M. Kane, *Electrical Safety in Health Care Facilities*, Academic Press, Inc., New York; 1975.)

5-4-1.3 LINE VOLTAGE TESTING

The voltages should be measured from each hot terminal to neutral and every other hot terminal in the vicinity. It is possible for these voltages to differ by 0, 115, 200, or 230 V if three-phase power is present. These voltages will differ by 0 or 230 V if only single-phase power is present. Electricity should be supplied from the same power-line phase and distribution line, which will produce zero voltage between the hot terminals of the receptacles.

5-4-2 Receptacle Force Testing

The metal prongs of a plug are held in the receptacle with spring-loaded metal contacts. If the spring has adequate clamping force, good contact is made. But the contacts will wear, resulting in poor contact. The plug can now become partially dislodged from the receptacle by a pull on the cord or perhaps even by just the weight of the cord. This could then disconnect the ground while allowing power to be applied to the equipment. This could also expose the hot conductor in the gap between the plug and the receptacle.

The force required to remove the plug from the receptacle is a good measure of the quality of the contact. The force needed to remove a metal prong from each contact should be measured with a calibrated spring or similar commercially available instrument. The lowest permissible removal force is 10 ounces per contact.

5-4-3 Pin Configurations

It is unwise to plug an instrument that draws 50 A into a receptacle that is connected with wire suitable for supplying only 15A. Plugs and receptacles are configured to prevent this occurrence.

Figure 5-19 shows the most frequently used configurations for receptacles and plugs, together with their electrical ratings. Certain pin configurations appear to be identical, although they have different current ratings, for example, $5-15R$ and $5-50R$. In these cases the basic pin configuration is the same, but the physical size of the plugs is different. The pin or receptacle opening marked *W* for *white* indicates the neutral wire in a grounded distribution system while *G* for *green* marks the protective grounding connection.

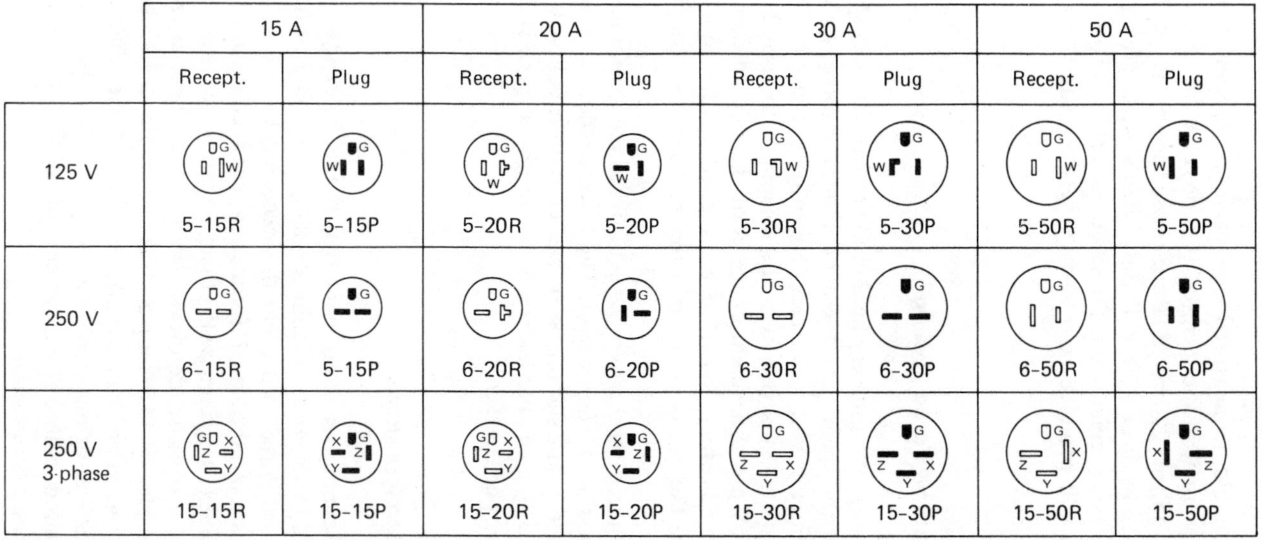

	15 A		20 A		30 A		50 A	
	Recept.	Plug	Recept.	Plug	Recept.	Plug	Recept.	Plug
125 V	5-15R	5-15P	5-20R	5-20P	5-30R	5-30P	5-50R	5-50P
250 V	6-15R	5-15P	6-20R	6-20P	6-30R	6-30P	6-50R	6-50P
250 V 3-phase	15-15R	15-15P	15-20R	15-20P	15-30R	15-30P	15-50R	15-50P

Figure 5-19 Pin configuration of the most frequently used plugs and receptacles. [From E. A. Pfeiffer, "Plugs and Receptacles in the Hospital: A Compendium for the Clinical Engineer," *Journal of Clinical Engineering*, 1 (1976), 46–55.]

5-4-4 *Grades of Plugs and Receptacles*

There are three basic grades of plugs and receptacles: general duty, heavy duty, and hospital grade. The Underwriters Laboratories, Inc., Standard for Electrical Attachment Plugs and Receptacles, UL 498, specifies the construction and necessary tests for each of these grades. One particular test is the overload test. It will be used to illustrate the differences in the three grades.

For a *general-duty receptacle*, a load of 150% of the rated current is applied, and a plug is inserted into and withdrawn from the test receptacle 50 times. The receptacle passes this test if there is "no electrical failure of the device or undue burning or pitting of the contacts." There are no special tests for performance of the ground contacts.

For a *heavy-duty receptacle*, a load of 200% of the rated current is applied, and the plug is inserted and withdrawn 250 times. There is an additional test which stresses the contacts and checks the contacts, including the ground, for damage or wear.

The *hospital-grade receptacle* is a very rugged device; its testing is also very rugged. The most severe of the several testing procedures as described by Pfeiffer (1976) is the "Abrupt Removal of Plugs Test." For this test, the receptacle is mounted vertically in a test fixture, as shown in Fig. 5-20, and a special test plug with replaceable brass power blades and grounding pin is inserted into the receptacle. A steel cable is attached to a steel rod projecting from the back of the plug approximately 2 in. from the plug face. The cable connects to a steel stem which has a strike plate attached at its opposite end. A cylindrical weight can slide over the stem. When dropped from a sufficient height, it hits the strike plate and abruptly removes the test plug from the receptacle. During this removal, the pins of the plug will normally be deformed and therefore must be replaced after each test. A test consists of eight abrupt plug removals, two in the first position, four after the receptacle has been rotated 180° with respect to the direction of the impact, and another two in the first position. After this sequence of tests, the receptacle has to be able to hold a test prong in the ground pin contact against a force of four ounces. It also has to be able to hold a standard two-wire plug with a pullout force of 3 lb. Also, there should be no breakage of the receptacle that interferes with its function.

For *hospital-grade plugs*, the most important test described by Pfeiffer is a series of strain-relief tests. A line cord of the minimum size likely to be used with the plug is installed in the strain-relief clamp of the plug. Its conductors remain unconnected. The cord is pulled in a number of different ways, including an abrupt-removal procedure similar to the one used for testing receptacles. The plugs pass the test when neither conductor, nor insulation, nor jacket are displaced by more than 1/32 in. during the test.

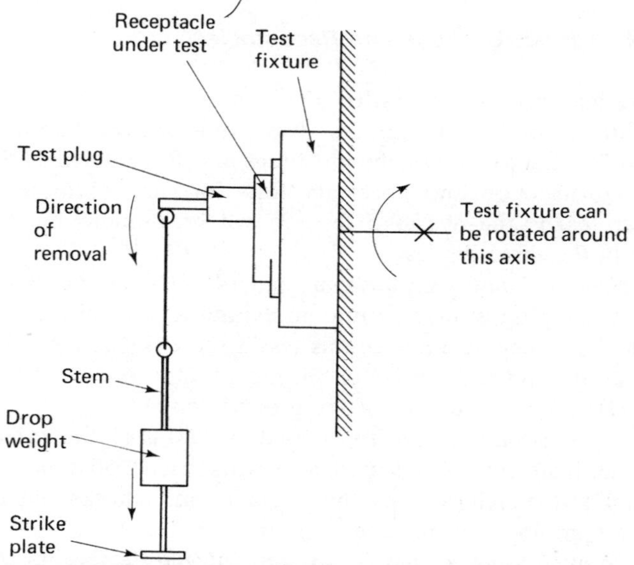

Figure 5-20 Abrupt-plug-removal test used by Underwriters Laboratories, Inc., for hospital-grade receptacles. [From E. A. Pfeiffer, "Plugs and Receptacles in the Hospital: A Compendium for the Clinical Engineer," *Journal of Clinical Engineering*, 1 (1976), 46–55.]

Plugs, receptacles, and cord connectors that pass these tests are hospital -grade devices. They are identified by the phrase *hospital grade*, which is visible during installation, and by a green dot visible after the device has been installed. These more rugged and reliable plugs and receptacles naturally cost more. Comparative prices for these receptacles are: general duty, $1.08; heavy duty, $3.59; and hospital grade, $4.95.

There are also many other special types of plugs and receptacles available. For example, there are isolated ground receptacles, locking plugs and receptacles, and explosion-proof devices.

5-5 GROUND-FAULT MONITORS

By periodically checking all receptacles one can be assured that no single fault will exist for longer than the interval between checks. But what about a hazard that occurs the day after the fault check? Continuous monitoring of all power receptacles using ground fault monitors will detect such failures. The schematic of a simple ground-fault monitor is shown in Fig. 5-21. If no fault exists, $I_1 = I_2$ and $V = 0$. But if a fault exists, $I_1 > I_2$, V does not equal 0, and an alarm sounds.

The purpose of an isolated power system is to have each conductor truly isolated from ground. Operating rooms with isolated power supplies

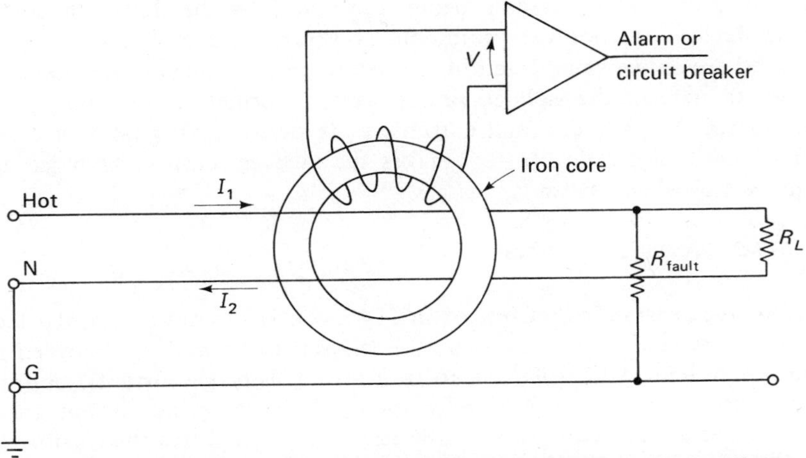

Figure 5-21 Ground-fault monitor: circuit for detecting current fault pathways between ground and either the hot wire or the neutral wire. If $I_1 = I_2$, then $V = 0$. If a fault exists, I_1 will not equal I_2 and a voltage will be induced in the coil.

are required to have ground-fault monitors to monitor the actual leakage current to ground. There are two common methods for doing this: current monitoring and voltage monitoring.

5-5-1 Current Monitors

Current monitoring (Fig. 5-22) senses current and thereby permits magnetic coupling to be used. The entire sensing circuit is then electrically isolated from the remaining system, thereby protecting the patient from the

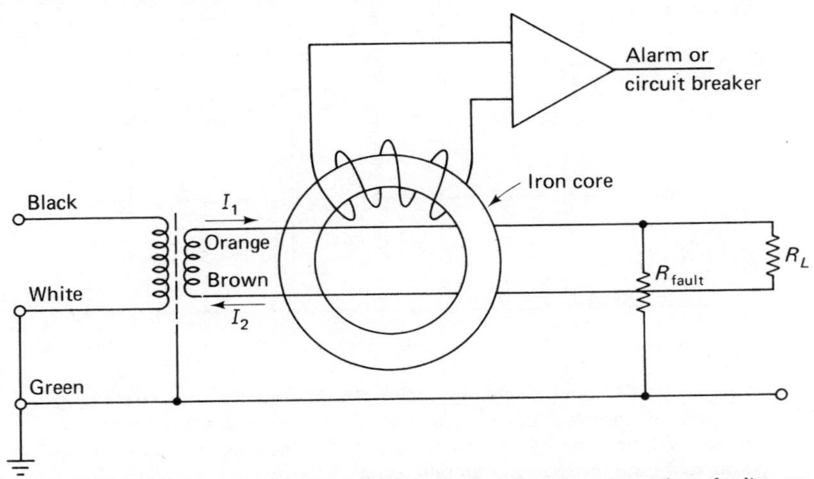

Figure 5-22 Current-sensing ground-fault monitor for detecting faults on isolated power systems.

possibility of hazard currents being introduced by the detection device. These devices are not very common, however, because the faults to be detected are only a small fraction of the total power-line currents. Depending on the size of the isolated power system, normal currents may be as high as 200 A, while the fault currents to be detected may be only 2 mA. This signal/noise ratio of 10^{-5} makes the devices sensitive to noise and requires critical adjustments.

5-5-2 Voltage Monitors

Voltage sensing, as illustrated in Fig. 5-23, is the most common technique for ground fault monitors. V_p is chosen to be midway between the two power lines, which is (due to transformer leakage currents) approximately ground potential. Normally, the meter indicates no current. In the presence of a fault, current will flow in the meter and trip the alarm.

One disadvantage of the voltage sensing system is that it has direct interconnection into the power and ground system. The direct connection permits the detector circuitry to introduce a small current into the ground system, which tends to increase any already existing hazard.

Impedances rather than pure resistances are usually used in ground-fault monitors, as illustrated in Fig. 5-24. The detector impedance **A** in series with the parallel combination of Z_1 and Z_2 is often designed to have a phase angle of 45° to ensure equal sensitivity to resistive and capacitive faults.

If the monitor of Fig. 5-24 is analyzed as a Wheatstone bridge circuit, it can be seen that there is a blind spot: there is a possible fault condition

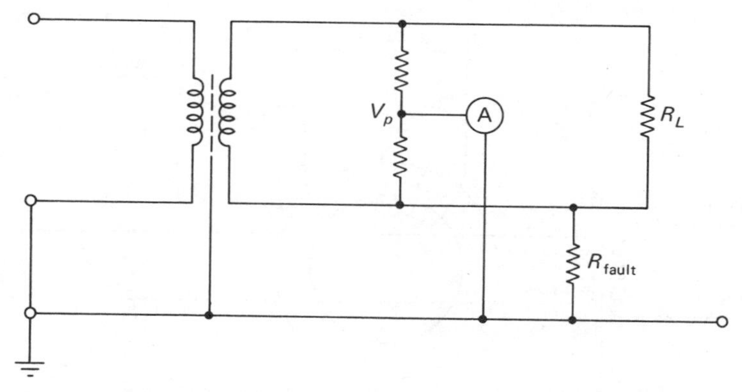

Figure 5-23 Voltage-sensing ground-fault monitor for detecting hazards on isolated power systems. When there are no faults, no current will flow though the ampmeter, A. If a fault exists, there will be a current path through the ampmeter.

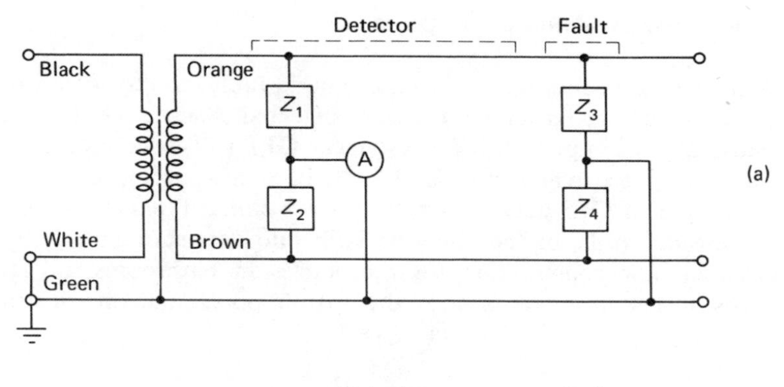

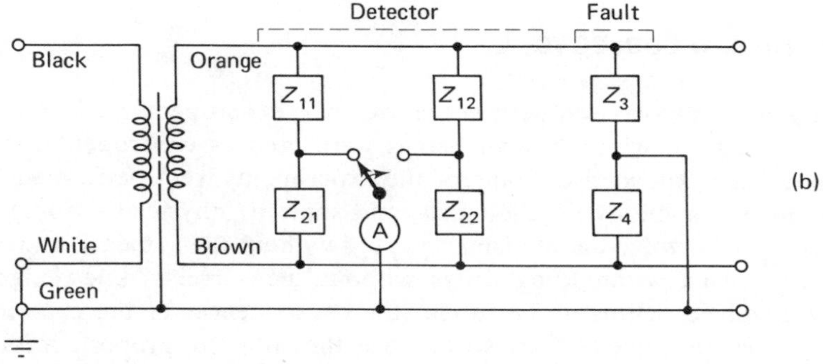

Figure 5-24 (a) Static and (b) dynamic voltage-sensing ground-fault monitors. The static monitor has a blind spot. If the ratio Z_3/Z_4 equals Z_1/Z_2, no current will flow through the ampmeter, A. In the dynamic monitor the switch alternates positions, so that the fault is compared to two different impedance ratios (Based on H. H. Roth, E. S. Teltscher, and I. M. Kane, *Electrical Safety in Health Care Facilities*, Academic Press, Inc., New York; 1975.)

that could go undetected. Namely, if there were two faults Z_3 and Z_4 and the ratio of Z_3 to Z_4 were the same as the ratio of Z_1 to Z_2, the bridge would be balanced and no alarm current would flow through **A** (see Fig. 5-24a).

The circuit shown in Fig. 5-24b eliminates the blind spot. The position of the switch varies as a function of time, so the monitor is called a dynamic detector. If the ratio of Z_{11} to Z_{21} is different than the ratio of Z_{12} to Z_{22}, then for any fault the bridge will be imbalanced in at least one of the two switch positions. A dynamic detector could be made with identical capacitors for Z_{11} and Z_{22} and identical resistors for Z_{21} and Z_{12}.

Line isolation monitors (ground-fault monitors) are permitted to supply 1 mA to the ground from the system being monitored. They are usually designed to trip the alarm if the total hazard current exceeds 5 mA.

5-5-3 Ground-Fault Interrupters

Rather than sounding an alarm, a ground-fault monitor could disconnect the power if a ground fault were detected. Such devices are called ground-fault interrupters (GFI). A typical GFI will disconnect power in less than 25 msec when a ground fault of 5 mA or more is detected. Various types of GFI may be substituted for circuit breakers, receptacles, and extension cords, or they may be built into new equipment. They are required in new construction for receptacles in bathrooms, in garages, outdoors, in pediatric wards, near therapeutic pools, and on construction sites.

5-6 HUMAN SUBJECTS

Many articles have been written that describe experiments performed 10 or 20 years ago in which human beings were used as experimental subjects without their knowledge. Some of the experiments were performed by the government, some by psychologists, and some by physicians. Some of the experiments involved administering drugs without the subjects' knowledge, some involved withholding drugs without the subjects' knowledge, and some involved deliberate lies about the consequences of the experiments. The gravest injustice of these studies was the failure to properly inform the subjects about the experiments.

Disclosure of such abuses led to policy statements guiding the treatment of human subjects. At first, voluntary guidelines, like the Declaration of Helsinki, were set. Later these guidelines obtained the weight of law, by publication in the *Federal Register*, May 30, 1974.

This regulation requires that all research grant applications submitted to the National Institutes of Health be reviewed by a committee at the submitting institution. The composition of this institutional review board (IRB) is precisely specified: the committee must consist of not less than five individuals with varying backgrounds, for example physicians, engineers, lawyers, housewives, and psychologists. The records of the committee must identify each member by name, earned degree, position or occupation, representative capacity, and other pertinent information. These regulations prescribe conflict of interest considerations and limit the number of institutionally affiliated people who may serve on the committee.

The committee is charged with determining if human subjects will be in danger, ensuring that the rights, welfare, and confidentiality of the subjects will be protected, conducting periodic reviews, and ensuring that informed consent will be obtained from the subjects.

Informed consent is the focus of considerable activity by institutional review boards. The Department of Health Education and Welfare's directions on this matter are as follows:

> *Informed consent* means the knowing consent of an individual or his legally authorized representative, so situated as to be able to exercise free power of choice without undue inducement or any element of force, fraud, deceit, duress, or other form of constraint or coercion. The basic elements of information necessary for such consent include:
>
> 1. A fair explanation of the procedures to be followed, and their purposes, including identification of any procedures which are experimental.
> 2. A description of any attendant discomforts and risks to be expected.
> 3. A description of any benefits reasonably to be expected.
> 4. A disclosure of any appropriate alternative procedures that might be advantageous for the subject.
> 5. An offer to answer any inquiries concerning the procedures.
> 6. An instruction that the person is free to withdraw his consent and to discontinue participation in the project or activity at any time without prejudice to the subject.
>
> Any institution proposing to place a subject at risk is obligated to obtain and document legally effective informed consent. No such informed consent, oral or written, obtained under an assurance provided pursuant to this part shall include any exculpatory language through which the subject is made to waive, or to appear to waive, any of his or her legal rights, including any release of the institution or its agents from liability for negligence.

Consent forms are difficult to write because they must explain technical equipment and procedures and yet be written in lay language.

This concern for human subjects has grown and expanded to include equipment being used to study, diagnose, and treat patients, as well as experimental human subjects.

5-7 MEDICAL DEVICE AMENDMENTS OF 1976

Consider these two scenes: a mother putting a bandage on her child's skinned knee, and a surgical team performing a heart transplant. What do they have in common? They are both using medical devices. The manufacture and sale of such medical devices is now controlled by the Federal Food and Drug Administration (FDA).

The law that gave this authority to the FDA, Public Law 94 295, defined a medical device as "an instrument, apparatus, implement, machine, contrivance, implant, *in vitro* reagent, or other similar or related article which is ... intended for use in the diagnosis of disease or other conditions, or in the cure, mitigation, treatment, or prevention of disease, in man or other animals or is intended to affect the structure or function of the body and which" is not a drug.

This law and the subsequent regulations published in the *Federal Register*[1] classify medical devices, establish marketing procedures, specify laboratory practices, and define good manufacturing procedures that must be followed by manufacturers of medical devices. Medical devices are classified into one of three categories:

Class I, General Controls.

Class II, Performance Standards.

Class III, Premarket Approval.

Simple devices, such as a bandage, a sponge, a sling, a bedpan, or a wood tongue depressor, are Class I devices which are subject to general controls. They must be labeled and manufactured under conditions spelled out in the regulations of good manufacturing practices. The manufacturers must be registered with the FDA and their facilities will be inspected.

More complicated devices, such as a closed-circuit television reading system, an eye-movement monitor, a biopotential amplifier, a tampon, or an electrocardiographic monitor, are Class II devices. They must comply with performance standards. Performance standards are being written by many organizations for different devices. University professors, IEEE groups, AAMI committees, industry groups, various medical committees, and the FDA itself are currently writing such standards. These performance standards specify a series of tests and specifications that must be met by the devices. These standards must include provisions to provide reasonable assurance of a device's safe and effective performance. They

[1] See the Department of Health, Education, and Welfare references. Reprints of these articles are available from your Congressman or from The Commissioner of Food and Drugs, Food and Drug Administration, 5600 Fishers Lane, Rockville, MD 20852.

prescribe permissible construction and components, and also describe appropriate labeling for proper installation and maintenance.

Devices such as pacemakers, cardiopulmonary bypass filters, or electro-anesthesia stimulators which are intended for use in supporting or sustaining human life, or are of substantial importance in preventing impairment of human health, or present a potential unreasonable risk of illness or injury are Class III devices. These devices are subject to general controls, they must comply with a performance standard, and in addition they are subject to extensive preclinical and clinical trials before they can be marketed.

The good manufacturing practice regulations deal with company organization, personnel, buildings, equipment, and records. There must be a written quality control manual. Written and oral complaints from customers and sales personnel expressing any dissatisfaction with the identity, quality, durability, reliability, effectiveness, or performance of a device must be reviewed by the quality control unit. The complaint and the action taken regarding the complaint must be kept on file. Personnel must be trained so that they understand the functions they perform, and those in contact with the medical device must be clean, healthy, and suitably attired. Buildings must have adequate space and suitable environmental conditions, such as lighting, ventilation, temperature, and humidity. Adequate cleaning and sanitation facilities must be provided. Equipment must be periodically inspected, calibrated, adjusted, and a maintenance schedule must be provided. Records must be kept of device history, quality control checks, and device design and specifications.

It is clear that many bioengineers will be working on projects proving compliance with FDA regulations, and on tests to meet the approval requirements of the FDA. The medical device amendments of 1976 have opened up new vistas for bioengineers.

REFERENCES

Association for the Advancement of Medical Instrumentation, *Safe Current Limits for Electromedical Apparatus*, An American National Standard. AAMI, Arlington, Va., 1978.

DALZIEL, C. F., and W. R. LEE, "Reevaluation of Lethal Electric Currents," *IEEE Transactions on Industry and General Applications*, IGA-4 (1968), 467–76.

Department of Health, Education, and Welfare, Food and Drug Administration, "Nonclinical Laboratories Studies: Proposed Regulations for Good Laboratory Practice," *Federal Register*, 41(225) (November 19, 1976), 51205–51230.

Department of Health, Education, and Welfare, Food and Drug Administration, "Medical Devices, Establishment, Registration and Premarket Notification Procedures," *Federal Register*, 41(173) (September 3, 1976), 37457–37465.

Department of Health, Education, and Welfare, Food and Drug Administration, "Medical Devices, Procedures for Investigational Device Exemptions," *Federal Register*, 43(93) (May 12, 1978), 20726–10757.

Department of Health, Education, and Welfare, Food and Drug Administration, "Manufacture, Packing, Storage and Installation of Medical Devices, Good Manufacturing Practices," *Federal Register*, 43(141) (July 21, 1978), 31508–31532.

Department of Health, Education, and Welfare, Food and Drug Administration, "Medical Devices, Proposed Procedures for Development of Standards," *Federal Register*, 43(143) (July 25, 1978), 32264–32271.

Department of Health, Education, and Welfare, Food and Drug Administration, "Medical Devices: Establishment of Procedures to Make a Device a Banned Device," *Federal Register*, 44(98) (May 18, 1979), 29214–29224.

GRASS, E. R., *Electrical Safety Specifically Related to EEG*. Bulletin X757C78. Grass Instrument Co., Quincy, Mass., 1978.

MEDICAL DEVICE ADMENDMENTS OF 1976, PUBLIC LAW 94 295, 94TH CONGRESS, S.510, MAY 18, 1976.

OLSON, W. H., "Electrical Safety," in *Medical Instrumentation*, ed. J. G. Webster. Boston: Houghton Mifflin Company, 1978.

OTT, H. W., *Noise Reduction Techniques in Electronic Systems*. New York: John Wiley & Sons, Inc., 1976.

PFEIFFER, E. A., "Plugs and Receptacles in the Hospital: A Compendium for the Clinical Engineer," *Journal of Clinical Engineering*, 1 (1976), 46–55.

ROTH, H. H., E. S. Teltscher, and I. M. Kane, *Electrical Safety in Health Care Facilities*. New York: Academic Press, Inc., 1975.

ROWLEY, B. A., *Towards an Understanding of Electrical Safety in Patient Care*. Texas Technical University, Lubbock, 1976.

SUMMERS, W. I., ed., *The National Electrical Code Handbook*, National Fire Protection Association, Boston, 1978.

Underwriters Laboratories, *Standard for Safety, Medical and Dental Equipment*. UL 544. UL Laboratories, Chicago, 1974.

PROBLEMS

5-1 One of the reasons that it is so easy to get shocked is that earth is used as a ground. If the secondary side of your local power step-down transformer were not grounded to earth, could this hazard be eliminated? Would it be advantageous to disconnect the neutral or ground wire from earth ground in your basement?

5-2 There are two major objections to the wiring schematic shown in Fig. P5-2. First, it is dangerous to the patient. Second, it will yield noisy signals. Explain each of these problems. The arm-to-arm resistance of the patient is 500 Ω, the ground circuit is composed of AWG No. 18 wire (6.4 mΩ/ft) and each mechanical connection has 5 mΩ of resistance. A diathermy machine is producing a stray magnetic field of 10^{-10} weber/m^2 at 2.5 GHz.

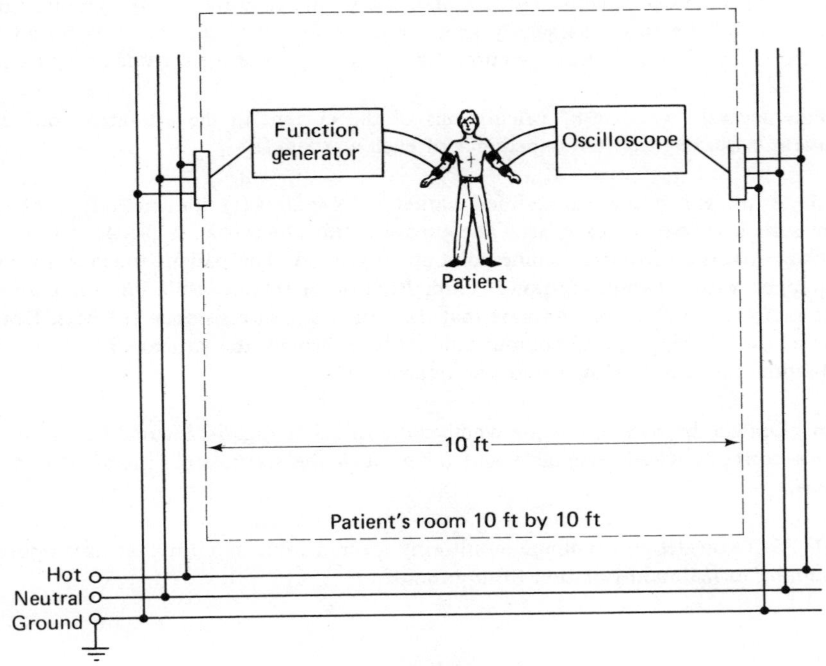

Figure P5-2

5-3 An engineer in contact with a grounded operating room floor is electrocuted when she touches the metal case of the isolation transformer. The capacitance between this case and the high voltage line is 0.7 μF. Assuming that the minimal fibrillating current is 100 mA and that her skin resistance is 1 kΩ, calculate the primary voltage.

5-4 An attendant was in the cardiac catherization lab preparing a patient for an upcoming exam. When the attendant was attaching ECG electrodes to the patient, he reached over to pick up an elastic strap on top of the dye injector (one hand was still pressing on the left arm electrode). The moment he touched the case of the dye injector, the patient went into ventricular fibrillation. You were called upon to investigate the incident. These facts were found:

1. The ECG monitor and dye injector were plugged into sockets on opposite sides of the room. A 100-ft piece of No. 12 copper wire (1.6 mΩ/ft) connected the ground terminal of the two wall sockets.
2. The contact resistance of the ground lead for the ECG monitor in the wall socket was 1 mΩ.
3. Both pieces of portable equipment were checked, with the following results:

   (a) the dye injector had less than 1 μA leakage current in the ground lead and the case was solidly grounded; (b) the ECG monitor was found to have 20 mA of leakage current between the left arm electrode and ground.

Make several "worst-case" calculations of the current in the attendant and the patient, assuming that the impedance of each was 500 Ω.

5-5 A patient has a saline-filled catheter ($R = 20$ kΩ) for measuring blood pressure inserted in her heart. The pressure transducer has a 3-MΩ, 0.001-μF leakage pathway from the saline solution to ground. The patient touches a radio equipped with a two-prong power cord. Because of frayed insulation, the case of the radio is at 120 V rms. Assume that the patient's skin resistance is 1 MΩ. Draw the equivalent circuit, and compute the current through the patient's heart. Is this situation dangerous? How could you improve it?

5-6 Draw a diagram indicating whether the LEDs of Fig. 5-17 would be on or off for a correctly wired receptacle and for each of the six possible miswired conditions.

5-7 Can you design a voltage-monitoring ground fault monitor that uses optical coupling to maintain isolation from ground?

6

LABORATORIES

It is important for bioengineers to have practical experience with biomedical instruments and their application to human beings. The most natural way to do this is with laboratories. The following is a list of laboratory sessions that have been found interesting and useful. A selection of labs from this list would complement the material of this text. They are not listed in any suggested chronological order. Often the selection of laboratories is based upon the equipment available. Many manufacturing companies will give large discounts on equipment, or even give free equipment for use in student laboratories. Kenneth C. Mylrea, of the University of Arizona, has prepared detailed laboratory descriptions for the set of biomedical instruments that he had available (personal communication). He may be willing to share these laboratory handouts.

6-1 ELECTROCARDIOLOGY

Students should attach the five standard electrodes to the limbs and the chest and record signals for each of the lead permutations, known as the 12-lead electrocardiogram (ECG). Differences between the lead configurations should be noted. Intrasubject variability of waveforms should be studied. An ECG amplifier, such as the Hewlett-Packard 8811A, is recommended but is not necessary.

If phonocardiology and pulse-wave pickup equipment is available, a subsequent lab should study the correlations between the heart sounds and electrical activity, the heart sounds and the carotid pulse, and the heart sounds measured by the electronic equipment and heard through a stethoscope.

References: Burton (1972); Hewlett-Packard (1972); Selkurt (1976); Reiser (1979).

6-2 CIRCUIT DESIGN CONSTRAINTS

The amplifier used for recording the electrocardiograms should be analyzed using an oscilloscope and/or a signature analyzer in conjunction with the circuit schematic. The constraints and special features demanded of the circuit design by physiological considerations should be studied. For example, students should study the patient lead devices used to limit fault currents, the spark gaps or neon tubes used to protect the electronics from defibrillator pulses, and the isolation circuits which ensure that there is no path to the power supply ground from the patient leads. The effects of varying the front-panel high- and low-frequency cutoff knobs should be measured for sinusoidal and step inputs.

References: Chaps. 2 and 5 of this book.

6-3 ELECTROMYOGRAM (EMG)

Surface electrodes are placed on a pair of antagonistic muscles, such as the biceps and the triceps muscles of the upper arm. The resulting signals are filtered (5 to 2000 Hz) and amplified. These signals can be displayed directly or may be rectified and integrated before display. The antagonistic behavior of the two muscles can be studied as the position of the arm or leg is varied or as weights are added.

References: Chap. 4 of this book; Heshler and Milner (1978).

6-4 ELECTROENCEPHALOGRAM (EEG)

Surface electrodes are placed on the skull and the measured electrical activity of the brain cells is filtered (1 to 50 Hz) and amplified. The alpha rhythm (8 to 12 Hz activity) can be seen when the subject is resting eyes closed. Opening the eyes or attending to some novel stimulus should make it go away. If an averager is available, visual evoked potentials can be recorded.

References: Regan (1972 and 1979); Lindsley and Wicke (1974); Chap. 11 of Geddes and Baker (1975); Barlow (1979); van der Tweel (1979).

6-5 ELECTRO-OCULOGRAM (EOG) AND PHOTOELECTRIC MEASUREMENTS

Four surface EOG electrodes are applied around the eye with a reference electrode on the earlobe. The resulting signals are filtered (0 to 100 Hz) and amplified. The independence of the horizontal and vertical eye-movement channels can be demonstrated and the functional relationship of peak velocity and duration to saccadic magnitude can be plotted. For a subsequent lab, a pair of spectacle frames can be modified to accommodate some photodiodes, and the photoelectric method of eye-movement measurement can be used. The effects of variable feedback can be studied, and the small eye movements and fine detail of the larger movements can also be seen.

References: Chaps. 2 and 3 of this book; Young and Stark (1963); Bahill, Clark, and Stark (1975); Bahill and Stark (1979).

6-6 CLINICAL ENGINEERING

Students should check all the receptacles in the area for correct wiring and sufficient contact resistance using a receptacle analyzer and a calibrated spring. The leakage current of some biomedical instruments (e.g., the electrocardiograph machine) should be measured in each of the four standard configurations of power on and off, hot and neutral, normal and reversed. This lab can be supplemented with a tour of a local hospital given by their resident clinical engineer.

Reference: Chap. 5 of this book.

6-7 RADIOLOGY

Because radiology equipment is so expensive, its use is best illustrated by a tour of a medical school's radiology department. The tour should point out computer-aided tomography (CAT) scanners, angiography equipment, and ultrasound equipment. This could be one of three or four hospital tours given during the semester.

6-8 BLOOD FLOWS

Real or simulated blood flows should be measured with one or more blood flow meters, and the operation and limitations of the devices studied.

References: *Medical Instrumentation* (1977); Chap. 8 of Cobbold (1974).

6-9 PRESSURE

Responses of a pressure-tip catheter should be compared to a hydraulically coupled pressure catheter. If a suitable pump is available, sinusoidal inputs may be applied. Otherwise, steps and sinusoids can be approximated by raising and lowering a beaker of water that provides hydrostatic pressure. An impulse can be approximated by pinching the water tube. A blood-pressure cuff should be used to measure the blood pressure of several subjects before and after mild exercise.

Reference: Chap. 7 of Cobbold (1974).

6-10 DIGITAL COMPUTER TECHNIQUES

Some physiological data (e.g., EOG recordings) should be fed through an analog-to-digital convertor and processed by digital programs: for example, fast Fourier transforms, digital filters, or cross-correlation programs.

Reference: Oppenheim and Schafer (1975).

6-11 ELECTRODE IMPEDANCE

The half-cell potentials of several different types of electrodes should be compared. The frequency and current dependent impedance of at least one set of electrodes should be measured. One electrode is placed at one end of a shallow trough filled with 1% saline solution. The second electrode is located at two distances from the first electrode. A function generator (producing a few hundred millivolts), a large resistor (many megohms), the two electrodes, and the water in the trough are connected in series. The voltage, magnitude, and phase across the electrode and the water and the voltage across the series resistor are recorded. The impedance of the electrodes and the water can be computed from these measurements. The effect of the water can be eliminated by using data with two different electrode separations. Thus the impedance of the electrode pair can be plotted as a function of frequency. The effects of current density can be observed by varying the series resistance.

An alternative and simpler setup is the electrode sandwich. Two electrodes are taped face to face with some electrode cream in between. This eliminates the water trough and makes the impedance calculations easier.

Reference: Chap. 2 of this book; Chap. 9 of Geddes and Baker (1975).

REFERENCES

BAHILL, A. T., M. R. CLARK, and L. STARK, "Dynamic Overshoot in Saccadic Eye Movements Is Caused by Neurological Control Signal Reversals," *Experimental Neurology*, 48 (1975), 95–122.

BAHILL, A. T., and L. STARK, "The Trajectories of Saccadic Eye Movements," *Scientific American*, 240 (January 1979), 108–17.

BARLOW, J. S., "Computerized Clinical Electroencephalography in Perspective," *IEEE Transactions on Biomedical Engineering*, BME-26 (1979), 377–91.

BURTON, A. C., *Physiology and Biophysics of the Circulation*. Chicago: Year Book Medical Publisher, 1972.

COBBOLD, R. S., *Transducers for Biomedical Measurements*. New York: John Wiley & Sons, Inc., 1974.

GEDDES, L. A., and L. E. BAKER, *Principles of Applied Biomedical Instrumentation*, 2nd ed. New York: John Wiley & Sons, Inc., 1975.

HERSHLER, C., and M. MILNER, "An Optimality Criterion for Processing Electromyographic (EMG) Signals Relating to Human Locomotion," *IEEE Transactions on Biomedical Engineering*, BME-25 (1978), 413–20.

HEWLETT-PACKARD, *ECG Measurements*. Application Note AN711. Hewlett-Packard, Palo Alto, Calif., 1972.

HEWLETT-PACKARD, *Phonocardiography*. Application Note AN732. Hewlett-Packard, Palo Alto, Calif., 1973.

LINDSLEY, D. B., and J. D. WICKE, "The Electroencephalogram: Autonomous Electrical Activity in Man and Animals," in *Bioelectric Recording Techniques*, Part B, *Electroencephalography and Human Brain Potentials*, ed. R. F. Thompson and M. M. Patterson. New York: Academic Press, Inc., 1974, pp. 3–83.

Measurement of Blood Flow, special issue of *Medical Instrumentation*, 11 (May–June, 1977).

OPPENHEIM, A. V., and R. W. SCHAFER, *Digital Signal Processing*. Englewood Cliffs, N.J.: Prentice-Hall, Inc., 1975.

REGAN, D., *Evoked Potentials in Psychology, Sensory Physiology and Clinical Medicine*. London: Chapman & Hall Ltd., 1972.

REGAN, D. M., "Electrical Responses Evoked from the Human Brain," *Scientific American*, 241, (December 1979), 134–46.

REISER, S. J., "The Medical Influence of the Stethoscope," *Scientific American*, 240 (February 1979), 148–56.

SELKURT, E. E. (ED.), *Physiology*. Boston: Little, Brown and Company, 1976, Chaps. 13 and 14.

VAN DER TWEEL, L. H., "Pattern Evoked Potentials: Facts and Considerations," Proceedings 16th ISCEV Symposium, Morioka, Japan, (1979), 27–46.

VAN DER TWEEL, L. H., O. ESTEVEZ, and J. STRACKEE, "Measurement of Evoked Potentials," in *Evoked Potentials*, ed. C. Barber. Baltimore: University Park Press, and also Lancaster, England: MTP Press Limited, 1980, pp. 19–41.

YOUNG, L. R., and L. STARK, "Variable Feedback Experiments Testing a Sampled Data Model for Eye Tracking Movements," *IEEE Transactions on Human Factors Electronics*, HFE-4 (1963), 38–51.

INDEX

V

W